Medical Physics
Volume III
Synapse, Neuron, Brain

Medical Physics

Volume III
Synapse, Neuron, Brain

A. C. Damask
Department of Physics
Queens College of the City University of New York
New York, New York

C. E. Swenberg
Armed Forces Radiobiology Research Institute
Bethesda, Maryland

1984

ACADEMIC PRESS, INC.
(Harcourt Brace Jovanovich, Publishers)

Orlando   San Diego   San Francisco   New York   London
Toronto   Montreal   Sydney   Tokyo   São Paulo

COPYRIGHT © 1984, BY ACADEMIC PRESS, INC.
ALL RIGHTS RESERVED.
NO PART OF THIS PUBLICATION MAY BE REPRODUCED OR
TRANSMITTED IN ANY FORM OR BY ANY MEANS, ELECTRONIC
OR MECHANICAL, INCLUDING PHOTOCOPY, RECORDING, OR ANY
INFORMATION STORAGE AND RETRIEVAL SYSTEM, WITHOUT
PERMISSION IN WRITING FROM THE PUBLISHER.

ACADEMIC PRESS, INC.
Orlando, Florida 32887

United Kingdom Edition published by
ACADEMIC PRESS, INC. (LONDON) LTD.
24/28 Oval Road, London NW1 7DX

Library of Congress Cataloging in Publication Data
(Revised for volume 3)

Damask, A. C.
  Medical physics.

    Vol. 3 by A. C. Damask and C. E. Swenberg.
    Includes bibliographies and indexes.
    Contents: v. 1. Physiological physics, external
probes.--v. 2. External senses.--v. 3. Synapse, neuron,
brain.
    1. Medical physics. 2. Biophysics. I. Swenberg,
Charles E. [DNLM: 1. Biophysics. QT 34 B155m 1978]
H895.D35   612'.014   78-205
ISBN 0-12-201203-8 (v. 3)

PRINTED IN THE UNITED STATES OF AMERICA

84 85 86 87   9 8 7 6 5 4 3 2 1

Contents

Preface ix
Acknowledgments xi
Contents of Previous Volumes xiii

CHAPTER 1 Introduction 1

References 6

CHAPTER 2 The Evolution and Morphology of the Brain

Introduction 7
Origin of the Two-Sided Brain 9
Origin of Vertebrates 10
Development of the Brain 12
Architecture of the Human Brain 17
The Triune Brain 20
Cephalization of the Species 23
Brain Asymmetry 26
Left-Handedness 29
Survival Value of Cerebral Dominance 31
Arrangement of Cortical Neurons 32
References 39

CHAPTER 3 **Electrical Properties of the Brain; EEG and Evoked Potentials**

Introduction	41
Electroencephalography	42
Computer Diagnosis of Electroencephalograms	48
Cellular Origins of Electroencephalogram Brain Waves	57
Evoked Potentials	62
Visual Evoked Potentials	63
Diagnostics of Children Using Visual Evoked Potentials	66
Diagnosis of Optic Neuritis Using Visual Evoked Potentials	69
Origin of Evoked Potentials	72
Auditory Evoked Potentials	73
Evoked Potentials as a Measure of Mental Chronometry	79
References	82

CHAPTER 4 **Chemical and Electrical Properties of Synapses**

Introduction	85
Kinds of Synaptic Connections	88
Chemical Synapses	90
Electrical Synapses	96
Properties of Postsynaptic Potentials and Currents at Chemical Synapses	98
Presynaptic Properties of Chemical Synapses	110
Exocytosis	121
A Molecular Model of Postsynaptic Responses	125
Drug Action at Synaptic Receptors	131
Central-Nervous-System Neurons and Drug Effects	135
Common Diseases Associated with Synaptic Function	137
References	142

CHAPTER 5 **Neuronal Integration and Rall Theory**

Introduction	145
Effects of Environment and Malnutrition on Dendrite Growth	149
Dendrite Electrotonus and the Linear Cable Equation	155
Special Solutions to the Cable Equation	160
Boundary Conditions at Dendritic Branch Points	164
Equivalent Cylinder Approximation	167
Properties of Somatic Potentials with Dendritic Synaptic Inputs	169
Compartmental Model	169
Shape and Time Behavior of Soma Potentials	170
Transient Passive Membrane Responses	176
References	181

CHAPTER 6 **Analysis of Membrane Noise at Synaptic Junctions**

Introduction	184
Analysis of Random Signals	187
Autocorrelation Function	189
Power Spectrum	192

CONTENTS

Determination of Power Spectrum (Spectral Density Function)	196
Thermal Noise	199
Shot Noise	201
Flicker Noise	202
Conductance Noise	202
Current Noise at the End Plate	203
Frequency Composition	208
Channels at the Neuromuscular Junction	212
Effect of Pore Structure	216
Single-Channel Events	219
Conductance Fluctuations in the Presence of Local Anesthetics	225
References	230

CHAPTER 7 New Techniques of Brain Studies: Autoradiography, Positron Annihilation, and Nuclear Magnetic Resonance

Introduction	232
Autoradiographic Determination of Regional Brain Metabolism	234
Dynamic Radiographic Studies of Brain Metabolism	237
Coincidence Counting of Positron Annihilation	238
Design Considerations of Positron Annihilation	242
Positron-Emission Measurement of Brain Metabolism	246
General Aspects of Nuclear Magnetic Resonance	248
Spinning Tops	250
Larmor Precession	252
Nuclear Moments	254
Effect of a Radio-Frequency Field	257
Population and Relaxation	260
Spin–Spin Relaxation	262
Spin–Lattice Relaxation	266
Spin–Echo Technique	268
Nuclear Magnetic Resonance Imaging	269
Chemical Shifts	277
Intracellular pH Measurement by ^{31}P Nuclear Magnetic Resonance	281
References	285

APPENDIX A	**Open-Channel Distribution Function**	289
APPENDIX B	**Rall's Branching Rule**	293
APPENDIX C	**Proof of the Equivalent Cylinder Approximation**	300
APPENDIX D	**Solution to the Cable Equation for a Cylinder with Sealed Ends**	303
APPENDIX E	**Autocorrelation Function**	307
APPENDIX F	**Fourier Coefficients**	311
APPENDIX G	**Fourier Transforms**	313

APPENDIX H	**Sampling Theorem**	316
APPENDIX I	**One-Dimensional Random Walk**	318
APPENDIX J	**End-Plate Current Spectral Function**	320
APPENDIX K	**Intracellular Trapping of 2-Deoxy-D-glucose**	324
APPENDIX L	**Statistical Properties of Single-Channel Events**	328
Index		333

Preface

This is the final volume of the planned three-volume set on medical physics. The purpose has been threefold.

The first is to assemble a body of knowledge in which physics is used in medicine, as well as in physiology and biology, where the topics are directed toward present or future use in medicine. A number of schools teach courses in medical physics, but most emphasize topics in which faculty members have some expertise. A course in physics, calculus, organic chemistry, etc., taken in any school has a common content, but this has been lacking in courses in medical physics. The attempt, therefore, has been to introduce a commonality of physical and mathematical techniques so that all who have taken such a course will have learned a minimum common amount of knowledge. No attempt has been made to be comprehensive, for the result would be encyclopedic. Instead, topics were chosen that illustrate the techniques. For example, Fick's laws of diffusion are derived in Volume I and used both in that volume in the interpretation of the nerve impulse and in the present volume for the diffusion of neurotransmitters at synapses. These laws of diffusion are equally useful in the interpretation of kidney function, which is not discussed in these volumes.

Second, an attempt was made to treat all topics at a common level; that level assumes the reader has had one year each of physics and calculus.

Although only first principles of physics are employed, some of the mathematical techniques are beyond first-year calculus. For these, appendixes have been added so that readers may understand what is being done without having to refer to other books on mathematics. With this approach, it is hoped that graduates in physiology and medicine who no longer take college courses will be able to improve their existing knowledge and be able to understand a significant fraction of recent research papers, many of which now use sophisticated mathematical techniques.

Third, these volumes are addressed to those scientists and engineers who, although they possess the mathematical skills, know very little of physiological processes. Therefore, each topic first covers just enough physiological terminology and function to enable these readers to understand something of what is going on. Appropriate references are given to the original literature in order to acknowledge the sources of the information contained within these pages and also to allow the inquisitive reader to broaden his knowledge. It is hoped that in some of this group of readers a degree of interest will be evoked, and they may consider contributing to the physics of medicine and physiology.

Attention is called to two matters. First, these volumes are addressed to the large segment of the scientific population that has minimal knowledge of biochemistry. Except when biochemicals are clearly related to a physical measurement, they are omitted. Thus, there is no attempt to write a complete discussion of topics, such as synapses. This has already been done in a number of excellent books. The chapter on synapses in this book is intended as a supplement to the chemical story to show where and how physics has been used.

The second matter is that of units. Although SI units are elegant and, by now, standard, much of the early work was reported in cgs units. If the writers converted all these, the reader, in referring to the original paper, would have to reconvert. We have largely reported work in the original units and, therefore, although the book lacks the elegance of consistent units, we feel it will be easier to use as a reference book.

Acknowledgements

We wish to acknowledge the helpful suggestions of the readers of various chapters of the manuscript while in no way holding them responsible for content or errors: Professor Walter Essman, Queens College, Chapter 4; Dr. Amy MacDermott, National Institute of Alcohol Abuse and Alcoholism, Chapters 2, 3, and 4; Dr. Michael McCreery, Armed Forces Radiobiology Research Institute, Chapter 6; Dr. Wilfred Rall, National Institutes of Health, Chapter 5; and Ms. Andrea Lunsford, Armed Force Radiobiology Research Institute, general reading of the manuscript for clarity and style.

Profound appreciation is expressed to Mrs. Marion Gaffga of Spring Hill, Florida, who expertly prepared and edited the typescript of this volume as well as those for Volumes I and II.

Contents of Previous Volumes

Volume I	PHYSIOLOGICAL PHYSICS, EXTERNAL PROBES
CHAPTER 1	Newtonian Fluid Flow: Respiration and Micturition
CHAPTER 2	Non-Newtonian Fluids: Mucus and Blood
CHAPTER 3	The Nerve Impulse: Action Potential and Transmission
CHAPTER 4	Muscle: Energy and Mechanism
CHAPTER 5	Bone: Mechanical and Electrical Properties
CHAPTER 6	Heart Motion: Electrocardiography and Starling's Law
CHAPTER 7	Circulation: Fluid Flow in Elastic Tubes
CHAPTER 8	Ultrasonic Probes: Scanning and Echocardiography
CHAPTER 9	Nuclear Medicine: Tracers and Radiotherapy
CHAPTER 10	Computerized Tomography: γ-Ray and X-Ray Brain Scanning
CHAPTER 11	Cryobiology: Cell Freezing and Cryosurgery
APPENDIX A	Chemical Thermodynamics
APPENDIX B	The Wave Equation
APPENDIX C	Binomial and Poisson Probability Distributions
APPENDIX D	Differential Equations
INDEX	

Volume II   EXTERNAL SENSES

CHAPTER 1   Introduction
CHAPTER 2   Gustation
CHAPTER 3   Olfaction
CHAPTER 4   Cutaneous Sensation
CHAPTER 5   Audition
CHAPTER 6   Vision
CHAPTER 7   Psychophysics
APPENDIX   Root Mean Square Deviation
INDEX

CHAPTER 1

Introduction

The existence of nerves within the bodies of higher organisms has been known since ancient times. In the second century Galen distinguished between sensory nerves and motor nerves. That electricity was somehow involved was not recognized until the latter part of the 18th century, when Galvani's experiments on the twitching of frogs' legs were interpreted correctly by Volta. A brief summary of the subsequent history of the electrochemical properties of neural conduction is found in many sources (e.g., Hubbard et al., 1969). For our purposes, only the discoveries of immediate relevance to the scope of this book will be mentioned.

The primary focus of this book is on neurons and their interactions. A neuron is a nerve cell. These come in many shapes and sizes. However, most of them are characterized by a *soma* (from the Greek word for body), and leading from this on one side is usually a structure of fine fibers much like branches on a tree, which bears the name *dendrite* (from the Greek word for tree). From the other side of the soma there is usually a single extended fiber, although it often has branches near its end, called the *axon* (from the Greek word for axis). An axon is often called a nerve fiber. Electrical signals between neurons are exchanged at small interaction sites on the body or on dendrites of the cell, called *synapses*. Neurons are distributed throughout the bodies of animals and are responsible for receiving and transmitting information. Some axons, such as those involved in transmitting the sensation of touch of the fingers to the brain, are as long as 1 m. Others, such as those in the retina of the eye, are quite short and go through many synapses from the sensory organ to the brain. A very high density of neurons exists in the brain. The human cortex is an integrated unit with an estimated 10^{10} neurons, some of which have 1000 or more synaptic connections with other neurons.

Much of our understanding and mapping of neurons and their interconnections comes from electrophysiological measurements in which very fine electrodes, usually glass tubes filled with an ionic conducting solution, are inserted into various parts of the neuron. When such an electrode is inserted into neurons of the brain, it is done knowing only the general location of the neurons under study because exposure of the neurons would harm or kill the animal. Often it is done blindly, measurements are taken, and then a small amount of dye is injected. The animal is then sacrificed, and microscopic examination of brain slices shows which neuron was measured. Repeat measurements are obviously impossible. More often motor neurons (motoneurons) in the spinal column are measured because they are usually larger than neurons in the brain and their specific function and location are more precisely known. With this prior knowledge, stimulation of a motor neuron, such as touching the paw of an experimental cat, may be achieved. The classic measurements on axonal conduction were done on the main axon of a squid, which is long, up to 1 mm in diameter, and may be removed from the squid and tested without alteration of its properties.

These electrophysiological measurements began about a half century after the experiments of Galvani, when better instruments, including the galvanometer, enabled DuBois-Reymond to discover the *action potential*. An electrode inserted into the axon of a neuron will show a negative potential of 40–90 mV relative to that outside the axon. If the axon is stimulated by a positive potential of about 30 mV applied internally, called a depolarizing potential, a sudden voltage pulse appears, which overshoots zero potential and becomes positive. This is illustrated in Fig. 1.1. The resting potential of -70 mV relative to the outside of the axon is seen on the left-hand side. The depolarizing voltage is applied to the nerve some distance from the recording electrode. The fiber subsequently develops a positive potential that lasts for about 3 msec, and then it recovers. The time between the stimulus artifact and the measured action potential is the time required for the action potential to travel from the site of depolarization to the recording electrode. If a second electrode is placed farther along the axon, it will record the action potential a few milliseconds later, undiminished in magnitude and unchanged in size.

INTRODUCTION

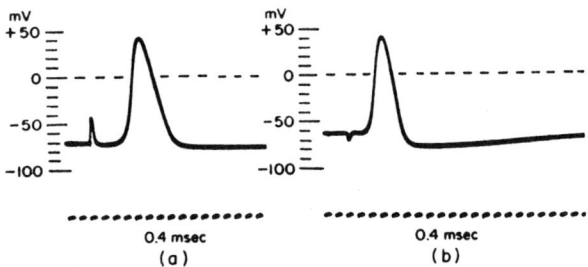

FIG. 1.1 Action potential of a squid axon: (a) in vivo; (b) isolated axon in seawater. The vertical scale is the internal potential in millivolts, the seawater outside being taken as the zero potential. [Unpublished work by A. L. Hodgkin and R. D. Keynes quoted by Hodgkin (1958).]

The velocity of the transmission of the action potential may be measured by the time of arrival versus distance. Intensity of a stimulus does not affect the size of the action potentials, only the frequency of their creation. Because of the recovery time required, action potentials within the same nerve fiber are limited to a maximum of about one every 3–4 msec.

The work of many skilled investigators went into the understanding of the action potential. The essence of the explanation is the following. The ionic composition inside all cells, including neurons, is different from the extracellular composition. A metabolic sodium pump within the nerve membrane maintains a deficiency of sodium ions and an excess of potassium ions in the nerve interior relative to their extracellular concentrations. When the nerve is stimulated, pores or channels in the cell membrane open and allow external sodium ions to enter, making the interior of the membrane locally positive. After a brief delay, a separate set of channels opens, allowing potassium to leave the cell, causing the interior potential to repolarize. The channels are open for only a short span of time and then close. The sodium pump restores the imbalance of ions produced by the action potential to that of the former resting potential. The change from negative to positive potential of a stimulated region of the nerve acts as a trigger potential to cause an action potential in the portion adjacent to the stimulated portion, and so on, thereby causing the action potential to travel along the nerve.

The calculation of velocity of this nerve impulse by Hodgkin and Huxley came within 10% of the experimental value for squid axons. This velocity increases as the square root of the diameter of the fiber. It is generally recognized, however, that, for survival of an animal, some nerve velocities have to be greater than this. The role of myelination of fibers was then studied. Certain nerve fibers have a myelin sheath around them except for exposed parts of the nerve about 1 mm apart, called *nodes of Ranvier*. It has been shown that the impulse jumps between these nodes. Such jumping permits faster travel of the action potential and has been named *saltatory conduction*. (See Vol. I, p. 79; Vol. II, p. 4.)

Parallel with the research on the electrochemical mechanism of nerve conduction were studies on what happened when this action potential reached an axon terminal. The most extensively investigated terminal region is the *neuromuscular junction* formed by nerve axons ending on muscle fibers. In the middle of the 19th century, Claude Bernard experimented with the South American arrow poison curare. He found that if it was applied to the neuromuscular junction, the muscle stopped contracting in response to nerve stimulation, yet the muscle could still be made to contract by directly stimulating it and the nerve fiber leading to the junction would still conduct normally. From these studies it was concluded that transmission from the nerve to the muscle was somehow blocked. This was early suggestive evidence that transmission of a stimulus from nerve to muscle (neuromuscular transmission) could be chemical in nature rather than electrical. Other experiments with other drugs followed, and there was a general recognition that a chemical was the mediator of neuromuscular transmission.

Loewi (1921) performed a classic experiment with two living excised frogs' hearts in separate aqueous solutions. He found that if he stimulated the vagal nerve to one of the frog's hearts, the heart rate slowed. He then extracted some fluid from it and injected it into the second frog's heart. The rate of the second heart slowed too. From this and similar experiments he clearly showed that transmission is chemical rather than electrical. Subsequent analysis showed that the responsible chemical in this experiment is acetylcholine (ACh).

Our present understanding of transmission at the neuromuscular junction is the following. Upon arrival of the action potential at the end of the nerve, the presynaptic nerve terminal, acetylcholine is released in little packets, *quanta*, containing approximately 10^4 molecules. Through normal passive diffusion the molecules pass across the gap to the postsynaptic terminal, which in a neuromuscular junction is called the *end plate*. At the postsynaptic membrane the neurotransmitters bind noncovalently to a special class of membrane proteins called receptors. This binding opens pores, thereby permitting the influx of sodium ions and the efflux of potassium ions, and possibly other ions, across the membrane as in the nerve itself. The impulse produced at the end plate can generate an action potential, which then spreads along the muscle fiber and causes it to contract (Vol. I, Chapter 4). From nearby sites in the end plate an enzyme is released, which destroys the effectiveness of ACh by hydrolyzing ACh into choline and acetic acid. The enzyme responsible for this is called acetylcholinesterase. The structural constellation associated with the transmission of a nerve impulse between two nerves or between nerve and muscle is called a *synapse*.

INTRODUCTION

An important area of synaptic study for the past 50 years has been to classify chemicals at synapses that are capable of effecting communication between cells. In the 1920s there was still a lively controversy between investigators about whether synaptic communication was chemical or electrical. The ideas of chemical synapse dominated research (and still do), although research of Fatt and Katz (1951) and others suggested that there are also electrical synapses. It has since been shown that both types of neuronal communication are common, so the early argument has ended in a draw. There is considerably less research on the electrical synaptic properties, however, possibly because of its late start. There are two distinctive features of electrical synapses worth noting. First, the gap of a synaptic junction is about an order of magnitude smaller in electrical synapses, 20 Å versus 200 Å for chemical ones. Second, there appears to be an ordered array of hexagonal structures spanning the gap made up of protein molecules called, for want of a better name, *connexin*. These conducting protein molecules create the electrical junction. There are obvious advantages and disadvantages to electrical synapses. The main disadvantage is that a graded response does not appear to be possible, i.e., a signal is either on or off. The main advantages are speed and temperature independence. In a chemical synapse there is a time lag, since the chemical transmitter must be released and diffuse across the synaptic gap. No such time lag exists in an electrical synapse. The temperature dependence of either transmitter release or diffusion can be disadvantageous for it is observed that cold-blooded animals such as fish have large numbers of electrical synapses. In cold-blooded animals this may be an adaptation to efficient operation. The word adaptation is used because there has been no observation of electrical synapses in protozoa, so electrical synapses are apparently not phylogenetically primitive. In mammals, however, electrical synapses are rapidly being found to be almost ubiquitous in the spinal column, hippocampus, hypothalamus, olfactory cortex, and retina. Clearly, much is still to be learned of these synapses.

Synaptic potentials, whether produced chemically or electrically, differ from action potentials; they are smaller in amplitude and have a longer time course. They may add, that is, the arrival of several small excitatory potentials may depolarize the postsynaptic nerve terminal until an action potential is created by the triggering of the sum of their effects. Some synapses make the postsynaptic nerve terminal more negative, *hyperpolarized*. These are called *inhibitory* synapses. It is more difficult to initiate an action potential in a hyperpolarized nerve terminal, for it is the sum of arriving potentials that determines whether or not an action potential is initiated.

As stated above, a synaptic signal may decrease the polarization of the postsynaptic membrane and thereby assist in the development of an action potential. This effect is called an *excitatory postsynaptic potential* (EPSP).

Other synaptic signals, however, may hyperpolarize (make more negative) the postsynaptic membrane and thereby make it more resistant to excitation. A postsynaptic response of this sort is called an *inhibitory postsynaptic potential* (IPSP). There is almost constant activity of a significant fraction of the neurons and their synaptic signals in the conscious brain, and ACh is not the only chemical transmitter involved. Statistical constructive interference of the sums of these activities gives rise to large potential waves, which are detectable not only on the brain itself but even on the scalp. These sums of spontaneous potential waves can be detected by electrodes on the scalp by a technique called *electroencephalography*. Another type of potential wave can be extracted from the total signal when an external stimulus is given, such as a sudden sound or light flash. These potential waves are called *evoked potentials* and are revealing much information on the neuronal processing of stimuli. Both these techniques are discussed in Chapter 3.

As will be shown in this book, there are two types of electrical behavior of the brain under intensive study, the microscopic and the macroscopic. Each must be understood individually before we can understand the collective behavior of the brain. Years of research are required to achieve even a small amount of reliable information, and the reader not in the field will have many questions. These questions have occurred to investigators, but at present they are simply unanswerable. The subsequent chapters present some of the techniques employed and some of the information determined. We wish we could present all the answers, but we cannot. If we could, however, scientific interest would fade. The late George Gamow once said, "I dread the day when science leaves the era of Columbus and Magellan and enters into the era of the National Geographic." The reader will see that this day is a long way off.

REFERENCES

Fatt, P., and Katz, B. (1951). An analysis of the end-plate potential recorded with an intracellular electrode. *J. Physiol. (London)* **115**, 320.

Hodgkin, A. L. (1958). Ionic movements and electrical activity in giant nerve fibers. *Proc. R. Soc. London, Ser. B* **148**, 1.

Hubbard, J. L., Llinás, R., and Quastel, D. M. J. (1969). "Electrophysiological Analysis of Synaptic Transmission." Arnold, London.

Loewi, O. (1921). Über humerale übertragbarkeit der Herznervenwirkung. *Pflügers. Arch. Ges. Physiol.* **189**, 239.

CHAPTER 2

The Evolution and Morphology of the Brain

INTRODUCTION

This chapter will deal with the brain and its interconnections with the spinal cord, which controls, among many things, the motor neurons (motoneurons) of the muscles. It will be in the nature of a brief orientation for the reader unfamiliar with the brain.

Almost everyone has seen pictures of the human brain; it appears to be very complicated, and indeed it is. The purpose of this volume, however, is not to treat the details, but rather the basic principles. The descriptions of the pathways of the brain will be kept simple and elementary so that the reader will understand the principles rather than all the details.

The brain is the sole organ concerned with thought, memory, and consciousness even though it is a relatively small structure, weighing no more than 1.5 kilograms. Figure 2.1 shows a view of the right cerebral hemisphere of a human brain. A prominent feature of the surface of the human brain is its convolutions. The crest of a single convolution is referred to as a *gyrus* and regions separating the various gyri are called *sulci*. The network of gyri and sulci has been found to be the same for all humans. This observation has

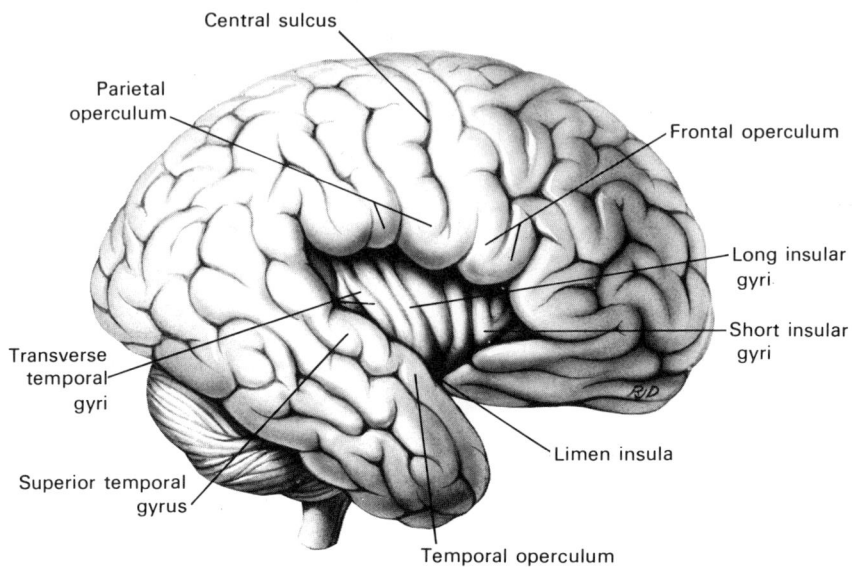

FIG. 2.1 View of right cerebral hemisphere of a human brain. [From Carpenter (1972) Copyright © 1972, the Williams & Wilkins Co., Baltimore.]

permitted the brain to be divided into six domains or lobes: (1) *frontal*, (2) *temporal*, (3) *parietal*, (4) *occipital*, (5) *insular*, and (6) *limbic*. A few of these are indicated in Figure 2.1. Additional names have been assigned to each convolution and sulcus in order that discussion among scientists of the various areas of the brain can take place. However, these finer structural details and terminology need not concern us.

In Vol. II, Fig. 4.4, general areas associated with the central location of the senses were indicated. In the human brain, which not only reacts to sensations but can think about them, even to the extent of recalling a fine meal tasted or a beautiful picture seen years before, it is evident that other parts must be connected in some way to these primary sense areas.

In the 1860s Broca showed that one part of the left hemisphere is associated with speech. Since then neurologists have mapped in great detail many of the neural pathways from the body to the brain and the reverse. Every detailed part of this system has a name or a number, and delving into neurological textbooks can be an unnerving experience.

To facilitate familiarization with the principal parts, we shall take a different approach and examine the brain in its most elementary forms as found in evolutionary stages. We shall also consider embryonic development, which to some extent repeats evolutionary stages. In this way the reader will follow the development from the simplest organism of the Cambrian period, about 500–600 million years ago, to the brain of *Homo sapiens*. A few zoological terms will, of necessity, be used, but these will be defined. Speculations of the evolutionary force for a dominant hemisphere and some facts of left-handedness will be considered, and finally some of the detailed arrangement of the neurons in the brain will be examined. In this way the reader will be prepared for the succeeding chapters on electrical phenomena of the brain and the action and chemistry of the synapses.

ORIGIN OF THE TWO-SIDED BRAIN

Fossils of bones in ancient rocks have been found of the most primitive forms of life but, unfortunately, there is little trace of the soft tissue (see, however, Haugh and Bell, 1980). Lacking that, examination of lower organisms that exist today must be relied on. One prominent group is the flatworm. *Acoela*, an order of small marine worms, shows stages of primitive transition. Three examples are shown in Fig. 2.2: (a) a diffuse network under the skin and a *statocyst* (a fluid-filled tube which determines equilibrium position as does the vertebrate vestibular organ); (b) a nerve net throughout the body with a primitive two-lobe brain; (c) an increase in brain size with longitudinal nerve strands.

It should be noted that in even the most primitive life forms the brain has two sides, one of which may be dominant in the sense that the limbs controlled by one side may be stronger and quicker than those of the other side. However, no evidence of asymmetry of function of the two sides has been observed. The role of two sides is obvious. Figure 2.3 illustrates schematically constant concentration contours of a chemical diffusing from a point source such as a food particle in water. If an organism has a single receptor with an associated neuron, as in that on the left-hand side, it could swim up the

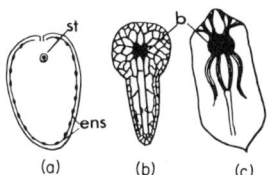

FIG. 2.2 Stages in evolution of brain and primary nerves among flatworms: (a) epidermal nervous system (ens) and statocyst (st); (b) bilobed brain (b) around statocryst and nerve net; (c) increased cephalization, loss of nerve net, and development of longitudinal nerves. [From Hodgson (1977).]

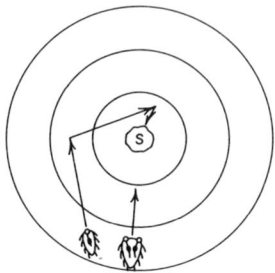

FIG. 2.3 Diagram illustrating evolutionary advantage of dual sensors. S: source diffusing taste-activating molecules with circles as contours of constant concentration. A primitive organism on the left-hand side with one sensor can swim up the gradient but course correction is broad. On the right-hand side a dual sensor organism can make constant small course corrections.

gradient but would have to make large course corrections. The organism on the right-hand side has two receptors, each coupled to a separate neuron. It can make constant small course corrections and thereby decrease the energy expended in reaching the source. In fact, the propelling cilia on such organisms move in concert, but each side is independent, much like rowers in a boat. The sensor on each side triggers a motoneuron on its side. Thus, Nature promptly rejected, if it ever existed, a single-sensor–single-neuron configuration. Since the two-sensor–two-neuron arrangement led to dual control of the motoneurons, further evolution of the two-sided brain was established.

ORIGIN OF VERTEBRATES

The fossil evidence of the origin of vertebrates is slim. Reliance must be placed on existing species and Haeckel's (1892) "biogenetic law." This so-called law proclaimed that individual development (*ontogeny*) repeats the history of the evolution (*phylogeny*). This law came about from the detailed study of embryos. In the earliest stages a mammalian embryo resembles a fish, complete with gills as in fish embryos. Romer (1967) pointed out that this law is only half true. New forms of the adult stages caused by environmental evolution are not piled on the embryos; they do not go through all stages. However, the early stages in the development of the brains of mammals appear to follow, within limits, those of lower orders of vertebrates. Careful anatomical studies of both existing species of lower orders and their larval stages, as well as embryonic stages of higher orders, are used to surmise the evolutionary history of vertebrates.

ORIGIN OF VERTEBRATES

The phylum *Chordata*, organisms with a central cord, is subdivided into four subphyla. In ascending order these are *Hemichordata, Urochordata, Cephalochordata,* and *Vertebrata*. The first three are referred to as protochordates or lower chordates. At some stage in their life history all forms possess a neural cord in a neural tube with a supporting rod known as the notochord (cord along the back). How did the chordates begin? The development of the larvae of hemichordates resembles the development of echinoderms (e.g., sea urchins). The phosphagen (phosphate) in the muscle of both protochordates and echinoderms includes a compound of creatine, whereas the phosphagen of all other invertebrates contains a compound of arginine (an amino acid). This suggests a common ancestry of the protochordates and echinodates.

Detailed studies of the larvae of the urochordate (sea squirt) have led to the following model. The ancestors of vertebrates resembled these larvae and lived on the surface of Cambrian seas. The sunlight caused a series of light-sensitive cells to develop along their backs, called a *neurectoderm*. Such a strip of nervous tissue is present even today in urochordate larvae. During the course of embryogeny this strip rolls up and sinks inward, with the bony notochord offering protection. This series of stages in urochordate larvae is shown in cross section in Fig. 2.4.

The development of the notochord and neural tube in vertebrates is currently being studied in considerable detail. For example, a sequence of critical moments in the development of the neural tube in chick embryos has been captured by Tosney (1982) with a scanning electron microscope. Figure 2.5a–c

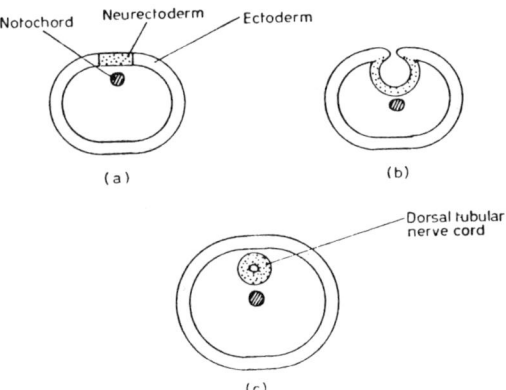

FIG. 2.4 Cross section of the formation of a dorsal nerve cord in a primitive chordate: (a) early embryonic stage; (b) the neurcetoderm rolls up and sinks inward; (c) a dorsal tubular structure is formed. [Reprinted by permission of Faber and Faber Ltd. from "The Brain, Towards an Understanding" by C. U. M. Smith (1970).]

shows three examples from the sequence. The formation process begins with a single plane of cells. At a given chemical signal the cells begin to migrate and then roll up to form the neural tube. As the two sides of the tube (Fig. 2.5a) come together (Fig. 2.5b), ectoderm cells move to the midline and fuse to cover the neural tube (Fig. 2.5c). Following this, they signal the growth of bony tissue. A brief survey, with key references, of current research in this area has been given by Thomson (1982).

Where and why did a change arise from the fixed position of the protochordates to the moving chordates? The embryo or larva of a fixed organism, such as the sea squirt, which clings to a rock, must find a safe place on the sea bottom to feed and to develop. The cilia of larvae of the lower orders permit some motion, but not much. A better development is the tadpole form with a muscular tail, nerves to control locomotion, and a notochord to stiffen the tail. With this development of locomotion a burst of evolution must have taken place with the notochord segmenting in a variety of ways to form vertebrae.

The lower organisms have a pharynx, a gut tube, gonads, and a simple nervous system. They are essentially feeding devices and little else. These are called *visceral* organisms. As locomotion developed, muscular control and more highly developed sense organs were added. The neurons associated with the senses moved forward, closer to the receptors. These later additions, muscles and improved sense receptors and transmitters, are in the outer tube of the body (i.e., outside the gut tube) and are called *somatic* structures. They have a separate nervous control system. As vertebrates evolved, they became a visceral–somatic organism with a welding into a single structure. However, the welding did not involve complete fusion, and the vertebrate can still be defined in terms of its visceral and somatic parts and of the associated control centers.

DEVELOPMENT OF THE BRAIN

The manner in which the brain develops may be observed by comparing embryos and by studying the anatomy of various life forms from primitive to primate. An understanding of both the evolution and embryogeny of the human brain simplifies recognition of its architecture. The human brain is quite complicated, and not all parts will be identified, but some terminology will be essential for the discussion.

The sensory receptors, other than those for cutaneous and balance sensations, are generally in the front (anterior) end of an organism, no matter how primitive. Furthermore, the sensory nerve cells related to these develop or migrate near the primary sensors. Thus, neural complexity began to

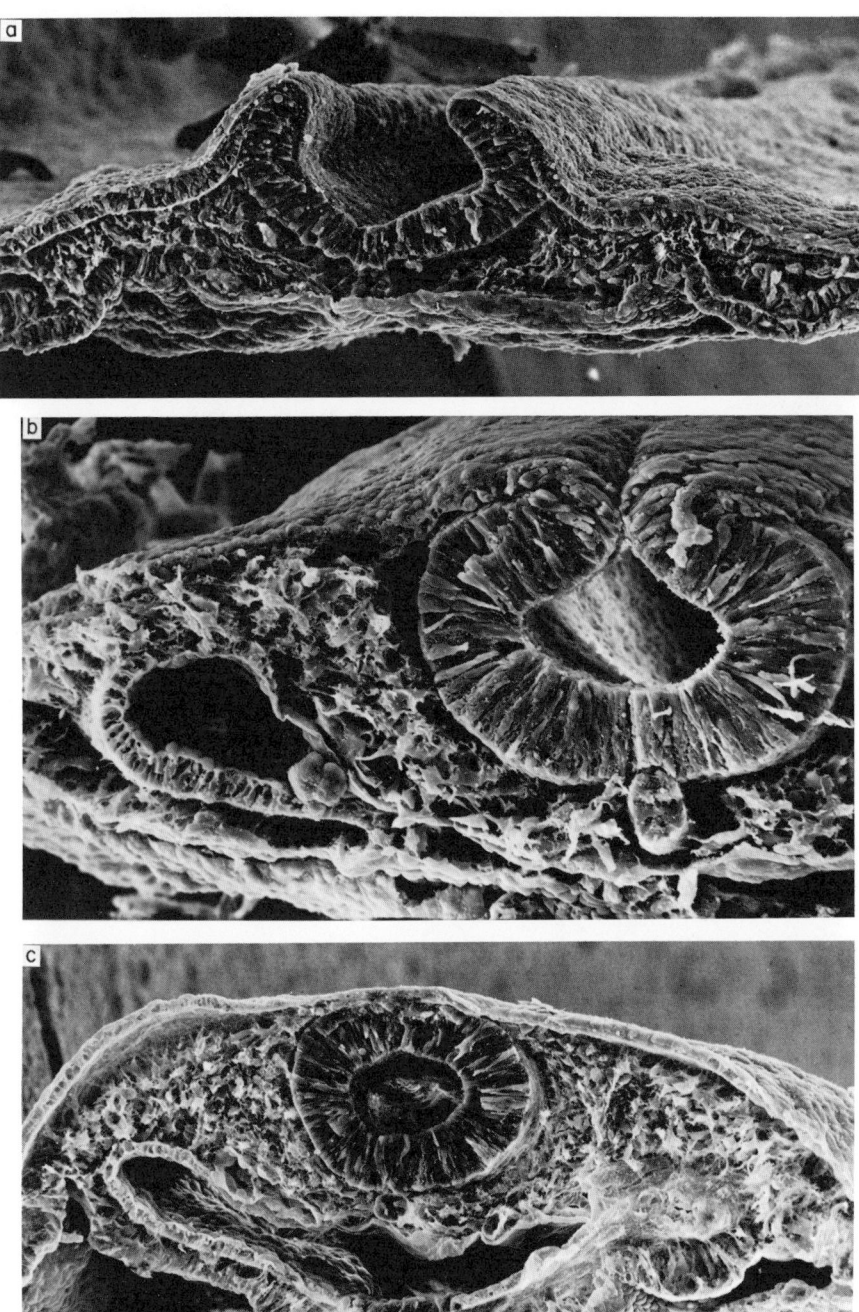

FIG. 2.5 Critical moments in the formation of the neural tube in a chick embryo: (a) early migration of cells (Tosney, 1982), (b) contact between two sides of (a) and (c) growth of epithelial cells that close the tube. [(b) and (c) courtesy of K. Tosney.]

expand the forward part of the nerve cord. Evidence from embryology and comparative anatomy indicates that the primitive brain of chordates developed not with one anterior expansion, but with three. These expansions, or in primitive forms merely bumps in the nerve cord, are called *prosencephalon* (forebrain), *mesencephalon* (midbrain), and *rhombencephalon* (hindbrain). These primitive sections of the brain processed sensory information and relayed it to motoneurons, which are large neurons in the brain stem and spinal cord with axons innervating muscle fibers. The spinal-cord material has both *white matter* and *gray matter*. Clusters of myelinated axons appear white, while clusters of cell bodies of information-processing neurons are gray.

The spinal cord is generally oval in the lower vertebrates and has a fluid-filled canal. However, as vertebrates developed this canal became small because large amounts of nervous tissue (gray matter) grew around it. In most higher vertebrates the gray matter has a bilateral symmetry, in an H shape like butterfly wings. Figure 2.6 shows this schematically in a left-side view of a cross section, with ant. and post. indicating anterior (front) and posterior (rear). Often, the anterior part is called *ventral* (toward the viscera) and the posterior part is called *dorsal* (toward the back). The ventral column is the seat of the cell bodies of the motor neurons of the spinal nerves, and their axons pass out of the spinal cord through ventral nerve roots. The number of these neurons will vary with the volume of musculature at that level of the body, with expected enlargements in regions of the limbs of land

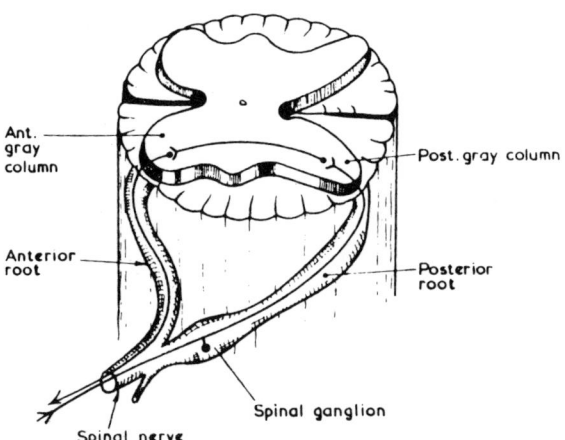

FIG. 2.6 Schematic of the cross section through the human spinal cord illustrating the butterfly-wing shape of the gray matter. Front of the body is to the left. *Anterior* is equivalent to *ventral* and *posterior* to *dorsal*. [From Starling and Evans (1962).]

DEVELOPMENT OF THE BRAIN

vertebrates. Most of the motor supply is to somatic muscles, but visceral axons are also present. The dorsal column of gray matter is associated with neurons through which impulses from the sense organs are relayed and distributed. These neurons have axons that ascend to the brain and descend to motoneurons, both somatic and visceral. The white matter of the cord, as stated earlier, is composed largely of myelinated fibers, both to and from motor neurons.

Returning now to brain development, the forebrain or prosencephalon turns downward as it increases in size, and becomes quite distinguishable because of a fold in the lower part, called a *ventral* (lower or toward the belly) *sulcus* (fold). (See Fig. 2.7a.) In the embryo the prosencephalon develops rapidly, and an upper indentation appears, called an *isthmus* because it narrows the growing brain at that position. The three bumps that are the forebrain, midbrain, and hindbrain are seen in Fig. 2.7b. Further development is shown in Fig. 2.7c. The hindbrain is now distinguished by two principal parts, the *myencephalon* (marrow brain or spinal cord) and the *metencephalon* (between brain). The midbrain remains as a single identifiable form, but the forebrain is subdivided into two parts, the *telencephalon* (anterior, paired part) and *diencephalon* (posterior part).

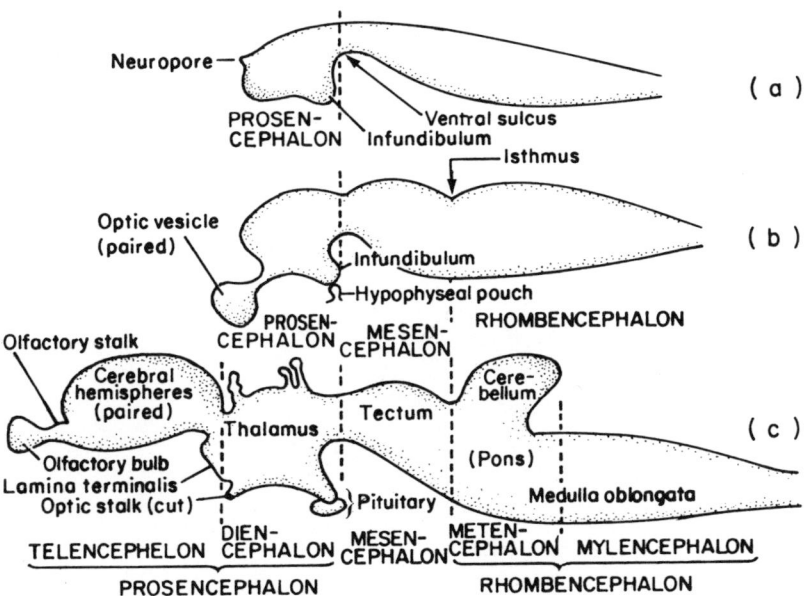

FIG. 2.7 The development of the principal brain divisions in vertebrates. [From Romer (1967).]

The above segmentations of the brain are not just landmarks. In the early stages of phylogeny three major sense organs appeared: eye, nose, and ear and lateral line (balance). In primitive vertebrates each of the three brain subdivisions is associated with one of these sense organs, and for each there develops a dorsal outgrowth of gray matter, i.e., sensory neurons. The three dorsal outgrowths (Fig. 2.7b) eventually become the *cerebrum* (brain), *tectum* (midbrain roof), and *cerebellum* (small brain, diminutive of cerebrum).

The original cavity in the spinal cord, which contains fluid, persists in the adult brain, but becomes extended, folded, and convoluted into *ventricles* (cavities), which are filled with fluid. A cavity, or *lateral ventricle*, is present in each of the cerebral hemispheres. These cavities connect with a middle third ventricle in the diencephalon. Within the upper part of the spinal cord, the *medulla oblongata*, is a fourth ventricle, which connects with the canal in the spinal cord. The spinal canal connects with these four ventricles, and all are filled with cerebrospinal fluid. In most of the brain the ventricles have thick walls of nervous tissue. There are two roof regions where the wall is thin, one at the junction of the hemispheres (telencephalon with the diencephalon) and the other in the roof of the fourth ventricle. In each of these areas is a highly folded region called *choroid plexus* through which there is an exchange of materials between the blood and the cerebrospinal fluid.

Note that the results of studies on embryos and primitive existing life forms have general agreement with available data from evolutionary studies. Some fossil remains are sufficiently well preserved that casts may be taken from their skulls to determine brain formation. One example of a *heterostracid*, a fish with bony head armor from 400 million years ago, has been laboriously reconstructed by Stensiö (1963). A diagram of the result is shown in Fig. 2.8. All the parts discussed above are seen to be present.

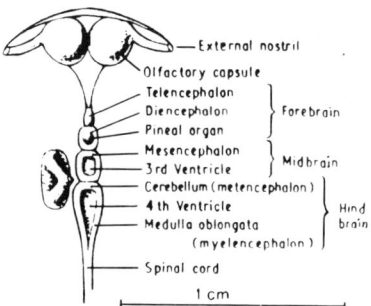

FIG. 2.8 A restoration of the central nervous system of the fossil *Heterostraci*. [From Halstead Tarlo (1965).]

ARCHITECTURE OF THE HUMAN BRAIN

Every detail and every layer of the brain has a name or numerical designation. We shall avoid these and give a general description of the architecture of the brain. Only the basics will be given as a guide to other topics in this volume and as a brief introduction for the reader who wishes to consult other references.

Lower vertebrates do not have the rolls of tissue (Fig. 2.1) in their cerebrum. Their skulls are large enough to contain their brains. The evolution of man and some other higher primates was so rapid, less than 30 million years, that the skull did not evolve rapidly enough to contain the enlarging brain. The neurons involved in evaluation of stimuli, thinking, and memory are gray matter that formed layers to create the *cerebral cortex*. The brain had to develop folds to increase the surface area. We shall deal with the sizes of brains in different species in a later section.

The general names used by physiologists for the regions of the human brain are given by the area of the skull under which they lie. These are the *frontal*, the *occipital* (rear), the *temporal* (the region of the temples), and the *parietal* (the two bones that form the roof and sides of the skull).

When the human skull is opened there are some obvious landmarks (Fig. 2.1). In Fig. 2.9 the lobes are labeled. Certain long folds give rise to surface sulci (folds or cavities). Deeper folds are called fissures. There is a pronounced fissure separating the temporal lobe from the frontal and parietal lobes. This is called the *fissure of Silvius*, or *Silvian fissure*. An extension of this is called the *lateral sulcus*. In the vertical direction the most prominent feature is the *fissure of Rolando*, generally called the *central sulcus*. On either side of this are two sulci named *precentral* and *postcentral* sulci. Rolls of brain tissue, called *gyri* (singular: gyrus) adjacent to these sulci are given names corresponding to their location.

Figure 2.9b shows the right half of a human brain sectioned medially. The dorsal part of the hindbrain, or rhombencephalon, becomes the *cerebellum*. The lower part swells and becomes the *pons*. These two are the main features of the metencephalon (Fig. 2.7), which is distinguished from the lower part of the *medulla oblongata*, the myencephalon. There is considerable knowledge of the function of these lower parts of the brain obtained by sectioning in live animals (see, for example, Guyton, 1971).

Neurophysiologists have inserted fine microelectrodes into neural cells in different regions of the brain and, by stimulating them electrically, have inferred a more detailed map of brain regions. The *reticular formation* (Fig. 2.9b) appears to be responsible for the wakefulness of the brain. Electrical stimulation of a cell in this system of a sleeping animal brings on immediate wakefulness. Injection of appropriate drugs or sectioning in this region can

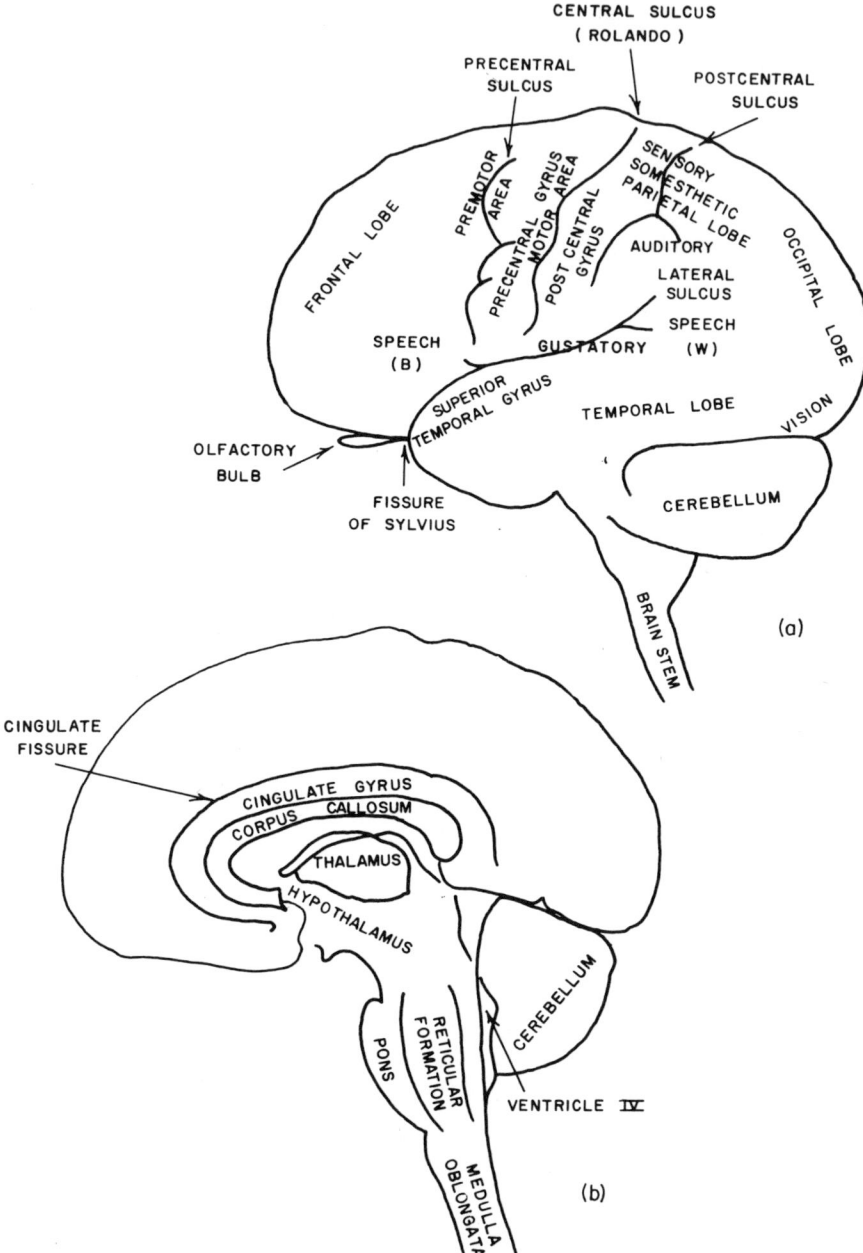

FIG. 2.9 Diagram of human brain: (a) surface of left hemisphere; (b) cross section of right hemisphere from longitudinal cut through center of brain.

put the animal to sleep. The *pons* contains fiber tracks that connect the medulla oblongata and cerebellum with upper portions of the brain. The cerebellum is the master control of motor skills in that it coordinates information from the muscles, cortex, etc., to make a well-executed intentional movement. For example, the cortex signals the desire to pick up a book and, with no further thought, the hand unerringly reaches out without overshoot and properly performs this function. These are learned mechanical skills, and even more complex ones, such as playing the piano, which can proceed without conscious thought, are apparently controlled by the cerebellum. Holmes (1939) studied the effects of brain wounds on soldiers of World War I. (Recall in Vol. II, Fig. 6.62, that he mapped out the striate cortex from patients.) He devised simple maneuvers of moving the hand in a zig-zag fashion between a vertical array of lights and photographed a blinking light on the finger of a patient. He found that the cerebellum controls the ease of such movement on the *same side* as the hand. That is, if one side of the cerebellum was damaged, the patient could not perform the exercise easily with the hand on the side of the injury. Patients reported that they had to stop and think at each light to plan the direction of muscle movement toward the next light.

The portions of the nervous system that control the visceral functions of the body are called the *autonomic nervous system*. Examples of visceral functions are arterial pressure, body temperature, and digestion. The *hypothalamus* is the major area in the brain that controls this system through the secretion of hormones. The hypothalamus has been called "the master organ of the hormonal processes." These functions are often called the vegetative functions of the body.

As the brain develops in both the embryo and the phylogenetic scale, the increase in size seems to be programmed by the *thalamus*, which came early in evolution. For each area of the cerebral cortex, there is a corresponding connecting area of the thalamus. Activation of any part of the thalamus causes an activation of a much larger portion of the cortex. Destruction of a portion of the cortex in the *somesthetic region* (the region of evaluation of muscular activity, touch, and other sensory properties) does not destroy a person's touch sensation, but it does destroy his ability to reason the shape of the object. From this and similar examples it is concluded that the cortex deals with the abstract, whereas the thalamus maintains all of the motor and sensory evaluations necessary for survival. This leads to the concept of the triune brain, which will be considered in the next section. However, before proceeding, it will be advantageous to refer to the map of the cortex.

The cortex map (Fig. 2.9a) was originally developed by Penfield and his associates (1950). They found that touching various areas of the exposed

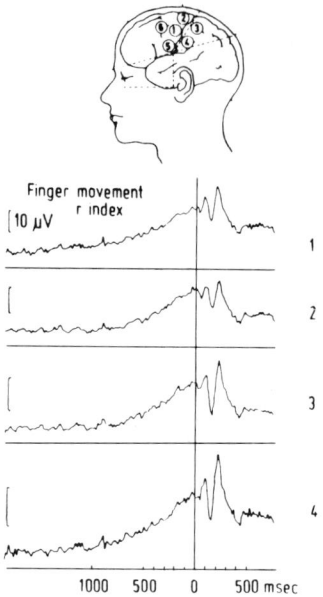

FIG. 2.10 Readiness potentials recorded at indicated sites on scalp in response to voluntary movement at will of right finger. Zero time is onset of movement. [From Deecke et al. (1969).]

cortex during brain surgery under local anesthesia produced either physical or mental reactions. It should be noted, however, that such areas are neither specific nor precise. That is, the indicated areas are those of generalized response, but the type of response is diffuse and cannot be located precisely. For example, the premotor and the motor areas have a relation: the former gives rise to the will or desire to cause movement while the latter receives the impulse to signal the movement. This has been demonstrated by Deecke et al. (1969), who placed electrodes on the scalp of a subject who would move an index finger at irregular times. The actual movement would be recorded most strongly by the electrode over the motor area for that finger. However, a diffuse "readiness potential" was observed throughout the motor and premotor area for up to 2 sec before the actual movement (Fig. 2.10).

THE TRIUNE BRAIN

Neurophysiologists have worked for a century in attempting to understand the neural connections in the body and brain. Most of these have been mapped out either by stimulating a cell electrically and observing the response or by severing a nerve and observing the subsequent path of its degeneration.

THE TRIUNE BRAIN

What has emerged as the most difficult and least understood portion is that of the *cingulate gyrus* (Fig. 2.9b). Within it, the *corpus callosum* has been determined to be the internetwork between the left and right hemispheres; however, below the corpus callosum are the thalamus, hypothalamus, and other structures.

Man has apparently inherited the structure and organization of three basic types of brain named by McLean (1973) *reptilian, old mammalian*, and *new mammalian* (Fig. 2.11). These three basic brains are quite different in both structure and chemistry (they absorb different stains), but all intermesh and function together in what McLean calls the *triune brain*. The collective response of the three together is greater than that of the sum of the parts.

The reptilian brain consists in general of the reticular formation, the midbrain, and the basal ganglia, which constitute its forebrain.

Observation of reptilian behavior shows a dominance of repetitive behavior, e.g., the return of sea turtles to the same place to lay eggs. The basic behavior of reptiles includes genetically constituted forms of behavior such as selecting homesites, establishing territory, display in mating selection, hunting, and breeding. The reptilian brain seems to be controlled by established precedent, and it has been suggested that this part of man's brain may be the cause of his desire for established rituals such as ceremonies and religions. Such repetitiveness does have survival value for reptiles. Birds are closest on the evolutionary tree to reptiles. Although no experiments on reptiles have been performed, the imprinting of newborn birds on a dummy instead of the mother can precipitate a sequential acting out of genetically constituted forms of behavior.

In the evolutionary development of lower mammals a cortex appears and becomes a "thinking cap" to the reptillian brain. This cortex is in all mammals and is called the *limbic* (border) lobe. The *neocortex* of neomammalians

FIG. 2.11 The three basic brain types that form the triune brain of human beings: reptilian, paleomammalian (old mammalian), and neomammalian (new mammalian). Each type has distinctive structural and chemical features and is integrated through evolution into the total brain. [From McLean (1973).]

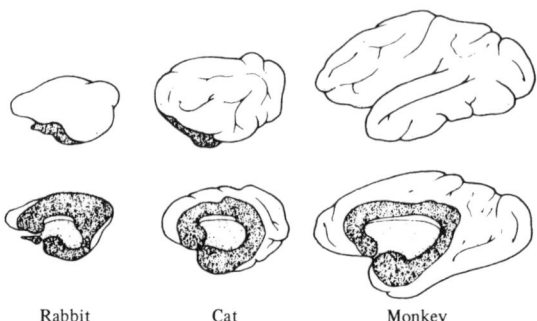

FIG. 2.12 Schematic views of side and cross sections (as in Fig. 2.9) of the brains of three mammals. Dark areas represent the limbic system. [From McLean (1973).]

varies with species, but the limbic system is common to all. Three examples are shown by the shaded areas in Fig. 2.12. The upper figures show the neocortex, which essentially encloses the limbic system, while the lower figures are the cross-sectional views.

The lower part of the limbic system seems to be related to emotional feelings and behavior that ensure self-preservation. Observations of patients with limbic epilepsy (where the center of the discharges lies within the limbic system) have shown that at the beginning of an epileptic discharge they experience elemental feelings of hunger, thirst, suffocation, and nausea as well as intense emotional feelings of terror, fear, anger, rage, etc. During electrical stimulation of the forward (anterior) region of the lower limbic system in dogs and cats, similar forms of behavior are elicited. When this part of the brain is excised in wild monkeys, they lose their sense of fear, eat all manner of objects (nuts and bolts, etc.), develop bizarre sexual behavior, and experience other changes that would be detrimental to survival in a natural environment. If the posterior part of the lower limbic system is electrically stimulated, the experimental animals develop enhanced grooming reactions and, in males, penile erection. This part undoubtedly is connected with courtship and reproduction. Note that of all the areas stimulated by neurosurgeons in the human cortex, erotic sensations were never elicited. The connection of the anterior and posterior parts of the lower limbic system suggests to some psychiatrists the cause of erotic stimulation experienced by some when wrestling or fighting with the opposite sex (McLean, 1973).

The upper part of the limbic system increases in relative size with increasing levels in the phylogenetic scale. The opossum, for example, has survived for 130 million years with little more than the reptilian and limbic brains, and it is largely guided by a fine olfactory sense. The upper part of the limbic system in higher vertebrates bypasses the olfactory system and has been shown to be dominated by vision. Thus, stimulation of vision, for example, with mirrors, in a higher-order animal, such as the monkey, brings on sexual display. McLean has suggested that this appealing to visual stimuli may be the reason primitive societies use phallic symbols as totems. Children also are generally uninhibited in genital display; in adulthood in civilized societies these feelings are suppressed, and the nonsexual neocortex is trained to dominate.

CEPHALIZATION OF THE SPECIES

Most of the controls of visceral and somatic mechanisms are located in the brain rather than in the spinal cord. It is therefore expected that the larger the animal, the larger the brain. The neocortex is associated with information processing, and one would expect a larger brain in the more intelligent animals. Both of these considerations have been examined to determine if there is a governing law.

Much is known about brain size versus body size for different species, including measurements from fossil fish, reptiles, and mammals. A lot of this has been summarized and analyzed by Jerison (1974, 1977). Representative data of brain weight versus body weight are plotted on a log–log scale in Fig. 2.13. Fossil data for reptiles and fish lie in the lower vertebrate grouping, and some early mammals, such as the ungulates (predecessors of the horse, camel, etc.), lie between the lower and higher vertebrates. If E is brain weight and P is body weight, it is seen that the average in each of the two groupings of Fig. 2.13 has a slope of $\frac{2}{3}$, that is,

$$E = kP^{2/3} \tag{2.1}$$

where k is a proportionality constant and is determined from the intercept on the ordinate.

Note from this figure that brain size does indeed increase with body size and, as Jerison points out, even the "small-brained" dinosaurs still fit within the lower group, albeit on the lower border of the dashed line. A vertical line through any body weight represents an increase in evolutionary scale from lower to upper brain size. Since a certain mass of neurons is required simply to control the body, the excess must be in the cortex and therefore in reasoning ability. If lines with slopes of $\frac{2}{3}$ are drawn through each point of interest, they will all be parallel but have different $\log k$ intercepts. Therefore, $\log k$ (or k itself) may be considered as a cephalization index, that is, the degree of brain development above basic body needs of the lowest animal of that weight.

Why should brain weight be proportional to the $\frac{2}{3}$ power of body weight? A comprehensive summary of data and hypotheses has been given by Gould (1971, 1975), the details of which are beyond the scope of our considerations. However, we shall briefly consider the concepts involved.

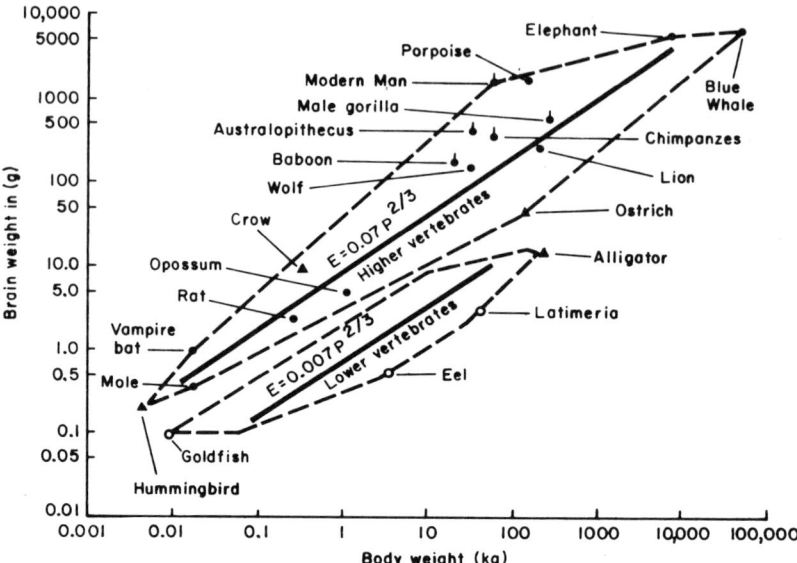

FIG. 2.13 Plot of brain weight versus body weight on log–log scale. The two solid straight lines have slopes of $\frac{2}{3}$ with respective intercepts 0.07 and 0.007 on the brain weight (E) ordinate. Fossil reptiles (not shown) lie within the dashed enclosure of lower vertebrates. [From Jerison (1974).]

First, we write the general form of Eq. (2.1) by setting the exponent $\frac{2}{3}$ equal to α. The logarithmic form of this equation is then

$$\log E = \alpha \log P + \log k \qquad (2.2)$$

Huxley (1932) proposed that the rate of growth of an organ should be determined by a growth factor and its size at any particular time. The growth rate would be rapid when small and decrease as the adult size was achieved. For a brain size E with a growth rate a at any time t, he suggested the differential equation

$$\frac{dE}{dt} = aE \qquad (2.3)$$

For body size P and growth rate b, a similar equation would be

$$\frac{dP}{dt} = bP \qquad (2.4)$$

Dividing Eq. (2.3) by Eq. (2.4) and rearranging yields

$$\frac{dE}{E} = \frac{a}{b}\frac{dP}{P}$$

Let $a/b = \alpha$ and integrate to obtain

$$\log E = \alpha \log P + \log k$$

This is the same as the empirical result of the data [Eq. (2.2)].

Consider now the implication of $\alpha = \frac{2}{3}$. For a spherical brain and body of similar density, Eq. (2.1) can be written, first in terms of volume V and then of length L,

$$V_E = kV_P^{2/3}$$
$$L_E^3 = k(L_P^3)^{2/3} = kL_P^2 \qquad (2.5)$$

Thus, the brain's volume increases as the surface area of the body. There are numerous hypotheses for this, and the debate continues.

One can gain insight into some of the debatable issues from an elementary consideration. It will be shown later in this chapter that the organization of cortical neurons in the brain is essentially vertical. That is, if one probes in a downward direction from most points, responses to and from the same part of the body are elicited, although the higher the layer, the more abstract the effect. Reference to Fig. 2.13 also indicates, at least for higher vertebrates, that there is a correlation between size and intelligence. Thus, as the brain grows radially, an increase in surface layers, and hence neural complexity, would occur. It is therefore not unreasonable that the brain need not grow in volume at the same rate as the body to achieve higher intelligence. What is not satisfactory with this explanation is the vertical increase in intelligence indicated in Fig. 2.13 with increasing k. Note that in Eq. (2.5) k has the dimension of length. It is conceivable that this implies that, for a constant body weight, intelligence increases with the radius of the brain.

There has been considerable criticism of the Huxley formulation, most notably that the time rates of growth Eqs. (2.3) and (2.4), if applicable at all, are certainly applicable only in the embryonic stage. For example, the size of the human brain relative to body weight is largest at birth. Furthermore, the relative sizes of each organ of all species change with time, each obeying a separate logistic, or S-shaped, curve. A careful review of the criticisms and alternative formulations is given in an appendix to the Dover reprint edition of Huxley's book (1972), and Gould (1971, 1975) has recently discussed this problem. However, the most disturbing new contribution to this problem is a recent examination of the data by Martin (1981). He shows that the $\frac{2}{3}$ rule of Fig. 2.13 is not generally valid when newer data are considered and, therefore, all the earlier arguments have been based on questionable data. A quotation from Kostitzin (1939) is appropriate here:

> It should be noted that nothing is as deceptive as the beautiful agreement between the observed and calculated values. A series of observed values is, in fact, equivalent to a narrow band rather than to a curve, and in this band can be traced a number of curves, corresponding very well with the conditions of the problem within the period of observations, but showing real divergence beyond these limits.

BRAIN ASYMMETRY

Somewhere along the evolutionary ladder the brain developed some asymmetrical (or *lateralized*) characteristics. An indication of this is the left-handed fraction of the population, and much early speculation was made of this, most of it wrong (Harris, 1980). We shall discuss some aspects of left-handedness in the next section. We wish first to consider even more dramatic aspects of brain asymmetry.

The first documented report of brain asymmetry was by Broca (1861), who did postmortem examinations of eight patients with *aphasia* (inability to articulate speech) and found that each had lesions in the left third frontal convolution of the brain. This portion of the brain is called Broca's area and is indicated in Fig. 2.9a as SPEECH (B). His findings initiated many subsequent investigations of the brain, and it was found that lesions in the same area of the right side of the brain did not cause aphasia, although there is an occasional person with Broca's area on the right side. One of these later studies by Wernicke in 1874 described cases of aphasia in which the temporal lobe of the left hemisphere was involved. This area is labeled in Fig. 2.9a as SPEECH (W). His patients, in contrast to Broca's, had difficulty in understanding the spoken word, and he called this region the *Wortschatz* (word store). Patients with lesions in this region had aphasia because they had difficulty in drawing on their store of words. In addition to these areas, others have been identified with speech because they involve lip, tongue, and jaw motion, as well as voice control, both volume and pitch, and feedback from the ear.

In the literature of hemispheric dominance of speech, frequent reference is made to the *amytal test*. In this test the patient is given an injection of sodium amytal in either *carotid artery*. (These are two arteries on both sides of the neck, which arise from the aorta and are the main conduits of blood to the brain.) The patient counts during and following the injection. If the injection is on the same side as the speech center, there is an aphasic arrest, i.e., the patient stops counting for a minute or more. If the speech center is on the opposite side, the arrest time is only a few seconds (Wada and Rasmussen, 1960).

The discovery of asymmetry by Broca and others began a century of intensive brain exploration and subsequent mapping. One of the early findings was that the motor cortex controls the opposite (contralateral) side of the body, e.g., the left hand is controlled by the right hemisphere. General mapping of the control areas is shown in Fig. 2.9a, and a more detailed map is shown in Vol. II, Fig. 4.4.

The *corpus callosum* (Fig. 2.9b) was long thought simply to hold the brain together. In some severely epileptic patients who suffered incessant seizures, however, it was severed cleanly by surgeons under the assumption that it was the path through which a seizure in one hemisphere transmitted its destabilizing impulses to the other hemisphere. The assumption proved to be correct. The result was even better than predicted, for there was diminution of seizures in both hemispheres. Apparently there is reciprocal excitation of the two hemispheres.

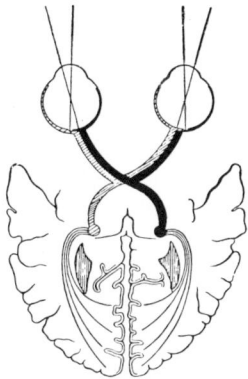

FIG. 2.14 The visual pathway showing partial decussation in the chiasma. Note that all impulses from one visual field pass to the same side of the brain. [From Starling and Evans (1962).]

These patients were studied for activity of the two sides of the brain, each independent of the other, by Sperry and his associates, and some of the results will be briefly summarized.[†] It is seen in Fig. 2.14 that part of the optic chiasma is severed in this operation so that everything from the right visual field of both eyes will be transmitted entirely to the left visual cortex and vice versa. Suppose that an array of different objects is on a table and that the subject's right eye is covered and his right hand is out of sight under the table. If the name of one of the objects is flashed on a screen, the left hand will reach

[†] R. W. Sperry received the Nobel Prize in 1981 for this work.

out and unerringly pick it up. This is because the left visual field projects onto the right visual cortex and the right hemisphere controls the left hand. The subject, however, cannot describe his action nor name the object. The left hemisphere, which contains the word storage and speech centers, has received no information from the right hemisphere, because the pathway of information exchange in the corpus callosum has been severed.

Experiments involving abstract concepts can be performed by the right hemisphere. For example, if the words "an object to light a fire" are flashed on the screen to the right hemisphere, the subject will retrieve a match with his left hand from an assortment of objects. However, when asked about his performance he cannot explain why his left hand is holding a match. Therefore, such a patient cannot be considered to have been conscious of his action. An early summary diagram by Sperry (1970) of the activities of the two hemispheres is shown in Fig. 2.15. (Stereognosis is the ability to recognize objects by touch.) For the majority of individuals, the left hemisphere is believed to process information in the phonetic, sequential, and analytic form. Language draws heavily on these and is therefore located in the left hemisphere. The right hemisphere has such skills as perception of form, spatial relations, abstract concepts, and some aspects of music.

The left hemisphere is considered by some to be the dominant hemisphere because it has an awareness of the conscious self. This somewhat unwarranted conclusion is based on the right hemisphere's inability to express itself in language; it therefore cannot disclose any experience of consciousness. The same might be said of animals, but the inability to communicate does not per se prove lack of awareness.

What is the situation with animals? Do they have a hemispheric asymmetry even though they cannot speak? We shall address the evolutionary basis for such asymmetry in the next section, but first we may consider some experiments on animals low on the evolutionary scale that communicate vocally, namely, birds. There are two important functions of song in birds. One is to attract mates, and the other is to define and defend territory. Note that these behaviors are basically reptilian in nature, as pointed out in the discussion of the triune brain. Therefore, the bird song has been under evolutionary control, perhaps for as long as birds have existed. Experiments on some birds, such as canaries, have shown that the vocalization is under the control of a region in their brains called the *hyperstriatum*. Nottebohm (1971) has shown that damage to the left hemisphere of the brain in this region strongly affects the ability to produce song, whereas damage to the right side does not. There is even morphological asymmetry in the form of a nucleus found in the left hyperstriatum but not in the right. This provides strong evidence that lateralization of the brain was an early evolutionary development.

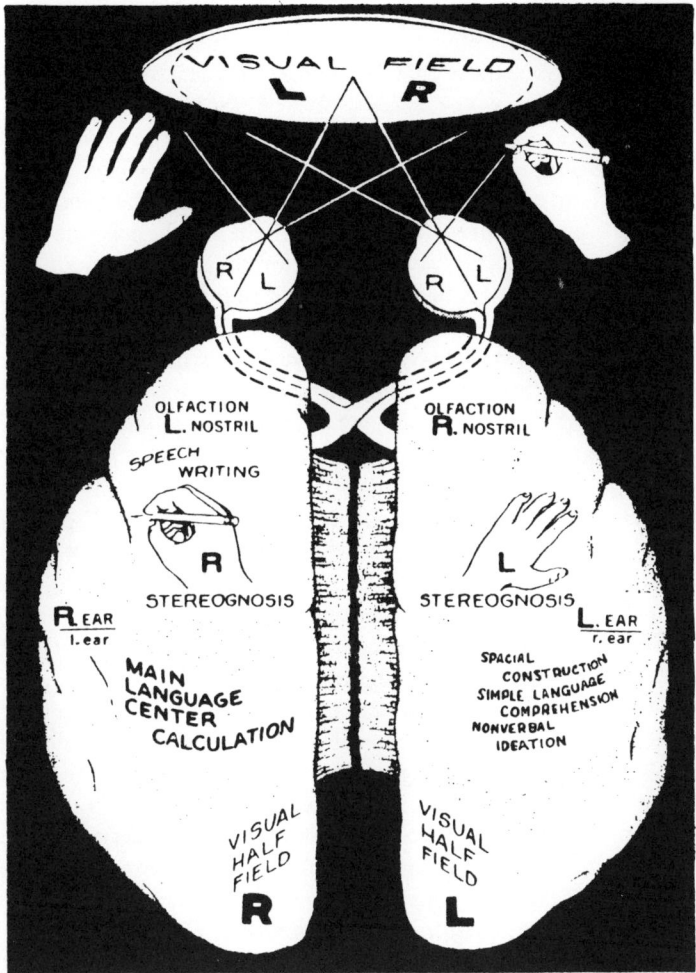

FIG. 2.15 The way the left and right visual fields project onto the right and left visual cortex. Also shown are the sensory inputs from the left limbs to the right hemisphere and vice versa. The dominance of each hemisphere in special abilities is also indicated. [From Sperry (1970).]

LEFT-HANDEDNESS

The earliest written record of left-handedness seems to be in the Bible (Judges 20:15, 16), which states that out of 26,000 soldiers, 700 were left-handed. This percentage, 2.7%, is lower than present-day estimates of 10–15%. It should be noted, however, that the verse in Judges says that of these 700, "every one could sling stones at an hair breadth, and not miss."

Therefore, this percentage compares favorably with modern data on the strongly left-handed percentage (see below). Careful analysis of cave drawings and chipped flint tools show that left-handedness existed to some extent in stone-age times.

As with any minority group, the preference of certain individuals for their left hand was thought to be unnatural and they were either persecuted or forced to use their right hands (Harris, 1980). Even the pejorative terms "sinister" and "gauche" arise from the Latin and French words for "left."

After Broca's discovery of the lateralization of the brain, much effort went into weighing various parts of the brain to determine morphological differences. Efforts were also made to determine from operations and autopsies if aphasia in left-handed patients was due to lesions in the right hemisphere. Generally it was not. Studies and controversy still continue, and conclusions are hampered by the lack of sufficient numbers for reliable statistics. There are some tentative conclusions, however. There is a very small number, about 2% of the population, of *strongly* left-handed individuals. These have their language center in the right hemisphere. The majority of left-handed individuals, however, are only partially left-handed. They are somewhat ambidextrous, i.e., there are varying degrees of left-handedness. This group seems to have stronger connections between the hemispheres of the brain in that there are potential language centers in the right hemispheres. Evidence for this is the following. If a right-handed individual develops aphasia from a stroke in the language center (Broca or Wernicke area) of his left hemisphere, he rarely recovers his power of speech. For left-handed individuals (other than the 2% with language centers in the right hemisphere), there is a good chance that with training they can recover from aphasia (Subirana, 1958), although there appears to be an age dependence for recovery. See Satz (1980) for a more recent review of the prognosis for recovery from aphasia.

At what stage of development does the hemispheric domination of speech occur? Careful postmortem morphological measurements of the size of Broca's area were made by Wada *et al.* (1975) on the brains of 100 adults and 162 infants. The ages of the adults ranged from 17 to 96 years, with a mean of 68, and of the infants from the 18th gestational week to the 18th postnatal week, with an average of 48 weeks after conception. They found the brains of 10 adults and 12 infants with the right side larger and 8 adults and 32 infants with equal development. The remainder had greater development of Broca's area of the left hemisphere. Within the statistical limitations of their sample, they found the left/right enlargement ratio to be 77% in infants and 75% in adults, and a reversal of dominance in infants of 16% and in adults 9%. There was no difference between male and female subjects. These data strongly indicate that the dominance is established in most cases prior to birth. The origin of right-hand dominance has been proposed by Calvin (1983). See p. 39.

SURVIVAL VALUE OF CEREBRAL DOMINANCE

Accepting the situation that in any group of organisms there will be variations (that is, if there is cerebral lateralization, there can be left-handers as well as right-handers), why is there cerebral dominance at all? What was its survival value in evolution?

Ernst Mach, the 19th-century physicist and pioneer in psychophysics, devoted much of his attention to this question (1914). Many modern ideas are rooted in his considerations. We shall just mention two.

In discussions with a military friend, Mach learned that soldiers in a snowstorm with no visible landmarks tend to wander in a cricle. This phenomenon is now well known. Mach suggested that this has evolutionary survival value in that the young in wandering away from the mother will tend to return. The slight dominance of the left hemisphere will make the strides of the right limbs be somewhat greater and cause the resulting path to be a large circle.

The other idea is reviewed by Corballis and Beale (1970) and by Webster (1977). It is best explained by reference to Fig. 2.16a, representing equal hemispheres, and Fig. 2.16b, representing a dominant hemisphere. In the upper parts of Fig. 2.16 stimuli S_1 or S_2 can be given to a foot or paw. Recognizing the left hemispheric control of the right side and vice versa, the different stimuli will cause responses R_1 and R_2, respectively. The stimulus may be,

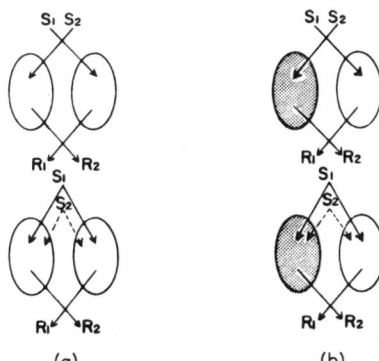

FIG. 2.16 The advantage of hemispheric dominance: (a) equal hemispheres; (b) unequal hemispheres. S_1 and S_2 in upper figures are touch stimuli, such as in walking, which cause respective unilateral responses R_1 and R_2. In lower figures S_1 and S_2 are stimuli that have a change in intensity, such as odor or illumination, which cause bilateral responses. In this situation R_1 and R_2, such as right and left limbs, would have equal response if both hemispheres were equal and the animal would not be programmed for a rapid response. If one hemisphere were dominant, a fight-or-flight reaction could commence reflexively with, for example, the right limb in right-handed animals. [From Webster (1977).]

for example, touch. A subconscious programming of the brain could develop for walking in that when one foot touches the ground the weight shifts to it and the other foot responds. Suppose, however, the stimulus is not unilateral like touch, but instead is a change of intensity, such as in olfaction or vision, i.e., the change in odor or reflection from a prey or predator, which would be a bilateral stimulus. This is indicated in the lower part of the figure. If both hemispheres are equal, the response will be equal in both legs. The brain cannot develop a coded reflexive response, since the intensity of nerve impulses to both legs will be equal. The survival value lies in a hemispheric dominance and therefore a limb preference. A reflexive code could be developed to eliminate time-consuming consideration; the response would always be with the leg controlled by the dominant hemisphere. Recall from the earlier discussion of damage to one side of the cerebellum that the patient had to consciously plan the muscular movements, which, not being reflexive, took considerable time to execute. Thus, the ability to program reflexes has high survival value, and a mechanism to facilitate such programming, such as a dominant hemisphere with the resulting right- or left-handedness, would be expected to have a very early evolutionary development.

ARRANGEMENT OF CORTICAL NEURONS

If one photographs the cross section of a portion of one of the gyri of the cortex of higher vertebrates under a microscope, distinct layers are seen. An example of human visual cortex is seen in Fig. 2.17. In this picture the broad inner gray part is "white matter," which was identified previously as arrays of nerve axons for transmitting or receiving signals. The layered, outer portion is the "gray matter" in which there is a high density of cell bodies.

What is visible under higher magnification depends on the method of staining. The pioneer work in this area was done by Ramón y Cajál (1909–1911), and later by Lorente de Nó (1949). Some stains reveal the *perikaryon*, or cell body of the neuron, while others are more selective for dendrites or axons. As discussed in Vol. I, Chapter 3, the axon is a large single fiber with branches that extends from the cell body of the neuron and transmits the action potential. The dendrites, on the other hand, are very fine fibers that make synaptic contact with the perikaryon, axons, or dendrites of other cells. Drawings from a microscope focused at different depths are shown in Fig. 2.18. The staining procedure used for these examples generally did not reveal the finer dendrites.

There are basically two categories of cells in the cortex, *pyramidal* and *stellate*, named for their approximate shapes. The pyramidal cells have four

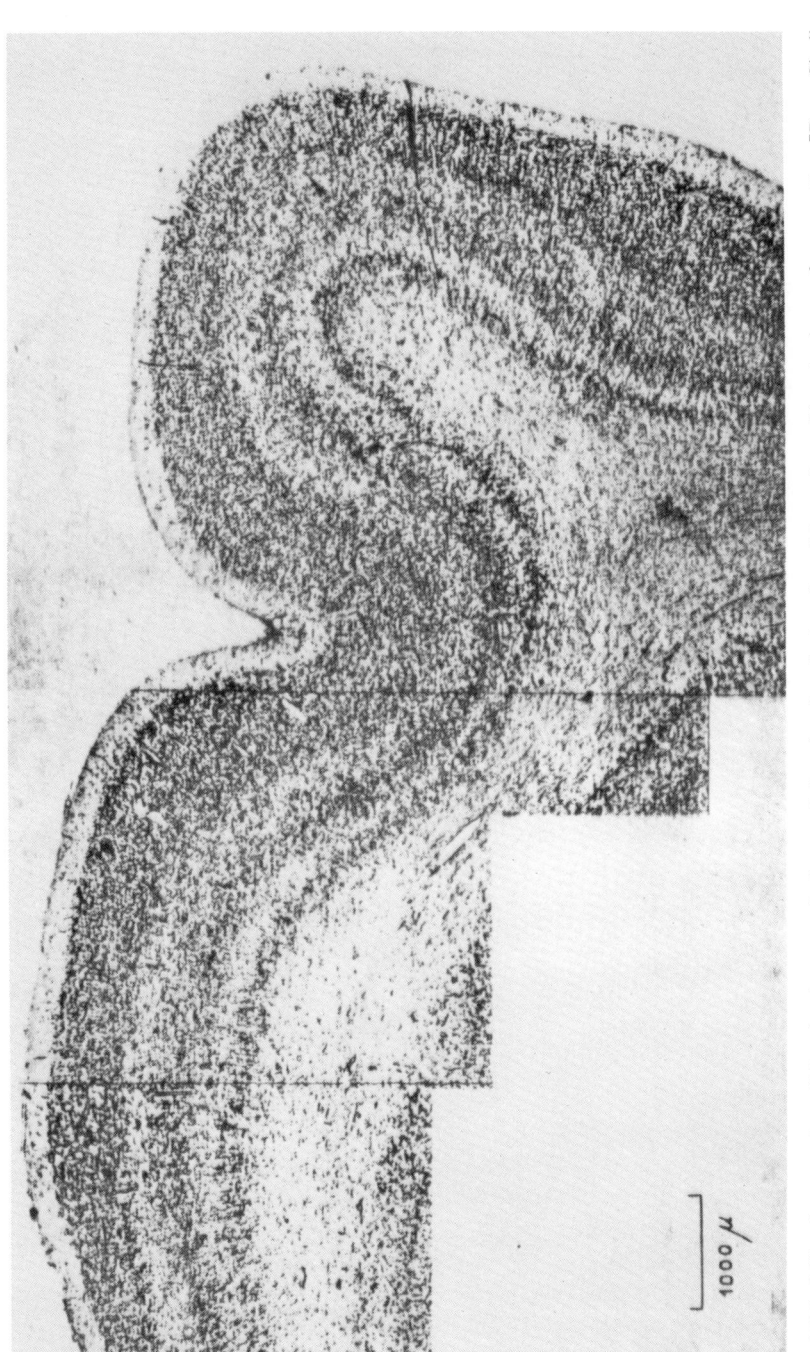

FIG. 2.17 Cross section of the visual cortex of a human being showing the interior white matter with outer layer of gray matter. [From Sholl (1956).]

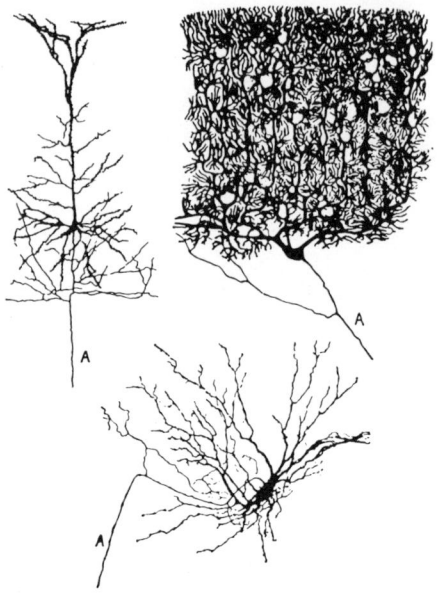

FIG. 2.18 Upper left: a drawing of a pyramidal neuron from the cerebral cortex; upper right: Purkinje cell of the cerebellum. Lower: motoneuron of the spinal cord. In each case A indicates the axon that is cut off. [From Rall (1962), based on drawings by Ramón y Cajál (1909–1911).]

basic types and the stellate three, although investigators of cell histology have claimed that no two cells are identical. The main difference between the two categories of cells are the length of the axons and the shape of the cell body. The axons of pyramidal cells are longer and dip down into the white matter. The axons of stellate cells terminate locally.

Nerve cells in the brain are surrounded by *glial* (glue) cells. They outnumber the neurons by about 10:1 and make up about half the bulk of the nervous system. Although they have been studied for over a hundred years, the only quantitative roles found for them so far are (1) they speed conduction of signals; (2) they assist brain nutrition; and (3) they have a role in regulating the extracellular ion concentration. They may also have a role in the life cycle of neurotransmitters. They apparently speed the conduction of signals by having their membrane potential lowered by an activation signal from neighboring neurons. It is peculiar that most brain studies continue as if the glial cells did not exist without even a justification for ignoring them, but that is the unfortunate state of knowledge. A review of the physiology and electrical behavior of glial cells is given by Kuffler and Nicholls (1976).

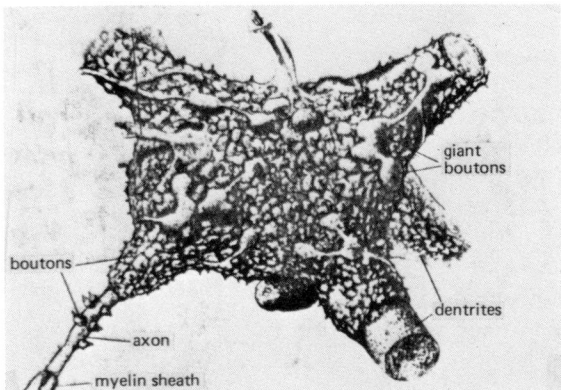

FIG. 2.19 A reconstructed drawing from a large number of serial electron micrographs of the body of a motoneuron of a cat, showing it completely covered with synaptic boutons. [From Poritsky (1969).]

Each neuron apparently leads its own biological life and does not form a syncitium with other cells as, for instance, cardiac cells do. Each receives and imparts information to other neurons by means of fine fibers that attach to the body, cell appendages, and dendrites of the other neurons. These attachments terminate in little knobs, or *boutons* (buttons), where *synapses* occur ("synapse" comes from the Greek *synapto*, to clasp tightly). The cell body and dendrites are covered with these synapses; estimates of the more dense numbers are in the thousands. A reconstruction of the synaptic boutons on the body of a motoneuron of a cat (from the spinal cord where the neurons can be quite large), taken from serial electron micrographs, is shown in Fig. 2.19. Diagrams of examples of the estimated 16,000 connections in a motoneuron of a rabbit are shown in Fig. 2.20. Note at this point the terminology *afferent*, which means leading into, and *efferent*, which means leading away from. Figure 2.20a shows a schematic arrangement of only a few of the afferent synapses around a perikaryon of a pyramidal cell and some of the initial segments of the dendritic trees. Figure 2.20b shows the diversity of regions of synaptic contact with a pyramidal cell, and on the right of this figure are some of the various shapes and contact positions of the boutons. Note in this drawing that none of the boutons actually touches the cell or its parts. There is a gap called the *synaptic cleft* through which a chemical diffuses upon electrical stimulation of the presynaptic nerve fiber. This will be discussed fully in Chapter 4. It should be noted at this point that not all synaptic contacts are by boutons. Within the spinal cord and medulla oblongata some contacts appear as the interweaving of mossy fibers and bear that name.

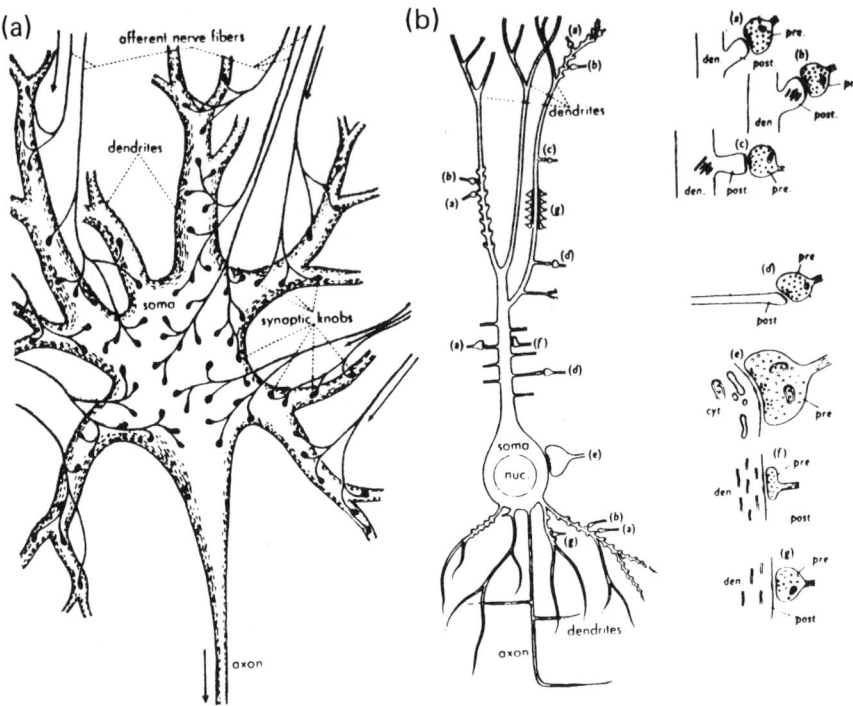

FIG. 2.20 Schematic (not to scale) of the general arrangement of synaptic boutons on a pyramidal cell: (a) general arrangement; (b) diversity of zones of contact on the cell with some of the types of contact of boutons with the cell. [From Hamlyn (1963).]

It has been estimated that there are about 10^{10} cortical neurons in the human brain (Shariff, 1953). If dendrites of each make thousands of connections with its surrounding neurons, the complexity of the brain and its possible varied responses can begin to be appreciated. Careful study of pyramidal and stellate neurons by Bok (1936) showed that the number of dendrites per cell range from about 20 to 80. The length of the dendrites ranges from 40 to 200 μm. The dendrites of different neurons can overlap considerably and make numerous synaptic connections. The ranges of such overlappings from typical neurons, drawn to scale, is shown in Fig. 2.21. This figure shows the axons as vertical lines with axonal branches. The solid triangles and circles represent the body of pyramidal and stellate cells, respectively, and the large circles are the extent of their dendritic fields. Considerable overlap between any cell and others is evident, particularly since

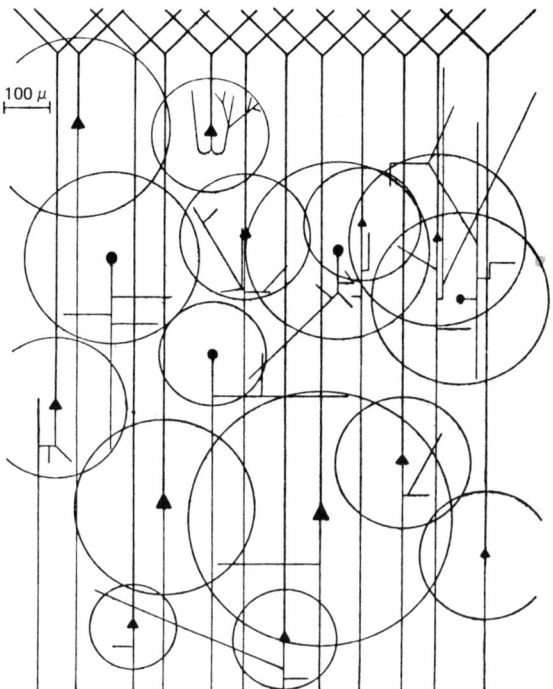

FIG. 2.21 Diagrammatic drawing of a number of cortical neurons with their axons, dendritic fields (large circles), and axonal branches drawn to scale. ▲: pyramidal cells; ●: stellate cells. Surface of the brain is at the top. [From Sholl (1956).]

this is a cross-sectional view and each circle must be considered as a sphere.

The number of fibers passing between the gray–white matter boundary is estimated to be about $100,000/mm^2$, of which 75,000 are efferent (going from gray to white matter) and 25,000 are afferent (Sholl, 1956).

The axons of Fig. 2.21 suggest that there is a columnar nature to cortical organization. This is shown in Fig. 2.22, which is a section of the visual cortex of a cat. The stain shows only a few of the neurons, so that the columnar arrangement is evident, with deep pyramidal cells at the bottom near the white matter. This columnar arrangement with layered cross-connections was mentioned in the discussion following Eq. (2.5), which contained an attempt to explain why brain growth was a function of body growth to the two-thirds power. In this model the columns represent the length and the layers (six in the case of cats and human beings) represent the surface area, which is proportional to the square of the length.

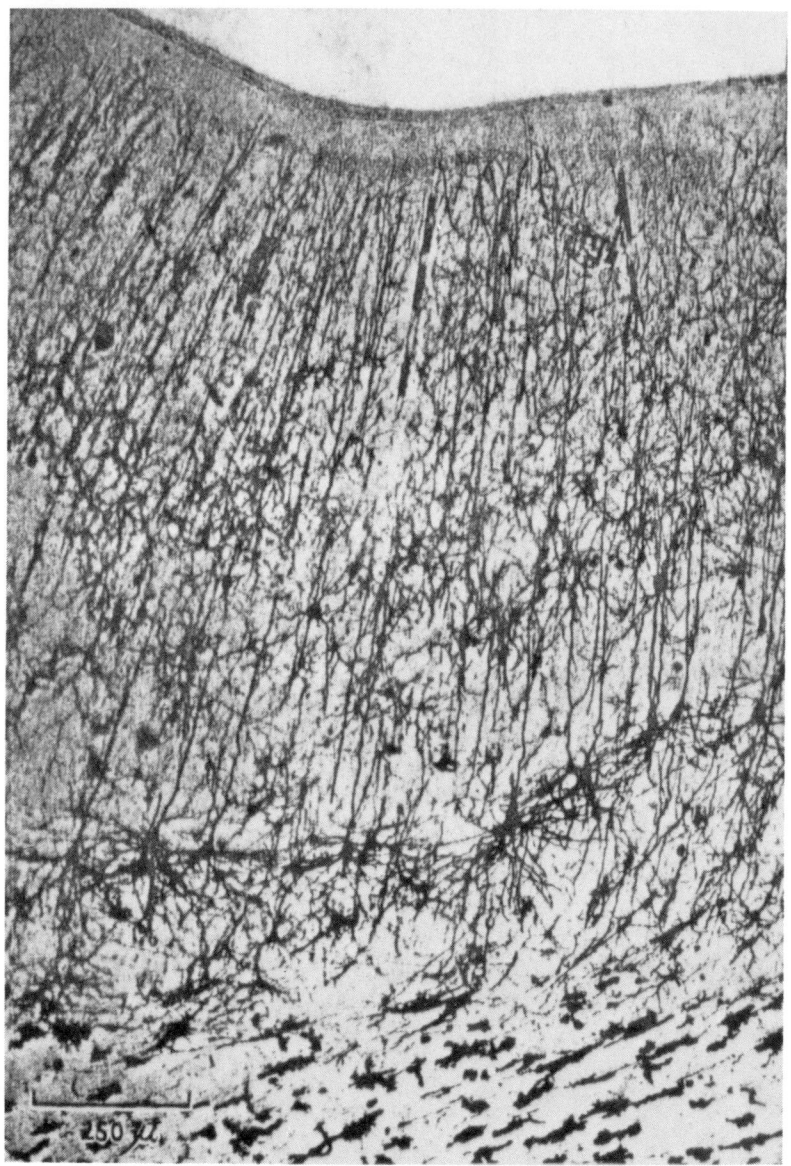

FIG. 2.22 A section of the visual cortex of a cat stained to reveal neurons. Note the columnar arrangement leading to the white matter at the bottom of the picture. [From Sholl (1956).]

Note added in proof: Calvin has suggested in his book, "The Throwing Madonna" (1983), that the predominance of right-handedness has been bred into the human species because of the asymmetrical location of the heart. It is known that the sound of a heartbeat soothes infants. When bipedalism developed, a mother would tend to carry an infant in her left arm against her heart to quell cries that might attract predators or frighten prey. With her right arm unencumbered, she could then throw stones at small creatures and, if successful, supply food for her children. The offspring of the successful throwers would, therefore, have a higher survival rate. It would be an interesting confirmation of this idea if research could show that four-footed animals have no preferred handedness.

REFERENCES

Bok, S. T. (1936). The branching of dendrites in the cerebral cortex. *Proc. Acad. Sci. Amst.* **39**, 1209.
Broca, P. (1861). Remarques sur le siège de la faculté du langage articulé, suivies d'une observation d'aphémie (perte de la parole). *Bull. Soc. Anat. Paris* **6**, 330.
Calvin, W. H. (1983). "The Throwing Madonna." McGraw-Hill, New York.
Carpenter, M. B. (1972). "Core Text of Neuroanatomy." Williams & Wilkins, Baltimore, Maryland.
Corballis, M. C. and Beale, I. L. (1970). Bilateral symmetry and behavior. *Psychol. Rev.* **77**, 451.
Deecke, L., Scheid, P., and Kornhuber, H. H. (1969). Distribution of readiness potential, premotion positivity, and motor potential of the human cerebral cortex preceding voluntary finger movements. *Exp. Brain Res.* **7**, 158.
Gould, S. J. (1971). Geometric similarity in allometric growth: A contribution of scaling in the evolution of size. *Am. Nat.* **105**, 113.
Gould, S. J. (1975). Allometry in primates, with emphasis on scaling and the evolution of the brain. *Contrib. Primatol.* **5**, 244.
Guyton, A. (1971). "A Textbook of Medical Physiology," 4th ed. Saunders, Philadelphia, Pennsylvania.
Haeckel, E. R. (1896). "History of Creation." Appleton, New York. (English translation of 1892 German edition.)
Halstead Tarlo, L. B. (1965). Psammosteiformes (Agnatha): A review with descriptions of new material from the lower Devonian of Poland. I. General part. *Paleontol. Pol.* No. 13.
Hamlyn, L. H. (1963). An electron microscope study of pyramidal neurons in the Ammon's Horn of the rabbit. *J. Anat.* **97**, 189.
Harris, L. J. (1980). Left-handedness: Early theories, facts, and fancies. *In* "Neuropsychology of Left-Handedness" (J. Herron, ed.), p. 3. Academic Press, New York.
Haugh, B. N., and Bell, B. M. (1980). Fossilized viscera in primitive echinoderms. *Science* **209**, 653.
Hodgson, E. S. (1977). The evolutionary origin of the brain. *Ann. N.Y. Acad. Sci.* **299**, 23.
Holmes, G. (1939). The cerebellum of man. *Brain* **62**, 21.
Huxley, J. S. (1932). "Problems of Relative Growth." Allen & Unwin, London. Reprint (1972). Dover, New York.
Jerison, H. J. (1974). "Evolution of the Brain and Intelligence." Academic Press, New York.
Jerison, H. J. (1977). The theory of encephalization. *Ann. N.Y. Acad. Sci.* **299**, 146.
Kostitzin, V. A. (1939). "Mathematical Biology" (T. H. Savory, trans.). London.
Kuffler, S. W. and Nicholls, J. G. (1976). "From Neuron to Brain." Sinauer Assoc., Sunderland, Massachusetts.
Lorente de Nó, R. (1949). Cerebral cortex: Architecture, intracortical connections, motor projections. *In* "Physiology of the Nervous System" (J. F. Fulton, ed.). Oxford Univ. Press, London/New York.

Mach, E. (1914). "The Analysis of Sensations." Open Court Publ. Co., London.
McLean, P. (1973). "The Triune Concept of Brain and Behavior." Univ. of Toronto Press, Toronto.
Martin, R. D. (1981). Relative brain size and basal metabolic rate in terrestrial vertebrates. *Nature (London)* **293**, 57.
Nottebohm, F. (1971). Neural lateralization of vocal control in a passerine bird. I. Song. *J. Exp. Zool.* **177**, 229.
Penfield, W., and Rasmussen, T. (1950). "The Cerebral Cortex of Man." Macmillan, New York.
Poritsky, R. (1969). Two and three dimensional ultrastructure of boutons and glial cells on the motoneuronal surface on the cat spinal cord. *J. Comp. Neurol.* **135**, 423.
Rall, W. (1962). Theory of the physiological properties of dendrites. *Ann. N.Y. Acad. Sci.* **96**, 1071.
Ramón y Cajál, S. (1909–1911). "Histologie du système nerveux de l'homme et des vertébrés." Maloine, Paris.
Romer, A. S. (1967). "The Vertebrate Body," 3rd ed. Saunders, Philadelphia, Pennsylvania.
Satz, P. (1980). Incidence of aphasia in left-handers: A test of some hypothetical models of cerebral speech organization. *In* "Neuropsychology of Left-Handedness" (J. Herron, ed.), p. 189. Academic Press, New York.
Shariff, G. A. (1953). Cell counts in the primate cerebral cortex. *J. Comp. Neurol.* **98**, 381.
Sholl, D. A. (1956). "The Organization of the Cerebral Cortex." Methuen, London.
Smith, C. U. M. (1970). "The Brain, Towards an Understanding." Capricorn Books, New York.
Sperry, R. W. (1970). Perception in the absence of neocortical commissures. *Res. Publ.—Assoc. Res. Ment. Dis.* **48**, 123.
Starling, E., and Evans, L. (1962). *In* "Principles of Human Physiology" (H. Davson and M. G. Eggleton, eds.), 13th ed. Lea & Febiger, Philadelphia, Pennsylvania.
Stensiö, E. (1963). The brain and the cranial nerves in fossil, lower craniate vertebrates. *Skr. Nor. Vidensk.-Akad.* [*Kl.*] *1: Mat.-Naturvidensk. Kl.* [N.S.] **13**.
Subirana, A. (1958). The prognosis in aphasia in relation to cerebral dominance and handedness. *Brain* **81**, 415.
Thomson, K. F. (1982). Reflections on the neural crest. *Am. Sci.* **71**, 72.
Tosney, K. (1982). The segregation and early migration of cranial neural crest cells in the avian embryo. *Dev. Biol.* **89**, 13.
Wada, J. A., and Rasmussen, T. (1960). Intercarotid injection of sodium amytal for the lateralization of speech dominance: Experimental and clinical observations. *J. Neurosurg.* **17**, 266.
Wada, J. A., Clarke, R., and Hamm, A. (1975). Cerebral hemispheric asymmetry in humans. *Arch. Neurophysiol.* **32**, 239.
Webster, W. G. (1977). Territoriality and the evolution of brain asymmetry. *Ann. N.Y. Acad. Sci.* **299**, 213.

CHAPTER 3

Electrical Properties of the Brain; EEG and Evoked Potentials

INTRODUCTION

Cells comprising the nervous system are not quiescent. The potential across their outer membranes is continuously altering in magnitude. In some neurons this electrical activity arises from inherent metabolic processes contained within the cell. For other neurons this electrical activity is caused by the continuous flow of information they receive due to small electrical potential signals of other cells. Through corresponding changes in their own membrane potentials they send electrical signals to other cells. The electrical changes for both types of neurons can be probed either with microelectrodes inserted in the cell (intracellular recording) or placed just outside of one (extracellular recording).

An electode placed directly on the brain will register a variety of waves with a magnitude in a normal subject of 0.1–5 mV. The frequency range of 1–14 Hz is considered the most significant for study and diagnosis. These same waves can be measured by placing the electrode on the scalp, although the insulating effect of the skin, skull, and dura mater will reduce the signal by a factor of 10 or more. A record of these waves is called an *electroencephalogram* (EEG). A statistical study of EEG and single-neural-cell discharges has shown that the EEG results as a random sum of the individual cell potential changes. This analysis will be discussed later in some detail.

When a stimulus is presented to a sensory organ a small potential change occurs in the area of the brain primarily responsible for its processing. This potential change is called an *evoked potential*. It is very small compared to the EEG, and the resulting waveform must be both added and averaged from many such stimuli, e.g., flashes of light or clicks of sound, and the background EEG subtracted. The importance of evoked potentials as a diagnostic tool for both the sensory organ and the neural processing of the stimuli will be discussed.

ELECTROENCEPHALOGRAPHY

Electrical activity of the brain has been known since the 19th century, when Caton (1875) demonstrated voltages from the exposed brains of animals. The voltages obtained from electrodes placed directly on the surface of the brain range from 100 to 500 μV, although smaller waves with voltages of only 20–30 μV are also seen.

Only about 20% of the cortical surface of the brain lies on its external convexity, i.e., adjacent to the skull. Between the brain and the exterior of the skin of the scalp lies 2–3 cm of dura mater, skull, and scalp. This thickness of both dielectric and ionic conducting material reduces the observed brain potential to about $\frac{1}{10}$ or less of its direct contact value.

The medical world was startled by a series of papers by Berger (1929, 1931, 1932 a,b, 1934, 1936) that described his discovery of electroencephalography (EEG). Electrical contact was made to the scalp, first by pads wetted with salt solution and later by silver electrodes held on with collodion. The output was amplified and recorded on a strip-chart recorder. Although his studies were supplemented by the work of other investigators during this period, it was the work of Adrian (1934) and of Adrian and Matthews (1934) that finally convinced the medical world of the validity and importance of Berger's work.

With the development of rapid chart recorders and amplifiers in which potentials at higher frequencies can be read, the measurements are now much more inclusive than the original ones of Berger. It is not our purpose to show diagnostics with EEG. However, to help establish a picture of the electrical activities of the brain, even though their origins are not completely understood, we shall present some basic findings.

The EEG generally discussed and used in diagnostics is that recorded from

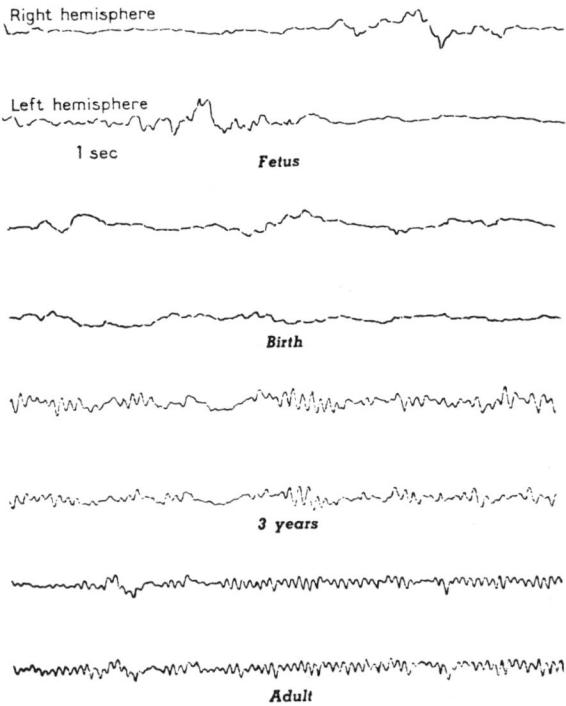

FIG. 3.1 Electroencephalograms from human beings at different ages. [From Tuchmann-Duplessis *et al.* (1975).]

adult human beings because the brain wave patterns do not reach the descriptive state usually employed until around age 13–14. This can be seen in Fig. 3.1, in which the waves of the two hemispheres of younger human beings lack both the rhythm and synchrony of adults. At birth the most noticeable waves have a frequency of about 3–4 Hz. At age 2–3 years a rhythm of 6–7 Hz appears and has greater amplitude. It is well organized in the occipital regions, but less so in the frontal regions. Finally in the adult the normal 8–12 Hz rhythm appears and is measurable over the entire cerebral surface. From this point onward in the text, all EEGs discussed will be those taken from adults unless otherwise stated.

If a subject rests quietly with his eyes closed with electrodes R and L attached to the right and left *occipital* regions (Fig. 2.9a) of the scalp, the recorded voltage is in a series of waves of frequency 8–13 Hz. This is called the alpha rhythm or alpha waves. These are shown in the upper two traces of Fig. 3.2. This rhythm is suddenly blocked when the subject opens his eyes, as indicated in Fig. 3.2. If the room is darkened and the subject opens his

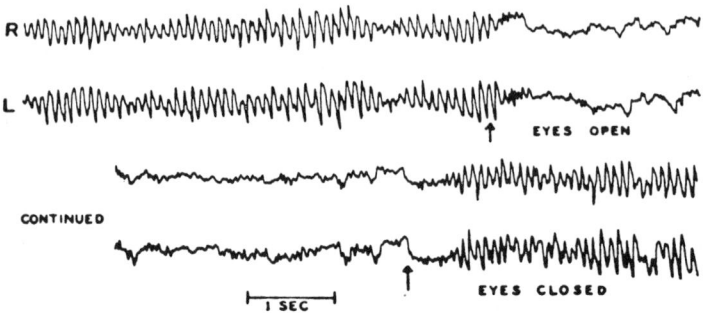

FIG. 3.2 Electroencephalogram from right (R) and left (L) occipital regions on the scalp of a subject. The alpha rhythm occurs when the eyes are closed and is blocked when they are open. [From Penfield and Jasper (1954).]

eyes, the same effect occurs. Trying to see has the same effect as seeing (Adrian and Matthews, 1934). It is attention rather than visual stimulus that blocks the alpha rhythm. A patient solving simple arithmetic problems with his eyes closed has an unaltered alpha rhythm, but a difficult problem, requiring attention, blocks the rhythm. Auditory stimuli in general do not block the alpha rhythm. It has since been shown that with training, i.e., biofeedback, a subject may achieve some voluntary control over his alpha rhythm.

The characteristic rhythms change with variations in the state of consciousness of a subject. These are shown in Fig. 3.3, in which the first two are the same as those shown in Fig. 3.2.

If electrodes are placed in the *parietal* region near the area of muscular control, i.e., the precentral and postcentral gyri of Fig. 2.9a, there is a different resting rhythm called the beta rhythm, with a characteristic frequency greater than 13 Hz. A muscular movement such as clenching the fist interrupts the rhythm at the initiation of the movement, but the rhythm is restored while the fist is clenched. This is illustrated in Fig. 3.4. Note that for this figure the electrodes were in direct contact with the brain. With the absence of the skull and dura mater the voltages are about 10 times larger than those shown in Figs. 3.2 and 3.3. Unclenching the fist has the same effect during its initiation. Thinking about the movement has no effect, but getting ready to move even though the movement does not take place does interrupt the rhythm (Jasper and Penfield, 1949). The potential change in preparing to move was shown in a more detailed study in Fig. 2.10.

The various periodicities of the waves recorded in EEG are given the general names in Table 3.1.

In normal persons the alpha rhythm is dominant in the occipital region of the relaxed brain, and the degree of consciousness can be observed from the patterns of Fig. 3.3, although these are altered with drugs. Brain death,

ELECTROENCEPHALOGRAPHY 45

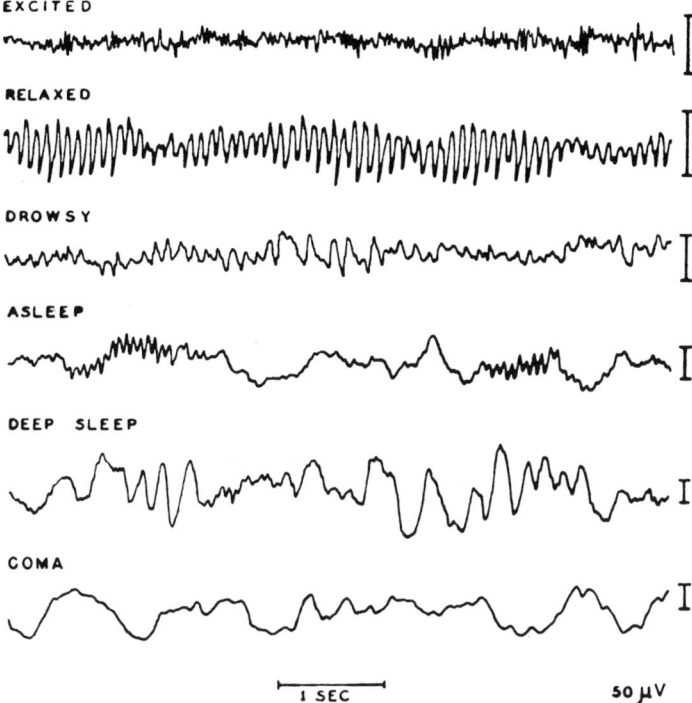

FIG. 3.3 Characteristic electroencephalograms during variations in states of consciousness. Scale indicated by bars are each 50 μV. [From Penfield and Jasper (1954).]

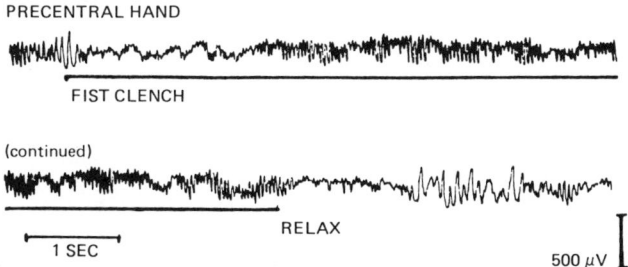

FIG. 3.4 Electrocorticogram (electrode in direct contact with the brain) from the precentral hand area in a human being during voluntary movement, clenching the fist and relaxing. The relaxed rhythm in this region, faster than the alpha rhythm, is called the beta rhythm. Note that it is blocked when the fist is first clenched but it returns during sustained contraction. It is again blocked during the initial relaxation movement. [From Penfield and Jasper (1954).]

TABLE 3.1

Classification of EEG Rhythms

Frequency (Hz)	Name
<4	delta
4.0–8.0	theta
8.1–13.0	alpha
>13.0	beta

a term of increasing legal usage, would be evident from the absence of any rhythm measured from the cortex, although there may be rhythm in the brain stem.

The depth of sleep is also determinable by EEG, as shown in Fig. 3.3, from the magnitude of the slow, or delta, wave. The depth of sleep is most easily measured by the duration of a sound that is necessary to arouse the subject. If the depths are given grades 1–4, where 1 is drowsiness or quiet wakefulness and 4 is deep sleep, it is found that the prominence of the delta wave is proportional to the depth of sleep. The normal pattern of sleep, as measured over a period of many hours by EEG, is one of variations in depth. This is shown in Fig. 3.5. Whenever the depth is smallest, the shaded areas, dreaming occurs. This can be determined by waking the subjects during this period. It is also determined by an effect called *rapid eye movement* (REM). This measurement is obtained from electrodes, attached near the eyes, which record voltages associated with movements of eye muscles.

Abnormalities of EEG can be recorded from any area of the scalp. They may be diffuse or irregular, and they may be observed from a single recording electrode. If all electrodes but one record normal rhythm and that electrode

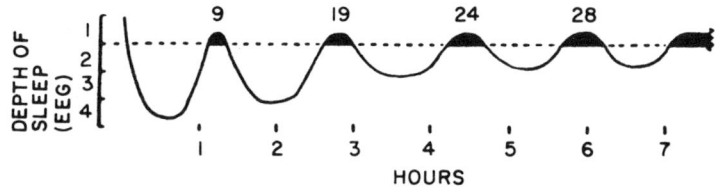

FIG. 3.5 A schematic of the variations in depth during a typical night's sleep. At Stage 1 level dreaming occurs which is accompanied by rapid eye movements (REM), as indicated by the shaded areas. The numbers above these areas are the average lengths of these dream periods in minutes. [From Dement and Wolpert (1958). Copyright 1958 by the American Psychological Association.]

exhibits a higher-voltage, slower wave, it is generally concluded that there is a lesion, tumor, or abscess in that region. Diagnosis by EEG alone, however, is not sufficient. Whereas *glioma*, a tumor of glial cells, invades the substance of the brain and causes an abnormal EEG, an arterial tumor (*angioma*) or membrane tumor (*meningioma*) generally will not. EEG is useful in conditions of acute cerebral abscess in which brain damage is rapidly progressive. It should also be noted that 10–15% of normal, healthy people exhibit EEG abnormalities.

The word epilepsy is from the Greek *epilepsia*, meaning to seize or take hold of, and is now used to denote a group of central nervous system disorders which have in common the spontaneous occurrence of seizures associated with the disturbance of consciousness. EEG records abnormal brain waves during such seizures. Some of these are illustrated in Fig. 3.6. Generally there is a spike, which indicates an abnormal discharge of many cells synchronously. The contrast with the upper two normal patterns is evident. Epileptic

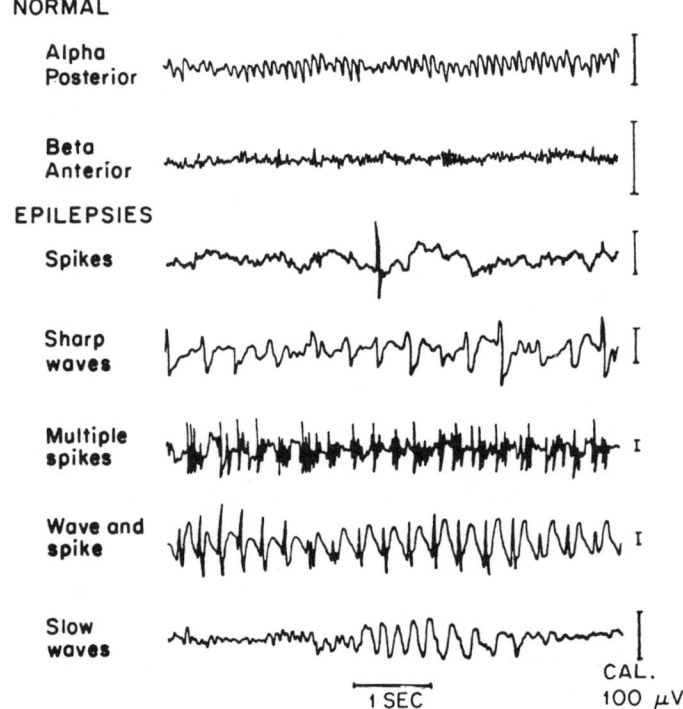

FIG. 3.6 Examples of a few types of epileptiform discharges in comparison with normal alpha and beta rhythms. Note the high voltages of the discharges. [From Penfield and Jasper (1954).]

discharges can originate in a variety of locations for a variety of reasons. For example, the brain may be damaged by hypoxia at birth or during delivery. Head injuries may also be the cause of epilepsy; nearly one-third of the people who sustain injuries of the brain through penetrating missile wounds later develop epilepsy. Tumors, infections, cerebrovascular diseases, and metabolic disturbances are also causes. Withdrawal from long usage of any habit-forming drug may give rise to an epileptic seizure, and an alcoholic admitted to a hospital for any reason, with the associated withdrawal from alcohol, may have a seizure. These have been called "rum fits."

In the days when surgery was the only possible cure, detection and location of the origin of the disturbance was required prior to surgical intervention. However, anticonvulsant drugs have been developed that generally control epilepsy without surgery. Brain surgery is now used only when the cause of the epilepsy, such as a tumor, must be removed for other reasons.

COMPUTER DIAGNOSIS OF ELECTROENCEPHALOGRAMS

Abnormalities of EEGs are useful to the diagnostician since they are a noninvasive tool for detecting a variety of neurological disorders, comas, and sleep stages. In a normal EEG, recording data from some 8–16 electrodes are recorded simultaneously on a strip chart for 10–20 min, resulting in 20–40 m of paper write-out. The physician has to look through these data for abnormalities. There is a general trend in developed countries for health care for all, and lower cost is achievable with increased automation. It is therefore of considerable interest to eliminate extraneous data that would consume a physician's time.

In Fig. 3.3 we saw that the EEG pattern for a sleeping person consists of a background activity, called a "stationary" or "stationarity," e.g., the alpha rhythm. This may be interrupted from time to time by sharp "spikes," of possible epileptic origin, or by short runs of another frequency. The physician will look for stretches of abnormal activity and describe them. In doing this he spends about the same time looking at the background activity, which contains little information, as he does with the abnormalities.

The concept of computer reduction of data, or even diagnosis, is an appealing one. Much research has gone into this area, but there is a sense of incompleteness of approach even if a specific goal is achieved. In general, with the help of a Fourier transform, we can define a power spectrum, that is, the energy distribution (proportional to the square of the wave amplitude) as a function of frequency (see p. 196). The frequency analysis of the waves and their changes can be made for different conditions. For a review of the mathematics of this technique, see, for example, van der Gon and Strackee (1966).

One of the recent methods, being developed by Bodenstein and Praetorius (1977) and by Praetorius *et al.* (1977), attempts to include all data and mimic the examination of a physician. This approach is summarized below.

A typical EEG consists of stationary rhythms, e.g., alpha rhythms, which are 2 sec or longer in duration. These may be interrupted by diagnostically more significant elements such as short periods of delta or theta waves. Superimposed on these relatively slowly varying patterns are sharp waves (lambda waves), or spikes. The above investigators have devised a computer technique for extracting and storing the spikes and dividing the stationary rhythmic periods into segments for storage and recall. The advantage of this method is that the time structure of the signal can be preserved and both the segments and spikes can be represented by a small number of parameters. Identical segments can be lumped together, and an algorithm for the entire EEG in terms of the stationary segments and transients can be written with the possibility that from these computer models diagnosis can be eventually achieved.

They have shown that an EEG power spectrum of a stationary segment, e.g., alpha rhythm, can be represented by a small number of parameters. An example of a three-peak power spectrum (ordinate in relative arbitrary units) fitted by a function with differing numbers of parameters is shown in Fig. 3.7. It is seen that very little is gained in accuracy in going from 10 to 20 parameters. Thus, an easily storable number, 10, can be used to describe a segment. Note that in this example the original spectrum is not given, only that there are three peaks in the power spectrum.

The general method used is as follows. The signal value of the EEG at each point in time can be expressed as a weighted sum of the values of a fixed

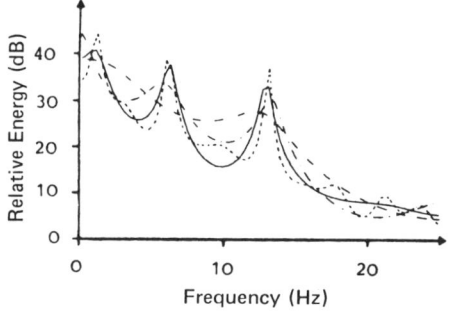

FIG. 3.7 Parametric representation of an EEG power spectrum containing three peaks, p, the number of parameters used, and their corresponding forms. It is seen that 10 parameters are sufficient. (Power spectrum of original data not shown.) ———: $p = 4$; —·—·—: $p = 6$; ———: $p = 10$; ---: $p = 20$. [From Praetorius *et al.* (1977).]

number of immediately preceding points plus a value equal to the difference between a prediction of the value based on preceding points and its actual value. This difference is called the prediction error. The prediction errors are determined on a computer by passing the data through a program (called an autoregressive filter) whose coefficients are derived from the power spectrum of the EEG data. The values of these coefficients are determined from the first 2 sec of the EEG data. The program compares the power spectrum with itself displaced in time (autocorrelation, see Chapter 6). If there is no change in the power spectrum of the EEG, no change is observed, the prediction error is accordingly small, and the filter is adapted to the signal. When a new spectral component appears in the EEG, the power spectrum shows a peak corresponding to the frequency of the new component, the prediction error is large, and the autoregressive filter is no longer optimal for the signal. The deviation of the filter from optimal adaptation can therefore be derived in terms of the deviation of the prediction error, which these investigators call *spectral error measure* (SEM). They establish, from comparison with typical EEG data, a threshold of SEM above which a new pattern has appeared. A typical example is shown in Fig. 3.8. The top trace is a portion of an EEG record of a sleeping child. The eye can discern the segments marked by the vertical lines, but the SEM (lower trace) also detects the pattern boundaries.

In order to analyze and store segments properly, large transients such as spikes must be treated separately. Whenever a large transient occurs in the EEG record, as shown for example in Fig. 3.9 (upper trace), the prediction error (middle trace) exceeds threshold limits (dashed lines, middle trace), and spikes are removed and stored for later use. The EEG stripped of these spikes is shown in the lower trace. Thus, by allowing the main signal to be slowly varying, three things are accomplished: (1) the segmentation process can proceed undisturbed; (2) the power spectrum features can be stored for later reconstruction and display, using only the 10 parameters; and (3) the position and magnitude of the spikes can be stored for recall and display.

The power spectrum of three segments of the EEG of a sleeping child are shown in Fig. 3.10 with the relative filter coefficients on the bottom as the stored information of each segment.

How good is all this? With the coefficients of each segment an EEG simulation can be reconstructed for display. Typical EEG traces (upper) and computer reconstruction (lower) are shown in Fig. 3.11 for several cases. Although individual waves are not in one-to-one correspondence, if the whole trace is examined, it is seen that all the features are present, at least all the features customarily examined by a physician. The extracted spikes can be restored as indicated in Fig. 3.12, which represents the EEG simulation with spikes of an epileptic patient. This is an example, discussed earlier, of

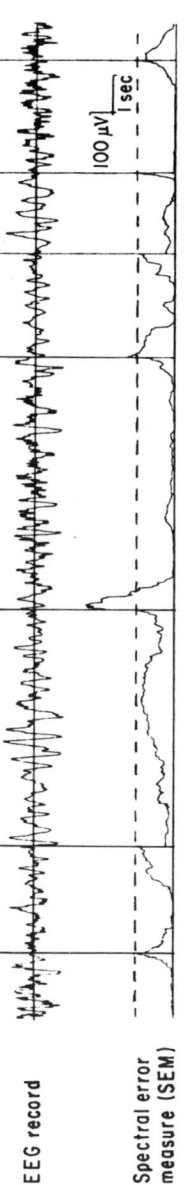

FIG. 3.8 Segment detection. Top trace shows an original EEG record containing 14-Hz spindles superimposed on a low-frequency background activity. When the spectral error measure (SEM) exceeds a threshold (dashed line) a pattern boundary is signalled (vertical lines). Data from a child age 8, sleep stage 3. [From Praetorius et al. (1977).]

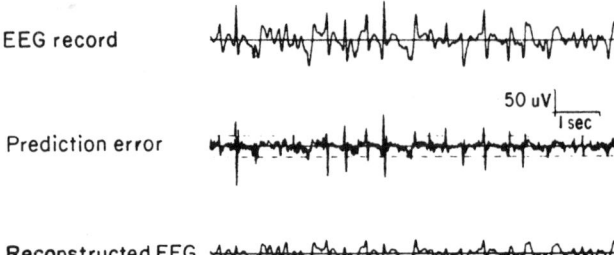

FIG. 3.9 Detection and removal of transients. Upper trace: original EEG. Middle trace: when error exceeds threshold limits (dashed lines) the values and locations of the errors are removed and stored for later use. Only the part within the dashed lines is used for processing of the SEM. Bottom trace: the remaining signal used for segmentation. [From Praetorius *et al.* (1977).]

how the 10-parameter representation of segments can be stored and simulated, with the spikes added in the reconstruction in their proper time and amplitude.

The ability to segment an EEG trace leads naturally to attempts to classify and possibly use pattern recognition in computer diagnosis. The eye notes comparable segments, and research is actively proceeding to develop

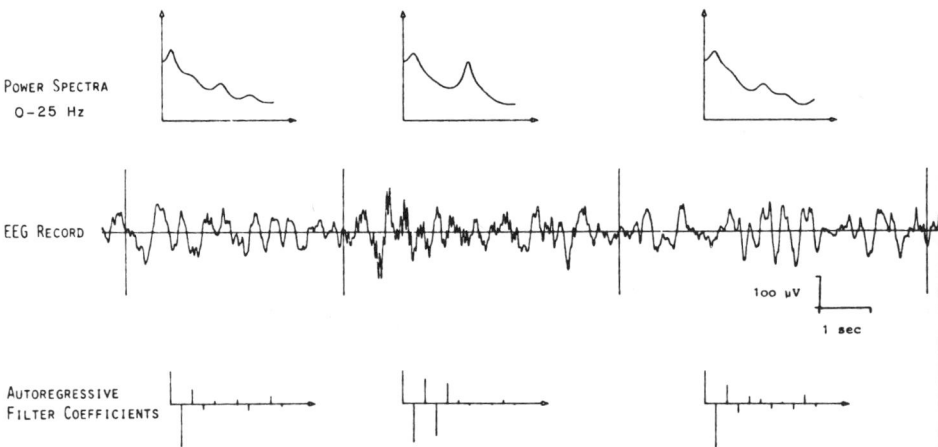

FIG. 3.10 The log power spectra of three segments of the EEG of a sleeping child with the corresponding 10 filter coefficients (arbitrary scale). Note that the 14-Hz component clearly seen in the middle segment shows up in the middle spectrum as the high-frequency component superimposed on the slower waves, but is much less pronounced in the other two. [From Praetorius *et al.* (1977).]

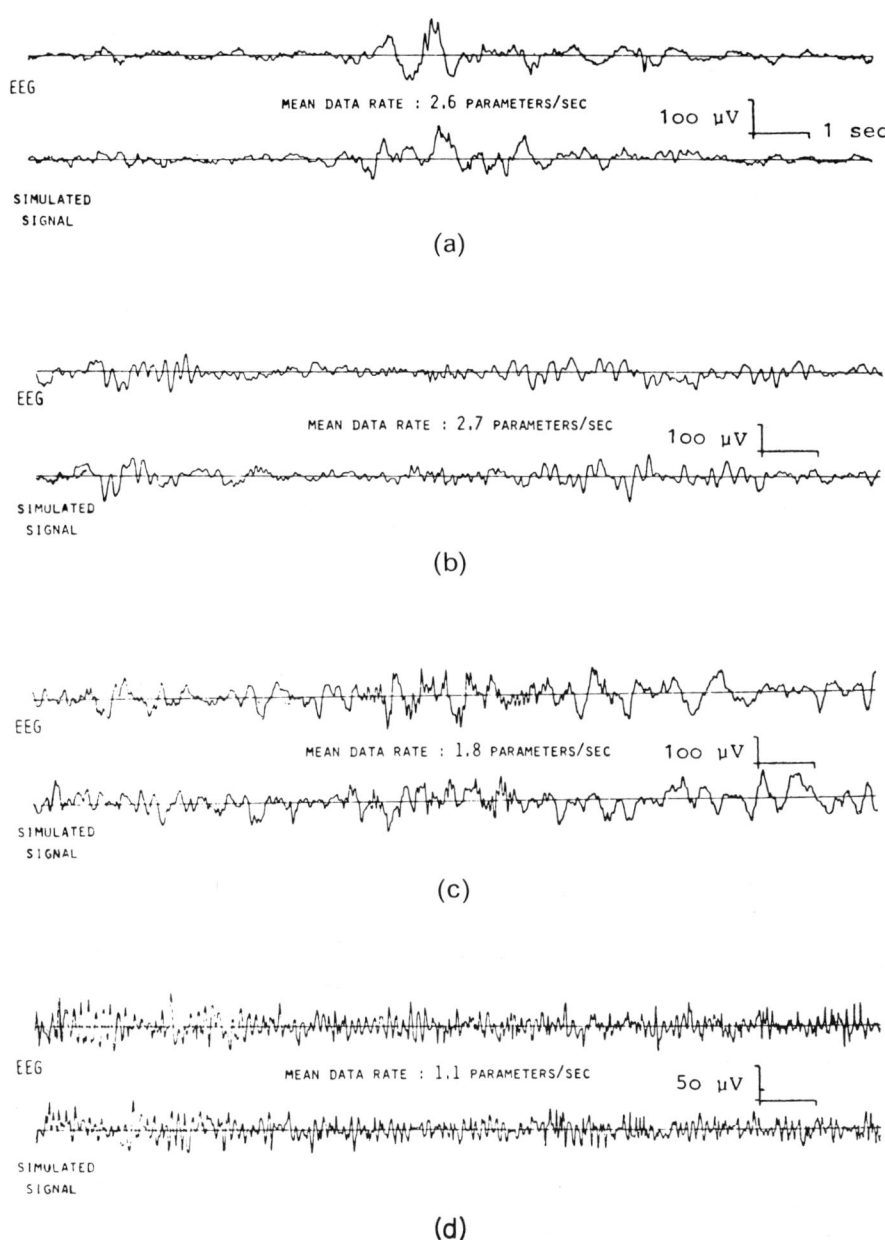

FIG. 3.11 Simulation of EEG records from stored features compared with original EEG. (a) Newborn, non-REM sleep; (b) child age 7, sleep stage 1; (c) child age 8, sleep stage 3; (d) adult, alpha EEG. [From Praetorius et al. (1977).]

54   3. ELECTRICAL PROPERTIES OF THE BRAIN

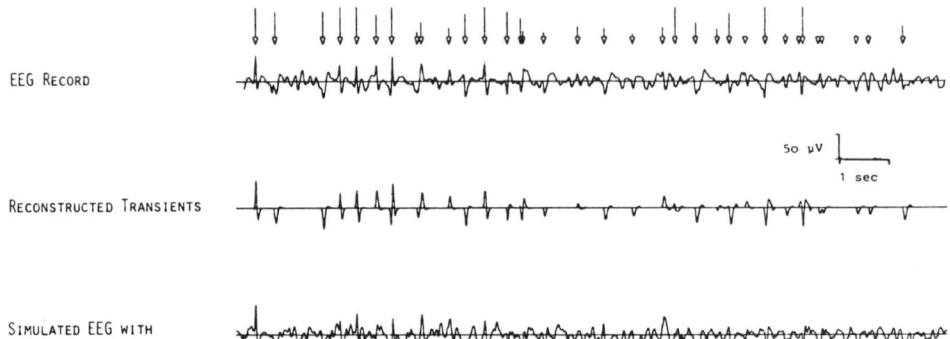

FIG. 3.12 Comparison of a stored simulated EEG and restored transients with original. Epileptic patient. [From Praetorius et al. (1977).]

appropriate statistical models for like segments so that the computer can reduce data from many like segments to a typically recognizable form. Two examples of the reduction of such feature sets from an algorithm by Ward (1963) are shown in Fig. 3.13. When a technique like this is perfected and clinically tested an enormous amount of data can be screened by computer with the neurologist's time being required for only special interpretations.

Other groups have developed digital computer techniques for fast Fourier transforms (FFTs) of EEGs. For example, John et al. (1980) developed a series of regression equations of fourth-order polynomials for the distribution of the relative power of the alpha, beta, delta, and theta bands of fronto-temporal, temporal, central, and parieto-occipital sources in resting (eyes closed) encephalograms of children as a function of their age. The regression equations yield the data of Fig. 3.14 for both left and right hemispheres of U.S. children, and the solid lines are data for Swedish children. It is seen that there is an orderly development with age of each component of the EEG and that there are also no apparent differences between country or culture. These investigators further assume that there are no differences from sexual, ethnic, or socioeconomic factors. Starting with this curve analysis method, Ahn et al. (1980) examined in the same way EEG data from normal children, those with learning disabilities, and children at risk for neurological disorders. They showed statistically what the probability is for a chance deviation from the prediction, and, in general, they found a very low level of such deviations for normal children from the predicted values of Fig. 3.14. In contrast, children at neurological risk or with learning disabilities showed a high number of deviations from the predicted values. The investigators suggested that measurement of these EEG parameters may provide a quick,

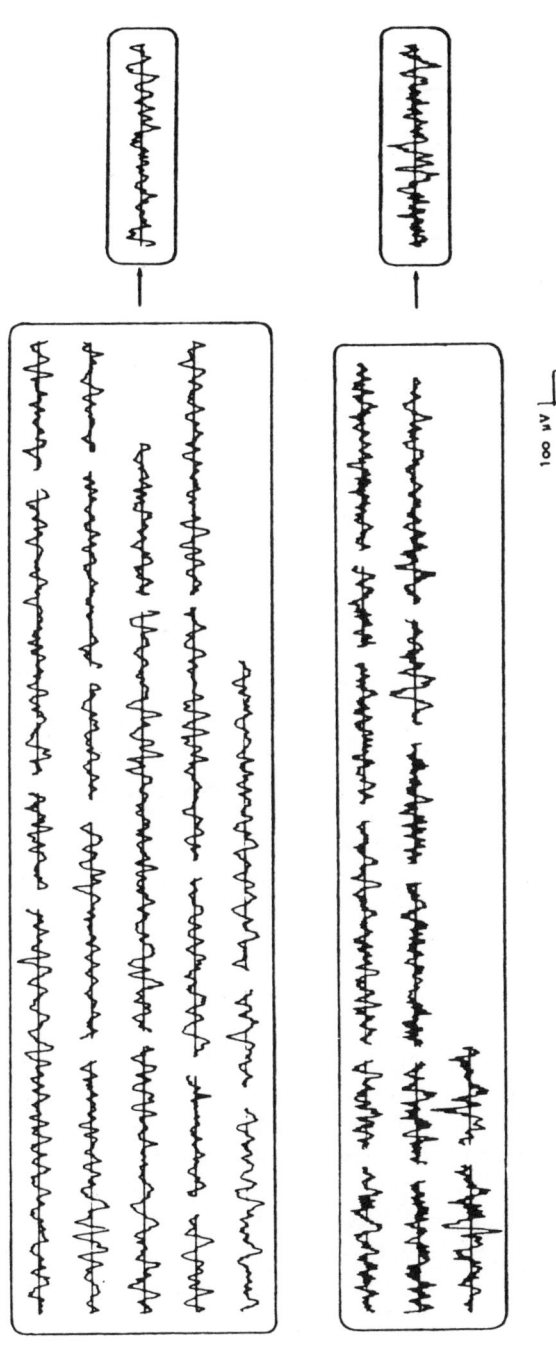

FIG. 3.13 Clustering of common feature segments from two portions of an EEG. On the right-hand side are statistical simulations of the features of both portions, based on an algorithm of Ward (1963). [From Bodenstein and Praetorius (1977). © 1977 IEEE.]

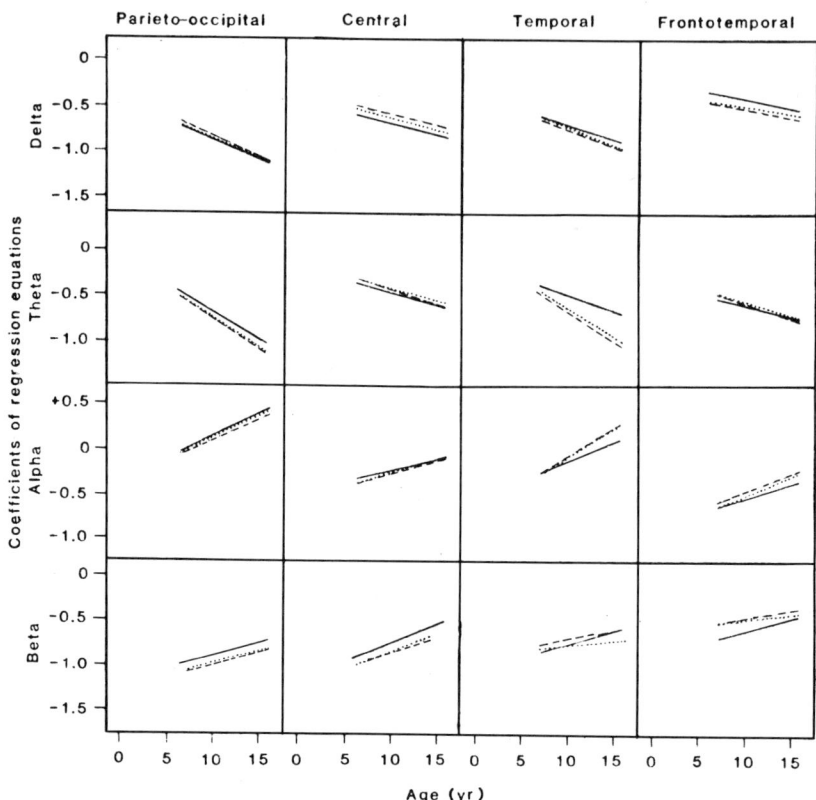

FIG. 3.14 Coefficients for statistical regression equations of fourth-order polynomials of the four main brain waves of normal U.S. and Swedish children as a function of their age. —: Swedish; ⋯: U.S., left hemisphere; ---: U.S., right hemisphere. [From John *et al., Science* **210**, 1255–1258 (1980). Copyright 1980 by the American Association for the Advancement of Science.]

reliable, and economic method of examining children who have consistent learning difficulties or who are considered at risk of brain dysfunction. While negative findings of deviation from the predictions of Fig. 3.14 may be considered inconclusive, positive findings would justify referral for more exhaustive examination.

An additional visual technique has been developed that enables a neurophysiologist to examine EEG data for brain-behavioral characteristics. This method has been called BEAM, for *brain electrical activity mapping* (Duffy *et al.*, 1979a). The outputs from 20 standard EEG electrodes are used. In addition, evoked potential (EP) signals are used. These are discussed later in this chapter. A computer interpolates instantaneous voltage values between the electrodes by a probability-mapping technique so that a map of the complete electrical field of the brain is produced at about 4-msec intervals of the theta (4–7.75 Hz) and alpha (8–11.75 Hz) activity. These are displayed on a video screen and retained as 64×64 numerical matrices in a computer memory for subsequent analysis. A series of 128 images at 4 msec each takes only 0.5 sec, so both photographs of the screen and digital storage are required for later analysis. The most extensive initial studies have involved comparison between normal and dyslexic boys (Duffy et al., 1979b, 1980, 1981; Duffy, 1981). Areas in which the waves did not propagate smoothly are obvious in the photographs. As a first important finding, Duffy et al. were able to show that such regions in the brains of dyslexic boys (defined as those $1\frac{1}{2}$ or more years behind their classmates in reading ability) involved both hemispheres and are present both at rest and during test readings. These investigators were able to develop values for classification that successfully diagnosed 80–90% of subjects not involved in the initial development of the diagnostic criteria. Duffy and his associates have additionally shown that tumors and lesions also visibly interrupt the smooth propagation of brain waves. It is not yet known whether this technique has the depth of penetration for diagnosis of deep-seated pathological conditions. Nevertheless, it is a powerful new tool whose uniqueness in the diagnosis of dyslexia has been amply demonstrated.

It is clear from the examples discussed above that a number of research groups are converging on methods of reliable computer-assisted diagnosis from EEGs. It is not expected that the role of the neurologist will be eliminated, but such fast scanning of data will greatly increase his or her efficiency.

CELLULAR ORIGINS OF ELECTROENCEPHALOGRAM BRAIN WAVES

A number of hypotheses have been proposed to explain the origin of EEG brain waves. One of the earliest was that the brain waves are envelope waveforms of the sum of action potentials of aggregates of neurons firing in synchrony. While the time course of an action potential is about 3 msec and a brain wave is 30–300 msec, it is not impossible that a large number of brain cells could produce a waveform envelope. Other suggestions involve the summation of spike after-potentials. However, it was later shown that while anoxia (deficiency of oxygen) will eliminate the spike potentials in cortical neurons, the EEG waves remain. In order to account for this it was suggested that the EEG waves arise from shifts of resting potentials owing to a local change in cell membrane potential caused by release of a transmitter chemical (Purpura, 1953). None of these hypotheses has been demonstrated to be valid. We shall now review an alternative hypothesis that appears to have validity.

The recorded EEG potentials must arise either in the neurons or from a source external to them. Let us first consider the latter possibility. The tissue of the brain is connected by ionic solution to the rest of the body. Since the potential waves are not found in the body, they must arise in the brain. But where in the brain? Outside the glial and neural cells, the media are only resistive and capacitive, and no generator potentials have been located. When a glial cell is pierced by an electrode, it exhibits the usual negative potential, but is nonexcitable. Thus, the search for the potential sources must focus only on the neurons. These cells are known to exhibit action potentials as well as presynaptic and postsynaptic potentials.

Careful experiments by Elul (1968) revealed significant facts about the behavior of the potentials of neural cells. He inserted pairs of electrodes into the cerebral cortex of cats. By doing this he was able to measure the potential recorded on each relative to a distant reference point and, in addition, the potential between the pair. His experiments have the additional variable of distance between the electrodes, and therefore he could measure the activity of progressively smaller volume of brain tissue. Note that the density of nerve cells in the brain of a cat has been estimated to be about 30,000 cells/mm^3 (Sholl, 1956). If the earlier hypothesis that the brain waves were envelopes of faster potentials from small generators, i.e., action potentials, then as the distance decreased between the electrodes the activity of the individual sources would become more prominent at the expense of the summed activity represented by the larger brain wave.

Experimental results indicated that this was not the case. The recorded activity was found to be unchanged for distances between electrodes as small as 30 μm. That is, the activity recorded between them could not be distinguished from the gross EEG or from the activity between either electrode and its reference point. Typical results are shown in Fig. 3.15 for two electrodes 30 μm apart. The frequency contents of the single electrodes and of their

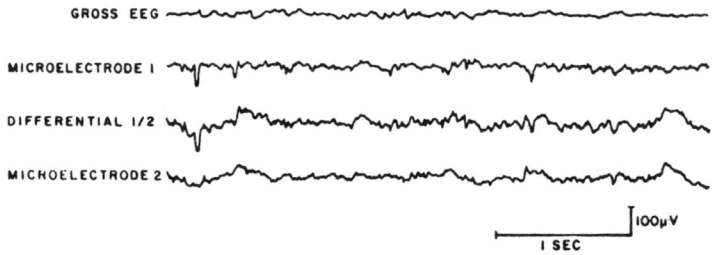

FIG. 3.15 Local wave activity between two microelectrodes placed 30 μm apart in the cerebral cortex. Frequency content in the differential record is comparable to that in records taken from each of the electrodes against a remote reference point, and each resembles the gross EEG. Cat, nembutal anesthesis, posterior suprasylvian cortex, 500-μm depth, steel microelectrodes. [From Elul (1968).]

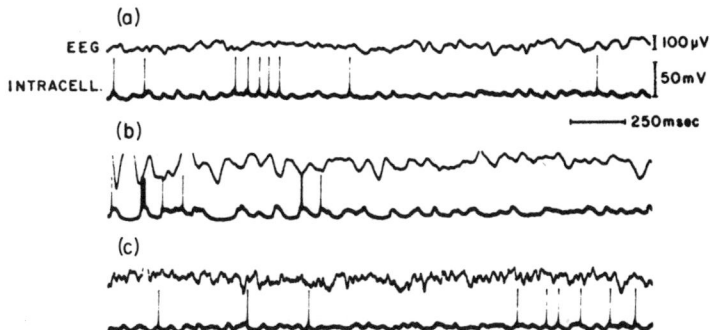

FIG. 3.16 Wave form measured with electrode penetrating a single cell (lower traces) compared with gross EEG (upper traces): (a) awake; (b) sleeping; (c) intensely aroused. (a)–(c) from same cell in posterior suprasylvian cortex, 750-μm depth, micropipette filled with KCl solution. Note that the peaks of the spikes (action potentials) are off scale. [From Elul (1968).]

difference resemble that of the EEG. The experiment also shows that the unitary generators cannot be larger than the 30-μm electrode separation. This restriction on size eliminates the possibility of sources larger than single cells.

Further exploration was made with a single electrode that penetrated a neural cell (Elul, 1964, 1967). Figure 3.16 illustrates a typical result from over 100 cells studied in cats. The upper trace in each part is the EEG, and the lower trace that of the cell. In (a) the animal is awake, in (b) it is sleeping, and in (c) it is intensely aroused. The typical spike (action) potentials are seen in the recordings from the neurons, but, more important, there is a 10–20-mV oscillation of the cell potential of the same frequency content as the EEG for each state of the animal; compare (b) and (c) in particular. Note, however, that the EEG trace is not a replica of the cell potential.

One may calculate the spectral intensity versus frequency (power spectrum) for the EEG and the cell. This is done by analyzing the data for frequency and amplitude of the individual waves. The intensity (see Vol. I, p. 206) is proportional to the square of the amplitude. The sums of the intensities of each frequency are then plotted versus frequency, as in Fig. 3.17. It is seen that the same shape is present for both the cell and the EEG. Note that Fig. 3.17 shows that in a cat's brain most of the spectral energy is in the 1–4-Hz band. For human beings, the EEG has been shown to have a maximum activity in the 8–12-Hz band (Saunders, 1963).

Elul considered the possibility that the generator source was outside the cell. If this is the case, then the potential just exterior to the cell membrane should be equal to or exceed that of the cell interior, because the signals within the cell would be a passive response. Figure 3.18 illustrates the wave activity recorded from an electrode placed just outside the cell. Its magnitude

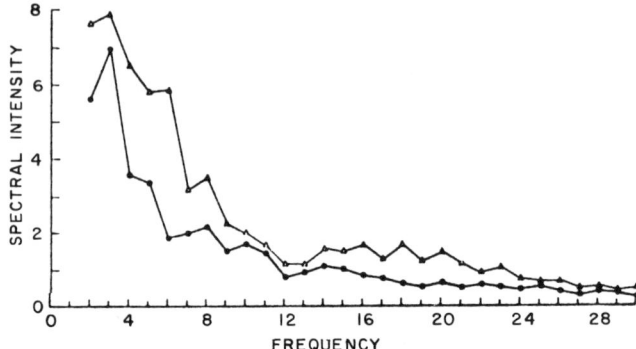

FIG. 3.17 A comparison of the EEG power spectrum, intensity vs. frequency, with the power spectrum of neuronal intracellular activity. △—△: EEG; ●—●: neuronal waves. [From Elul (1968).]

is about 1% of that measured inside the cell, as shown in Fig. 3.16. Thus, the conclusion is justified that the slow oscillations characteristic of EEG have an origin within the neural cells, but not from their action potentials.

Can the magnitude of the output of these small generators compare with that of the EEG? The experiments show that although the internal potential changes have a magnitude 100 times greater than that of the EEG, the potential drop across the cell membrane reduces the potential changes to the magnitude of the EEG. A second, more challenging question is why the individual outputs of these cells, assumed to be unrelated, do not effectively cancel and yield a zero output. Elul achieved a reasonable answer to this

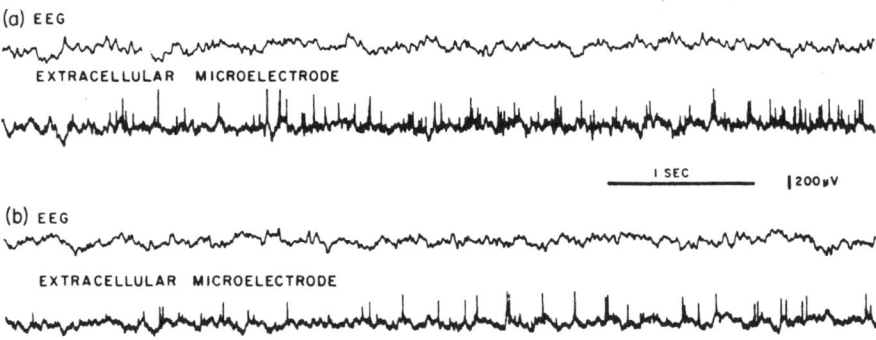

FIG. 3.18 Two examples, a and b, of extracellular recording of wave activity taken just outside a neuron compared with EEG. Note the change in voltage scale from Fig. 3.16. In Fig. 3.16 the intracellular base wave height is about 20 mV, whereas in this figure it is about 200 μV. In contrast to the intracellular recording (Fig. 3.16) the extracellular recording does not exceed the EEG. Posterior suprasylvian cortex, 500-μm depth, NaCl-filled micropipette. [From Elul (1968).]

question by a careful examination of the statistical behavior of an ensemble of the individual potential changes of the neurons. This can be seen by the following considerations.

The *central limit theorem* of statistics states that the sum of a large number of individual probability distributions always tends to assume a normal distribution regardless of the nature of the component distribution provided that (1) the individual distributions are either independent or at least not linearly related; (2) they possess a mean; and (3) they have a finite standard deviation. The measurements of the amplitudes of the cell potentials were referenced to a grounded electrode, and therefore the data can be examined statistically. No relation between the outputs from any adjacent cells was observed, so at least no linear relation appears to exist.

What is then measured statistically is the amplitude probability distribution for the EEG and the cell potentials. The details were obtained as follows. A 16-sec epoch was broken down into 400 samples/sec and, with a computer, the amplitudes were stored in 100 bins. A smoothed histogram of the amplitude probability was then plotted for both the EEG and the intracellular data. These are shown in Fig. 3.19 as the shaded areas. Normal distributions with standard deviations are plotted as the solid lines. The EEG data are seen to have a normal distribution and are in agreement with the earlier studies of this by Saunders (1963). The histogram for the intracellular recording from a single neuron is not a good Gaussian function. This deviation from Gaussian suggests that there may not be a completely linear independence of the cell generators. However, the results are close to a normal distribution, and this indicates that while total independence of the generators in different cells may not be likely in a system with such high connectivity as

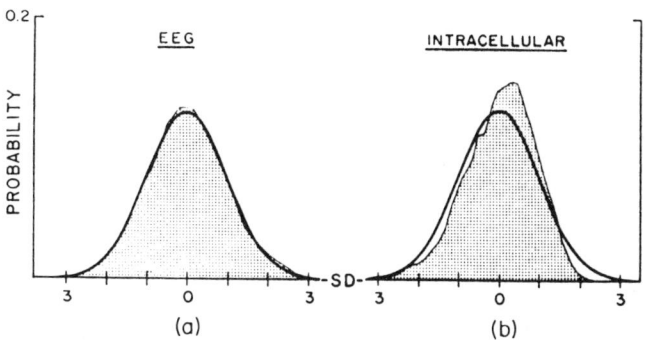

FIG. 3.19 Amplitude probability distributions of cortical EEG (a) and of neuronal wave activity (b) over the same period in the same preparation. Histograms were computed for a 16-sec epoch at 400 samples/sec. Normal distributions with corresponding standard deviations are shown by the solid lines. [From Elul (1968).]

the brain, they may be considered as such for a first approximation. On this basis Elul concluded that the EEG may be accounted for as the normally distributed output from a combination of the activity of many independent (or nonlinearly related) neural cell generators.

EVOKED POTENTIALS

A sensory stimulus is processed by specific groups of brain cells. In other words, they are activated by the stimulus. It is therefore not expected that the potential change from their collective electrical activity can be measured. Their signals at the scalp are representative of only a small fraction of the total number of brain neurons, and the signals are accordingly small compared to the total electrical activity of the brain as measured by the EEG. Typically, the EEG potentials have an amplitude between 5 and 100 μV, whereas the potential from a group of stimulated cells is usually no larger than 5 μV and is measurable as small as 0.5 μV.

Caton (1875), the discoverer of EEG on the exposed brains of animals, was also the first to report on electrical changes in the brain upon stimulation of a sense organ. He put one electrode of a voltmeter on the exposed surface of the brain and the other electrode on a cut surface on the neck as a reference electrode and, using a lamp as a stimulus to the eye, recorded a potential change.

It was mentioned earlier that there is about a factor-of-10 decrease in potential of brain waves in measuring from the scalp instead of from the exposed brain. The first investigator to demonstrate what are now called *evoked potentials* (EP) on the scalp was Dawson (1947). He stimulated the same wrist nerve many times on a human subject while recording the EEG. He assumed that while the EEG was basically random, the response from the stimulation would have a constant waveform. He photographed the EEG trace displayed on an oscilloscope screen following each stimulus. When he superimposed the traces, the random alpha rhythm summed to a heavy background above which, because of the summed amplitudes of the common wave forms, the EPs were visible.

Modern computer techniques now permit the summation of EPs to be exhibited directly by automatic subtraction of the alpha or beta backgrounds. A typical series of summations in response to light flashes is shown in Fig. 3.20, where it is seen that the magnitude of the common waveform of the EP increases with the number of stimuli. The alpha and beta random background has been subtracted. Other techniques have evolved. For example, a repetitive stimulus will cause a repetitive response and a series of filters that select only this fundamental plus higher harmonics, i.e., a

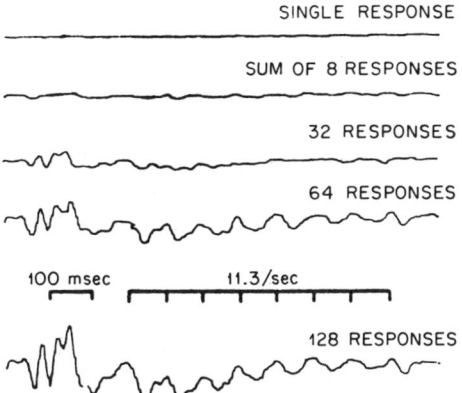

FIG. 3.20 Detection of evoked response in man by addition of single responses with EEG subtracted. A periodic flashing with intervals varying from 1.1 to 1.4 sec. Scalp electrodes, pariento-occipital linkage. [From Brazier (1960).]

Fourier analyzer, are used. For reviews of this latter technique, see Regan (1972, 1977b).

Research on *somatosensory evoked response potentials* (SERP) has been extensive (see, e.g., Sclabassi *et al.*, 1977). Some research has been reported on evoked potentials due to olfactory (Allison and Goff, 1967) and gustatory (Plattig, 1968) stimulation. By far the most extensive research and application to date of evoked potentials is that from visual and audio stimuli, and the subsequent discussions will cover some of the important techniques and results from these two.

The literature on *visual evoked potentials* (VEP) or *visual evoked cortical potentials* (VECP) and *auditory evoked potentials* (AEP) is extensive, and unfortunately much of the early work is contradictory. With the development of computers at reasonable cost and standardization of techniques, EPs are emerging as a useful medical diagnostic tool, although there are still some inherent difficulties, which we shall first illustrate by specific examples.

Visual Evoked Potentials

An often quoted simulation by Cigánek (1961) is a general scheme drawn for VEPs recorded from midline occipital–parietal electrodes. This is shown in Fig. 3.21, in which negativity at the electrode results in an upward deflection. The rhythmic after-discharge is shown on a different time scale. We shall see that this after-discharge is not of this periodic form, and the entire schema is rarely seen in experiments. However, it is a good starting point from which to examine experimental data.

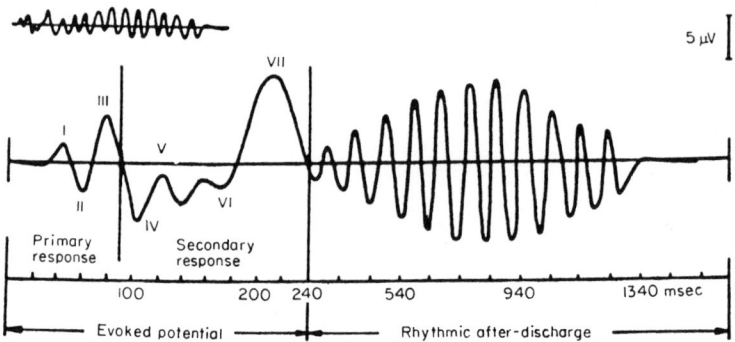

FIG. 3.21 Simulated schema for the human response to light flashes recorded from midline occipital–parietal electrodes. A different time scale is used after 240 msec for the rhythmic after-discharge. At the upper left the evoked potential and after-discharge are shown on a uniform time scale. Stimuli applied at $t = 0$. [From Cigánek (1961).]

The various peaks up to 240 msec are numbered. Some investigators label the peaks and valleys by letters, while others label the peaks as $P_1, \ldots, P_n$ and $N_1, \ldots, N_n$ successively as positive (P) and negative (N). Unfortunately, uniform standardization of all of the waves is unachievable (see Kinney, 1977; Shagass, 1972; Regan, 1972; Bergamini and Bergamasco, 1967). An example of five VEPs recorded from the same subject over a 2-month period is shown

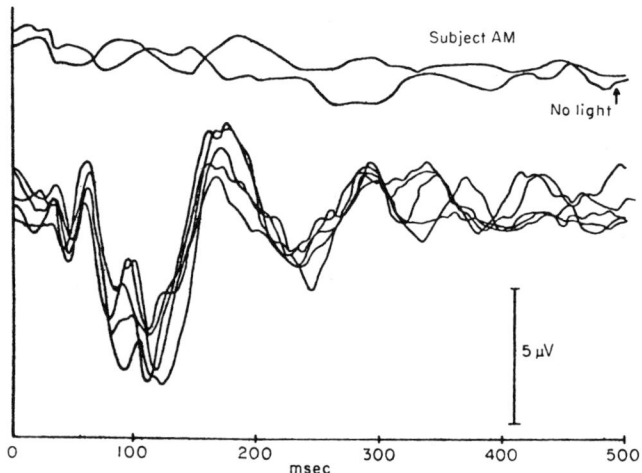

FIG. 3.22 Five VEPs recorded from the same subject in response to the same stimulus over a 2-month period. The light flashes at $t = 0$. Control runs with no light shown at the top. [From Kinney et al. (1972).]

EVOKED POTENTIALS 65

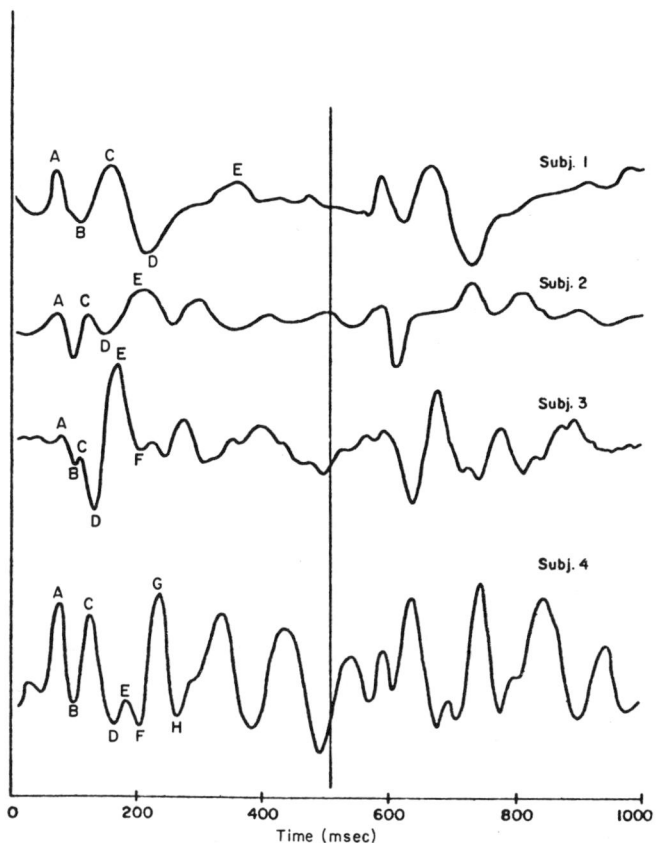

FIG. 3.23 VEPs from four individuals under identical conditions (flash rate of 2 sec^{-1}). [From Kinney and McKay (1971); also in Kinney (1977).]

in Fig. 3.22 and from four individuals under identical conditions in Fig. 3.23. Part of the problem in duplication of results on a single individual is electrode placement, and differences between individuals make achievement of common waveforms very difficult. An example of the varieties of VEP response from different positions is shown in Fig. 3.24. Note both the time difference in the occurrence of the peaks and the sign reversals of the potentials. Other artifacts disturb the patterns, such an acoustical or somatic stimuli during a VEP test.

In spite of these difficulties, however, certain peaks and valleys are essentially constant for both animals and man, and the time of occurrence after the stimulus, called *latency*, can be measured under a variety of conditions

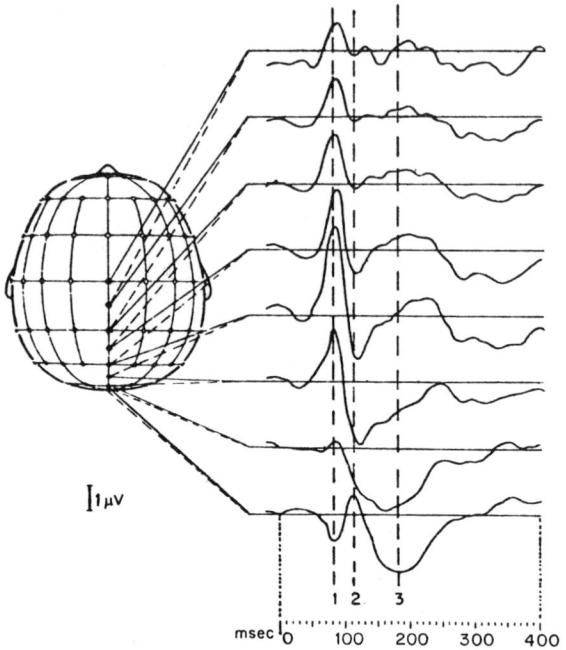

FIG. 3.24 Eight VEPs from the same stimulus recorded from different cortical locations. [From Kinney and McKay (1971); also in Kinney (1977).]

or physical states, e.g., illness and drug use. There are important applications of this technique to measurements on animals and young children who cannot vocalize the effects of various stimuli.

Diagnostics of Children Using Visual Evoked Potentials

A useful technique in VEP is a checkerboard-pattern projection in which the size of the squares can be varied. With this method corrective lenses for glasses have been fitted on small children. The following description of experiments explains the technique.

The VEP response is different for a pattern-field projection and a blank-field projection. This is illustrated on the left-hand side of Fig. 3.25 for two individuals. The solid line is from a blank field and the dashed line from a patterned field. If the blank-field VEP is subtracted from that of the patterned field, the resulting curve is the effect of the pattern. This is shown on the right-hand side of Fig. 3.25. Although the results for two individuals are not identical, strong similarities exist. In particular are the two latencies, the peak at about 90–100 msec. and the valley from 180 to 200 msec.

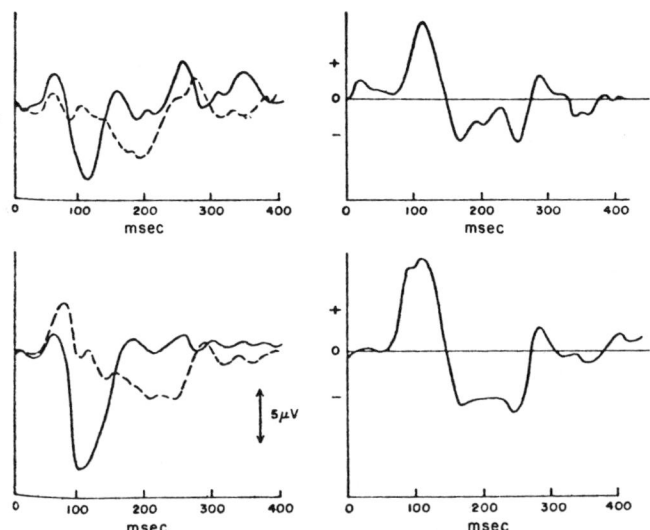

FIG. 3.25 Left curves: VEPs for two human subjects. Solid line: blank field projection; dashed line: pattern field projection. Right curves: effect of pattern field obtained by subtracting solid line from dashed line. [From Kinney and McKay (1974); also in Kinney (1977).]

There is a relation between visual acuity and the check size of the checkerboard pattern. That is, a pattern size below that of the resolving power of the eye will appear as a blank pattern and accordingly eliminate the pattern part of the evoked potential, the right-hand side of Fig. 3.25. There are obviously different degrees of blurring as the limit of visual resolving power is approached. Instead of changing the pattern size, lenses can be introduced between the viewer's eye and the pattern to produce an artificial blurring. This is shown in Fig. 3.26 for the two latencies with different check sizes. The upper curves are for the 90–100-msec latency and the lower curves for the 180–200-msec latency. It is seen that the percent loss of the pattern part of the evoked potential increases with blurring and the VEP is maximum for the small unblurred pattern.

The application of this technique is clear. Defective vision in a child can be optimally corrected by measuring the VEP from projected patterns while introducing a selection of lenses in his visual path. A method of determining proper corrective lenses for children by VEP that is even more rapid has been reported by Regan (1973). The fitting of glasses on small children is not yet in general practice, although VEP now permits this. Experiments in which growing monkeys have had their vision impaired in one eye have shown abnormalities in the development of the visual cortex (Hubel, 1979). It is therefore important that children with vision defects wear corrective glasses at an

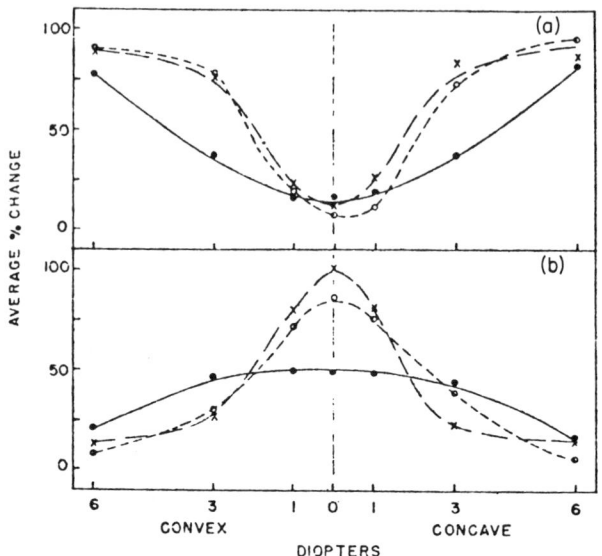

FIG. 3.26 Percentage change in magnitude of the response at the two latencies of Fig. 3.25 [(a) 90–100 msec; (b) 180–200 msec] as a function of blurring lenses for three different pattern sizes. Check size given in minutes of arc. Note from Fig. 3.25 that the first latency is negative, whereas the second is positive. For this subject, the second latency is the more sensitive to decreased visual acuity, decreasing from 100% to 50% for a check size decrease of 46 to 12 min of arc with no lens. ●—●: 46 min; ○—○: 20 min; ×—×: 12 min. [From Harter and White (1968).]

early age so that the cortex can develop normally even if glasses are not worn regularly at a later age.

Another important eye diagnosis on small children by VEP is for amblyopia. This is a condition of low visual acuity that is uncorrectable with lenses. In one form of amblyopia the infant starts to squint, and the squinting eye rapidly loses acuity. If left in this condition, the loss of acuity will be permanent. This condition can be reversed by occluding the good eye with a patch or a piece of transluscent paper. The problems in very young children are both to diagnose the condition as early as possible and to measure the progress of recovery with treatment. Regan (1977a) has been able to perform this measurement using VEP with a Fourier analyzer. In viewing a checkerboard pattern the amblyopic eye will perform nearly as well as the normal eye for large checks, but will perform poorly as the checks decrease in size. Figure 3.27 shows a plot of the VEP amplitude as a function of check size for the normal and amblyopic eyes. As treatment proceeds the extent of recovery is readily evaluated by repeating the measurements.

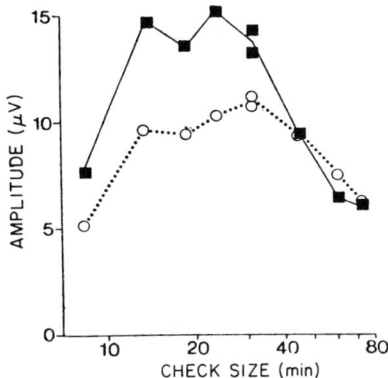

FIG. 3.27 VEP amplitude versus check size (in minutes of arc) for the normal and amblyopic eyes of a child who squinted during infancy. ■—■: normal eye; O ··· O: amblyopic eye; ···: noise. [From Regan (1977a).]

Color blindness is also measurable by VEP and can be detected in very young children (Regan and Spekreijse, 1974). In fact, VEP results for the spectral sensitivity of the eye are identical with those obtained by flicker photometry (Vol. II, Fig. 6.81). This is shown in Fig. 3.28, in which the crosses are psychophysical data from conventional flicker photometry and the solid circles are data obtained by VEP. The line is the curve established by the International Commission on Illumination.

Diagnosis of Optic Neuritis Using Visual Evoked Potentials

An early diagnosis of multiple sclerosis is usually difficult unless there is clinical evidence of involvement of the central nervous system at more than one site. This difficulty stems from the changing patterns of the disease with time in its propensity for spontaneous remission or reappearance. Both demyelination of nerve fibers and the presence of synaptic damage are common in people with multiple sclerosis. Standard neurological tests may suggest the onset of the disease, but further tests are desirable. Evoked potentials from stimulation of a nerve, for example, a vibrator applied to a finger (Sclabassi et al., 1977), have been used, but not always with consistent results. The disease can attack optic nerve fibers, and such damage is very sensitive to VEP. The resulting condition, optic neuritis, is usually painful and frequently results in loss of visual acuity of one or both eyes. Some patients develop this condition in only one eye, and the VEP of this eye can be compared with that of the good eye.

Halliday et al. (1972) used VEP from flash and checkerboard-pattern reversals on 17 healthy persons and 19 suffering from optic neuritis. They

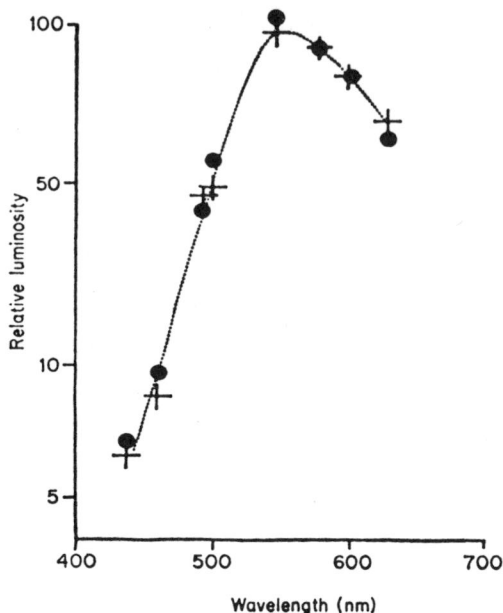

FIG. 3.28 Crosses: relative luminosities by flicker photometry. Circles: spectral sensitivity by evoked potentials. Solid line is the International Commission on Illumination (ICI) curve. [From Regan (1970).]

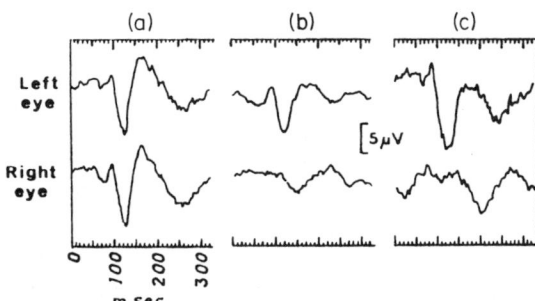

FIG. 3.29 Pattern-visual evoked responses from a midline occipital electrode from the left and right eyes of a healthy subject (a) and two patients who are recovering from acute attacks of optic neuritis in the right eye with onset 4 weeks (b) and 3 weeks (c) previously. Note both the smaller peak amplitude and increased latency in the affected eyes. [From Halliday et al. (1972).]

examined the latency of the 90–100-msec peak. This latency occurs following a very short flash, which they used. Their blank-flash-to-pattern-reversal detection instrument required 35 msec to change, and the latency from flash to pattern was therefore increased by this amount to about 120 msec. Figure 3.29a shows the latency (a negative peak) at 120 msec from a pattern flash for a normal individual. Figures 3.29b and c show the curves for two patients recovering from optic neuritis of the right eye, with the onset having occurred at 4 and 3 weeks prior to the measurements, respectively. Histograms of the latencies of this peak from a pattern-evoked response for the patients and healthy individuals are shown in Fig. 3.30. There was no observed correspondence between visual acuity and latency during recovery of the patients. That is, with subsidence of the neuritis visual acuity tended to recover but the latency did not.

In spite of the clear increases in the latencies in the pattern-field evoked responses of the patients, no increases of the latencies were observed following blank-field light flashes alone. The investigators suggested that different groups of optic fibers are responding to the two different types of stimuli. While the pattern-field response is thought to arise from central vision, the flash response may well involve the peripheral vision. Thus, it may

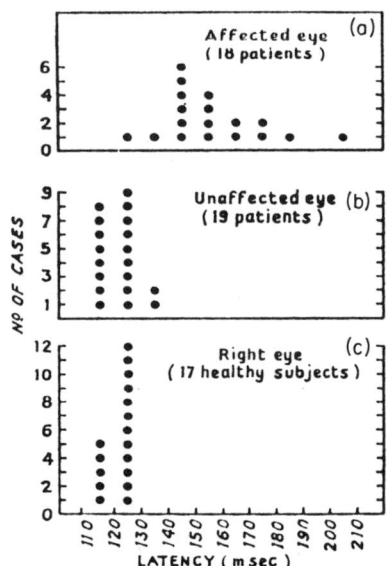

FIG. 3.30 Distributions of the peak latencies of the major positive component of the pattern VEP (a) from the affected eye of 18 patients with optic neuritis, (a) from the unaffected eye of 19 patients with optic neuritis, and (c) from the right eye of 17 healthy subjects. [From Halliday et al. (1972).]

tentatively be concluded that it is the central vision fibers that are primarily affected by optic neuritis.

It is tempting to suggest that partial demyelination of the optic nerve fibers causes a slowing of transmission, as indicated by the increased latency of 20–30% in Fig. 3.30. However, reference to Vol. II, Fig. 1.5, shows that the speed of impulse conduction in myelinated nerve fibers is faster than that in nonmyelinated ones of the same diameter only down to 1 μm. For smaller nerve fibers, nonmyelinated ones have a faster velocity. Furthermore, this figure shows that the change in conduction speed with diameter is large, that is, conduction speed in a myelinated fiber is linear with diameter, whereas in nonmyelinated fibers the speed varies as the square root of the diameter. Unless one knows the diameters and degrees of myelination of the fibers, very little can be said with any confidence. An alternative model is that a partial failure of the transmission at the synaptic junction has occurred and the reduced transmission of the signal at the synapses may be the cause of the increase of the VEP latency (Halliday *et al.*, 1972). See p. 76 for the results of audio evoked potentials on a patient with multiple sclerosis.

Origin of Evoked Potentials

What is the origin of the evoked potentials? It will be shown later in this section that most of the characteristic peaks in the auditory evoked potentials have been traced to their location in the auditory pathway. The visual evoked potentials are not so completely understood. However, it has been shown that single cells in the visual cortex in a cat exhibit action potentials with an average latency of the VEP. Fox and O'Brien (1965) inserted microelectrodes into the visual cortices of 40 cats and measured the time of the action potential after a visual stimulus for a large number of times (thousands). The oscilloscope traces were stored in a multichannel analyzer. If different neural elements sequentially activated relate to the sequential appearance of the various VEP waves, then individual cells should exhibit high firing probabilities in the time of one of the VEP waveforms. If, however, the firing frequency of individual cells does not correspond uniquely to any individual component of the VEP, but corresponds instead to a number of components, then this would suggest that the firing of each cell is related probabilistically to the VEP rather than having a one-to-one correspondence. Data are shown in Fig. 3.31. The upper curves show the distributions of the time of firing of the first spike after the flash, and the lower curves show the VEP. It is seen that there is a strong correlation between probabilistic firing and the VEP. The investigators give two possible interpretations, which are not yet resolved: (1) the VEP consists of summed asynchronous cell firings, with the cells in a given localized area having approximately the same probability of firing; or (2) the VEP may be compounded of electrical activity from a number of

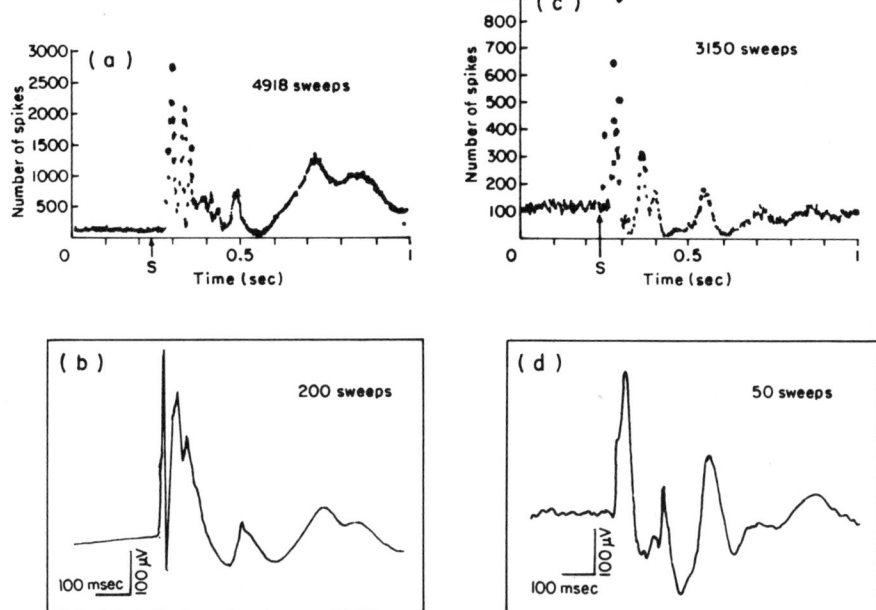

FIG. 3.31 Relation between the probability of firing of a single cell and evoked potential wave form. (a) Time distribution of spikes in a single cell in the visual cortex of a cat after stimulation by 4918 light flashes. (b) Averaged evoked potential recorded from the same microelectrode after cell death. (c) and (d) Example from another cell. [From Fox and O'Brien, *Science* **147**, 888–890 (1965). Copyright 1965 by the American Association for the Advancement of Science.]

sources, local or distant. The sources could originate from local potentials associated with electrotonic spreading within the dendrites. This is discussed in Chapter 5. It is not yet known with certainty where the origin of VEPs lies. However, experiments with audio evoked potentials, described in the next section, have conclusively demonstrated that different latencies of peaks correspond to synapses at different locations along the audio pathway. In order for the VEP experiment of Fox and O'Brien to be consistent with the audio results, there should be a close dendritic coupling between cells at the different points of the visual pathway. That is, activity at a later synapse may cause an earlier cell to fire. This hypotheses has not yet been demonstrated.

Auditory Evoked Potentials

The *auditory evoked potential* (AEP), or sometimes *audio evoked response* (AER), is quite different from VEP, with seven clearly identifiable peaks and a much shorter latency than that of the visual. Potentials are customarily

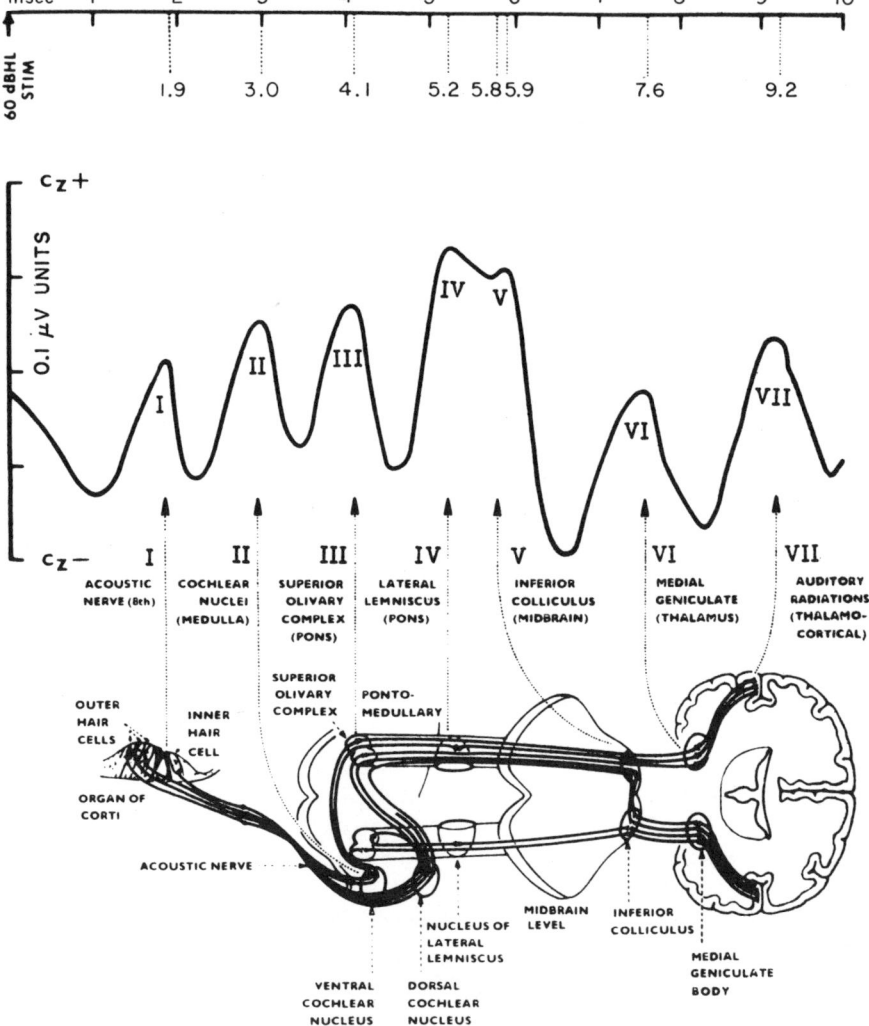

FIG. 3.32 Evoked auditory potentials from clicks 60 dB above hearing threshold shown in middle part of figure (C_z^+ and C_z^- are vertex positive and negative voltage deflections, respectively). Upper part of figure shows time of each peak after the click (called latency). Lower part of figure is the pathway of an auditory signal from ear through brain and the identification of each stage with corresponding potential peaks. [From Stockard *et al.* (1977).]

measured between a scalp electrode in the center midline, or vertex, of the brain and the reference electrode at an earlobe. Although square wave tones are sometimes used, the more common stimulus is a short click sound at a rate of about 10/sec. The stimulus is presented both to individual ears and as a general sound to both ears. The evoked response of normal human beings is shown in Fig. 3.32. It is seen that there are seven major deflections within 10 msec of the sound. These were first described by Jewett and his associates (1970) and Jewett (1970). Unlike VEP, the responses are independent of the subject or his state of arousal. That is, the responses can be elicited from a newborn or from a patient asleep or comatose. By inserting probes in the brains of experimental animals to determine directly the latencies of various nuclei of cells (Jewett, 1970) and by producing artificial lesions that eliminate some of the deflections (Lev and Sohmer, 1977), several of these deflections have been positively identified, as illustrated in Fig. 3.32. Thus, a powerful tool now exists for detection of some of the functional problems associated with portions of the brain. A few examples will be discussed.

The AEP amplitude and latency depend on the signal intensity. This is shown for a person with normal hearing in Fig. 3.33 for the left ear on the left-hand side of the figure and for binaural hearing on the right. The sound intensity in decibels is given by the numbers on the left-hand side. If one plots the latency of appearance of each of the peaks versus intensity, Fig. 3.34 results. Thus, a comparison of patient data with this standard yields a measure of degree of deafness and probable location of the cause. This tool is extremely useful in determining the degree of deafness in comatose emergency patients and infants, and has even been used to determine malingering in cases of industrial deafness (Heron, 1968).

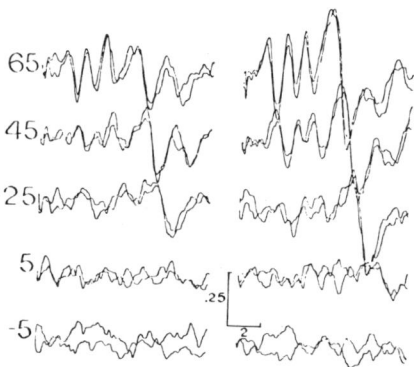

FIG. 3.33   AEP responses in a normal subject as a function of signal intensity (65 to −5 dB) with monaural (left) of binaural (right) presentation. Two separate runs shown for each. [From Starr and Anchor (1975). Copyright 1975 American Medical Association.]

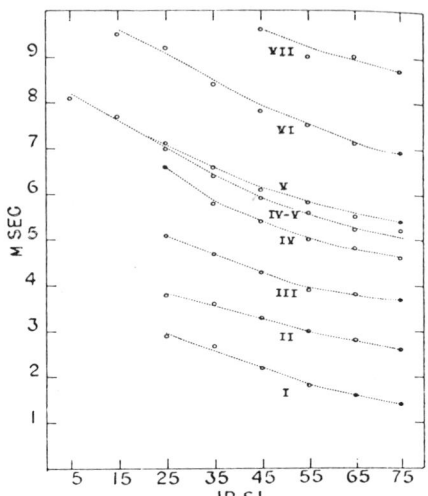

FIG. 3.34 Latency of each of the seven components as a function of signal intensity (decibel sound level) from curves such as Fig. 3.33. Data are averages from six subjects in response to monaural clicks. [From Starr and Anchor (1975). Copyright 1975 American Medical Association.]

Figure 3.35 shows the left-ear (upper) and right-ear (lower) AEP of a patient with multiple sclerosis, which involved the brain-stem structures. (The double curves are the superpositions of two separate tests taken in the same session.) Note that only wave I of normal latency was evoked. All others were decreased in amplitude and shifted in latency. The calibration lines in the figure represent 0.1 μV and 2 msec.

The upper curve in Fig. 3.36 is the AEP of a patient with a brain-stem tumor. (Note that the midbrain mentioned in the legend is part of the brain stem.) The lower curve is for a normal subject. Calibration lines are 0.25 μV and

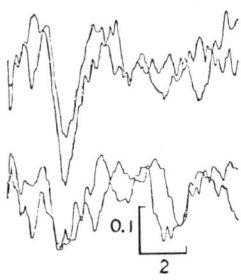

FIG. 3.35 AEP from a patient with multiple sclerosis with clinical involvement of brain stem structures. Monaural click at 75 dB on left (top) and right (bottom) ears. Note that wave I has normal latency (Fig. 3.32), but the remainder are reduced in amplitude and deviate substantially in latency from normal. [From Starr and Anchor (1975). Copyright 1975 American Medical Association.]

EVOKED POTENTIALS   77

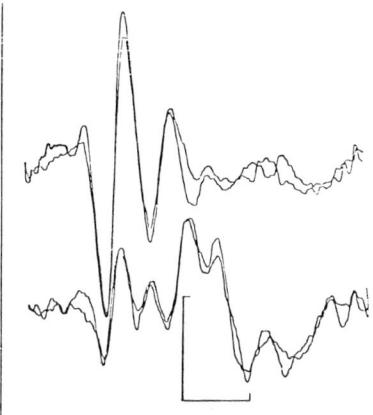

FIG. 3.36 AEP from patient with a tumor that destroyed the midbrain (top) compared with AEP of a normal subject (bottom). Note prolonged latency of second and third peaks and an almost total loss of resolution of higher peaks. Calibration lines are 0.25 μV and 2 msec. [From Starr and Anchor (1975). Copyright 1975 American Medical Association.]

2 msec. It can be seen that in the patient there is almost a total loss of components after the third upward deflection as well as a prolonged latency of the second and third peaks. Measurements on a patient with brain-stem glioma (cancer of glial cells) yielded the AEP of Fig. 3.37. Note the presence of only component I in the left-ear stimulation (upper curves) and no responses in right-ear stimulation (lower curves).

It is important to diagnose and correct deafness in young children and to distinguish it from autism. Rapin and Graziani (1967) first demonstrated what can be accomplished by AEP. The upper trace of Fig. 3.38 was unchanged in a 21-month-old child in the presence of a 109-dB tone at 500 Hz.

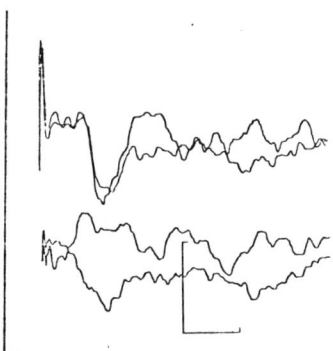

FIG. 3.37 AEP of left (top) and right (bottom) of patient with brain-stem glioma. Calibration lines are 0.25 μV and 2 msec. [From Starr and Anchor (1975). Copyright 1975 American Medical Association.]

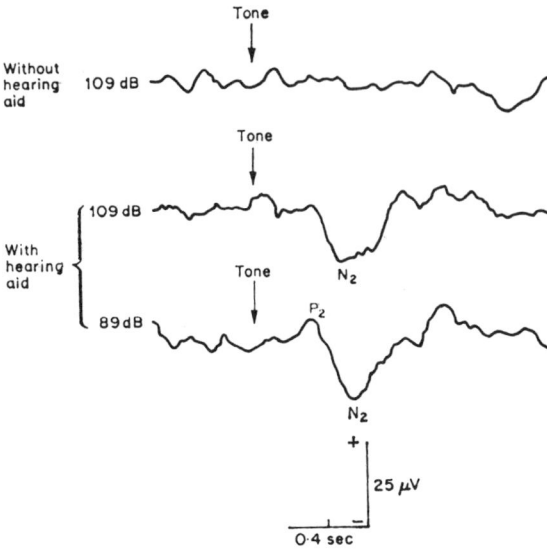

FIG. 3.38 AEP of 21-month-old girl taken with 500-Hz sound at indicated intensities. The upper trace remained unchanged with and without the sound. Lower traces show that a hearing aid facilitated AEP, and therefore hearing ability, at both 109 and 89 dB. [From Rapin and Graziani (1967).]

When a hearing aid was fitted to the child, a strong response was observed to the tone even when lowered by 20 dB. All three traces of Fig. 3.38 were taken on the same day.

The latency of both visual and auditory evoked potentials is initially large in newborn children but decreases significantly in a few weeks. This can be seen for VEP in Fig. 3.39, in which the mean latency for the second positive peak is plotted against age in days. The decrease is quite striking, and the reason is not known. A change in latency of the peaks in auditory evoked potentials also occurs with age in infants. This is illustrated in Fig. 3.40 by the solid lines on the three graphs at the left. The dashed lines are the latencies of infants who have a high risk of central-nervous-system damage. Although this illustrates the decrease of latency of waves I, III, and V with age, not much difference is noted between the latencies of normal and high-risk infants. On the right-hand side of Fig. 3.40 is shown the amplitude of the waves in normal and high-risk children. The differences are quite dramatic. The investigators of this effect, Salamy et al. (1980), have not yet been able to determine the cause of the differences in amplitude. They suggest differences in synaptic efficacy, neural synchrony, or simply a delay in development. An infant born at risk is predisposed to insults of the central nervous

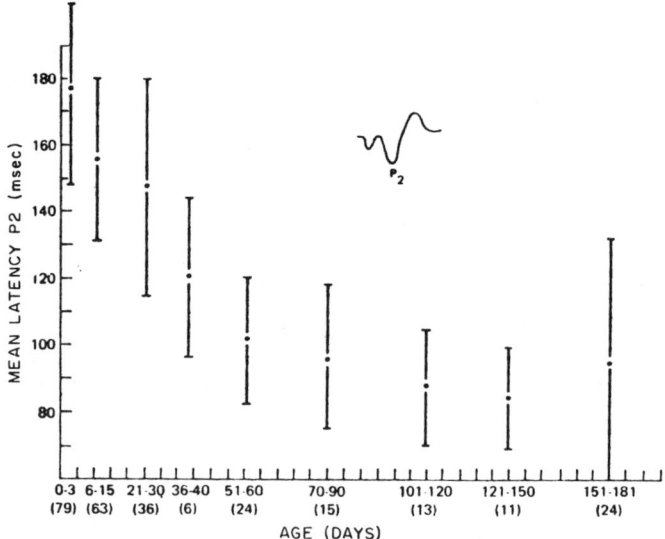

FIG. 3.39 Mean latency of the second positive peak of visual evoked potentials of newborn children as a function of their age in days. ( ): sample number. [From Ferriss *et al.* (1967).]

system, and these investigators also claim to have developed a method by which reversible versus permanent injury can be identified.

The technique for AEP is now so simple that a trained technician can perform it at a patient's bedside. AEP is rapidly becoming a standard clinical tool as more data are accumulated and correlated by disease, drugs, surgery, and autopsy.

Evoked Potentials as a Measure of Mental Chronometry

Early researchers had hoped that electroencephalograms could lead to an understanding of how the brain processes each stimulus through different steps up to a reaction time. We have seen how the form of EEG and the evidence that it is the sum of probabilistic firings of many cells precluded the identification of the behavior of a few cells. In essence, the difficulty was that of identifying a small signal in a large quantity of noise. However, researchers were able to subtract the noise and enhance the signal by multiple generation to sum the common signals of an evoked potential. Each delay, or latency, of an evoked potential could tentatively be assigned to the timing of a synapse, and thus a noninvasive diagnostic technique of brain damage or disease emerged.

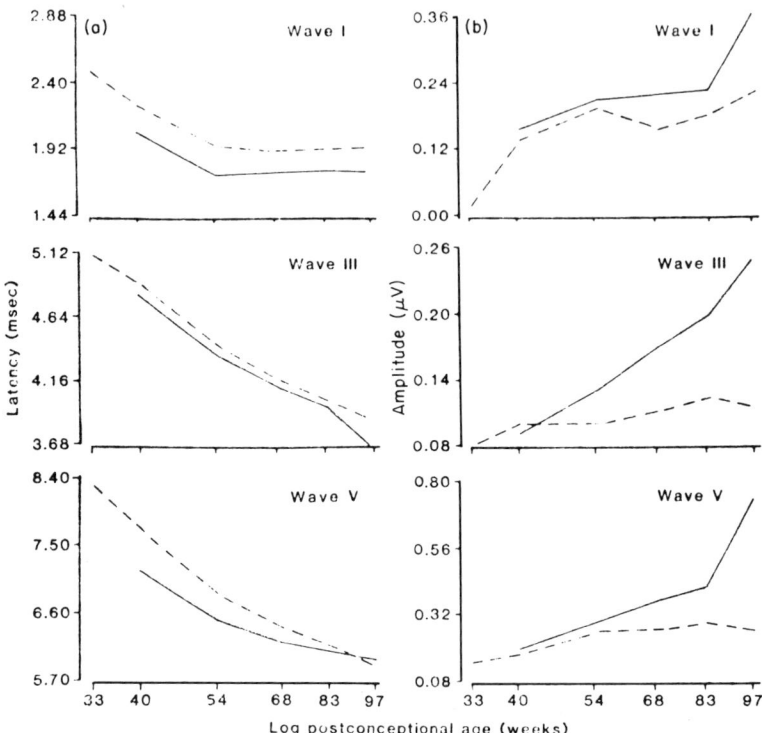

FIG. 3.40 (a) Latencies of AEP waves I, III, and V versus age for infants. (b) Amplitude of waves I, III, and V versus age for infants. Solid lines: normal infants. Dashed lines: infants with high risk for central-nervous-system damage. [From Salamy et al., Science 210, 553–555 (1980). Copyright 1980 by the American Association for the Advancement of Science.]

Electrical engineers have performed well in the design of evoked potential measuring devices, and the cost of these is now quite modest. Many departments of psychology and psychiatry now perform a variety of experiments with them and are exploring further steps of brain processing of stimuli by the introduction of new techniques.

One of these techniques is based on the discovery of what is called the P300 wave, which follows an auditory or visual stimulus with a latency of about 300 msec. This is long after the usual processing times described earlier. Its existence was first reported by Sutton et al. (1965), who found that its appearance was related to the uncertainty of a stimulus. The stimuli were either auditory clicks or flashes of light. There were two types of tests. In each a cuing stimulus would first be given followed by a random interval of 3–5 sec during which the subject was asked to predict the type of test stimulus that would be given. In one experiment the subject knew which would follow by previously being told the order of the stimuli. He would then be certain

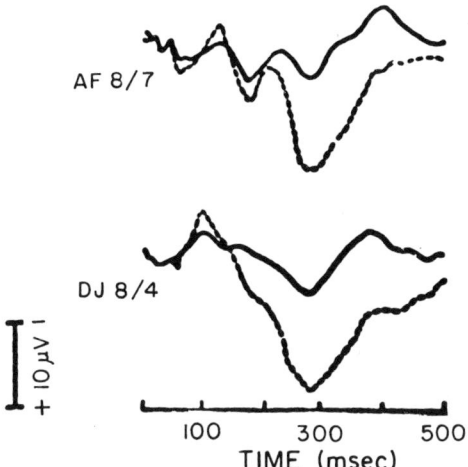

FIG. 3.41 Average wave form for certain and uncertain stimuli of auditory and visual stimuli for two subjects (AF and DJ). (Auditory stimuli presented one third of the times and visual stimuli two thirds of the times.) Solid lines: certain. Dashed lines: uncertain. [From Sutton et al., Science **150**, 1187–1188 (1965). Copyright 1965 by the American Association for the Advancement of Science.]

of the type of test stimulus. In the second type he would not know the order and would therefore be uncertain. A typical result of a series of tests on two subjects is shown in Fig. 3.41 for which the test stimulus was audio one-third of the time and visual two-thirds of the time. It is seen that the P300 latency (positive downward) is larger when the subject was uncertain. Other tests showed that the larger the uncertainty, the larger the P300 wave, and this was the result of all of the tests on all of the subjects.

Considerable study of the P300 latency has taken place. A major point is that although an uncertainty factor causes it, the latency time varies with a number of factors. The term P300 remains as an identification label, but 300 msec is not a fixed time for it. In fact, it is seen in Fig. 3.41 that an identification of the range of the entire positive peak may be more informative than the time of occurrence of the maximum, particularly if it is desired to superpose data as templates. This use of the entire epoch of the latency has been described by Kutas et al. (1977). This method has been used by McCarthy and Donchin (1980) to demonstrate that the P300 component of evoked potentials has its latency determined by processes involved in stimulus evaluation and categorization and is relatively independent of response selection and execution (reaction time). Some investigators now feel that the P300 is a measure of the degree of attentiveness of the subject (Truxal, 1983).

It seems clear that although much research has yet to be done, a technique has emerged in which the timing of the processing of sensory information into a reaction may be measured and evaluated in terms of the performance of a human brain both normal and abnormal. Audio–visual intersensory integration in children may be measurable, and specific learning disabilities such as dyslexia may be identified and remedied (see, e.g., Shipley, 1980).

REFERENCES

Adrian, E. D. (1934). Electrical activity of the nervous system. *Arch. Neurol. Psychiatry* **32**, 1125.
Adrian, E. D., and Matthews, B. H. C. (1934). The Berger rhythm: Potential changes from the occipital lobes of men. *Brain* **57**, 355.
Ahn, H., Prichep, L., John, E. R., Baird, H., Trepetin, M., and Kaye, H. (1980). Developmental equations reflect brain dysfunctions. *Science* **210**, 1259.
Allison, T., and Goff, W. R. (1967). Human cerebral evoked responses to odorous stimuli. *Electroencephalogr. Clin. Neurophysiol.* **23**, 558.
Bergamini, L., and Bergamasco, B. (1967). "Cortical Evoked Potentials in Man." Thomas, Springfield, Illinois.
Berger, H. (1929). Über das Elektrenkephalogramm des Menschen. *Arch. Psychiatr. Nervenkr.* **87**, 527.
Berger, H. (1931). Über das Elektrenkephalogramm des Menschen. *Arch. Psychiatr. Nervenkr.* **94**, 16.
Berger, H. (1933a). Über das Elektrenkephalogramm des Menschen. *Arch. Psychiatr. Nervenkr.* **100**, 301.
Berger, H. (1933b). Über das Elektrenkephalogramm des Menschen. *Arch. Psychiatr. Nervenkr.* **101**, 452.
Berger, H. (1934). Über das Elektrenkephalogramm des Menschen. *Arch. Psychiatr. Nervenkr.* **102**, 538.
Berger, H. (1936). Über das Elektrenkephalogramm des Menschen. *Arch. Psychiatr. Nervenkr.* **104**, 678.
Bodenstein, G., and Praetorius, H. M. (1977). Feature extraction from the electroencephalogram by adaptive segmentation. *Proc. IEEE* **65**, 642.
Brazier, M. A. B. (1960). Some uses of computers in experimental neurology. *Exp. Neurol.* **2**, 123.
Caton, R. (1875). The electric currents of the brain. *Br. Med. J.* **2**, 278.
Cigánek, L. (1961). The EEG response (evoked potential) to light stimulus in man. *Electroencephalogr. Clin. Neurophysiol.* **13**, 165.
Dawson, G. D. (1947). Cerebral responses to electrical stimulation of peripheral nerve in man. *J. Neurol., Neurosurg. Psychiatry* **10**, 134.
Dement, W., and Wolpert, E. A. (1958). The relation of eye movements, body motility, and external stimuli. *J. Exp. Psychol.* **55**, 543.
Duffy, F. H. (1981). Brain electrical activity mapping (BEAM): Computerized access to complex brain function. *Int. J. Neurosci.* **13**, 55.
Duffy, F. H., Burchfiel, J. L., and Lombroso, C. T. (1979a). Brain electrical activity mapping (BEAM): A method for extending the clinical utility of EEG and evoked potential data. *Ann. Neurol.* **5**, 309.
Duffy, F. H., Denckla, M. B., Bartels, P. H., and Sandini, G. (1979b). Dyslexia: Regional differences in brain electrical activity by topographic mapping. *Ann. Neurol.* **7**, 412.
Duffy, F. H., Denckla, M. B., Bartels, P. H., Sandini, G., and Kiessling, L. S. (1980). Dyslexia: Diagnosis by computerized classification of brain electrical activity. *Ann. Neurol.* **7**, 421.

Duffy, F. H., Bartels, P. H., and Burchfiel, J. L. (1981). Significance probability mapping: An aid in the topographic analysis of brain electrical activity. *Electroencephalogr. Clin. Neurophysiol.* **51**, 455.

Elul, R. (1964). Specific site of generation of brain waves. *Physiologist* **7**, 125.

Elul, R. (1967). Spike-wave relationship and processing of information in cortical nerve cells. *In* "Information and Control Processes in Living Systems" (D. M. Ramsey, ed.), p. 106 N.Y. Acad. Sci., New York.

Elul, R. (1968). Brain waves: Intracellular recording and statistical analysis help clarify their physiological significance. *In* "Data Acquisition and Processing in Biology and Medicine" (K. Enstein, ed.), Vol. 5, p. 93. Pergamon, Oxford.

Ferriss, G. S., Davis, G. D., Dorsen, M. M., and Hackett, E. R. (1967). Changes in latency and form of the photically induced average evoked response in human infants. *Electroencephalogr. Clin. Neurophysiol.* **22**, 305.

Fox, S. S., and O'Brien, J. H. (1965). Duplications of evoked potential waveform by curve of probability of firing of a single cell. *Science* **147**, 388.

Halliday, A. M., McDonald, W. I., and Mushin, J. (1972). Delayed visual evoked response in optic neuritis. *Lancet* **1**, 982.

Harter, M. R., and White, C. T. (1968). Effects of contour sharpness and check-size on visually evoked cortical potentials. *Vision Res.* **8**, 701.

Heron, T. G. (1968). Industrial deafness and summed evoked potential. *S. Afr. Med. J.* **42**, 1176.

Hubel, D. H. (1979). The visual cortex of normal and deprived monkeys. *Am. Sci.* **67**, 532.

Jasper, H., and Penfield, W. (1949). Electrocorticograms in man: Effect of voluntary movement upon the electrical activity of the pre-central gyrus. *Arch. Psychiatr. Nervenkr.* **183**, 163.

Jewett, D. L. (1970). Volume conducted potentials in response to auditory stimuli as detected by averaging in the cat. *Electroencephalogr. Clin. Neurophysiol.* **28**, 609.

Jewett, D. L., Romano, M. N., and Williston, J. S. (1970). Human auditory evoked potentials: Possible brain stem components detected on the scalp. *Science* **167**, 1517.

John, E. R., Ahn, H., Prichep, L., Trepetin, M., Brown, D., and Kaye, H. (1980). Developmental equations for the electroencephalogram. *Science* **210**, 1255.

Kinney, J. A. S. (1977). Transient visually evoked potential. *J. Opt. Soc. Am.* **67**, 1465.

Kinney, J. A. S., and McKay, C. L. (1971). "The Visual Evoked Responses as a Measure of Nitrogen Narcosis in Navy Divers," Rep. No. 664. Naval Submarine Medical Research Laboratory, Groton, Connecticut.

Kinney, J. A. S., and McKay, C. L. (1974). Test of color-defective vision using the visual evoked response. *J. Opt. Soc. Am.* **64**, 1244.

Kinney, J. A. S., McKay, C. L., Mensch, A. J., and Luria, S. M. (1972). Techniques for analyzing differences in VERs: Colored and patterned stimuli. *Vision Res.* **12**, 1733.

Kutas, M., McCarthy, G., and Donchin, E. (1977). Augmenting mental chronometry: The P300 as a measure of stimulus evaluation time. *Science* **197**, 792.

Lev, A., and Sohmer, H. (1977). Sources of averaged neural responses recorded in animal and human subjects during cochlear audiometry (electrocochleogram). *Arch. Klin. Exp. Ohren-, Nasen- Kehlkopfheilkd.* **204**, 79.

McCarthy, G., and Donchin, E. (1980). A metric for thought: A comparison of P300 latency and reaction time. *Science* **211**, 77.

Penfield, W., and Jasper, H. (1954). "Epilepsy and the Functional Anatomy of the Human Brain." Little, Brown, Boston, Massachusetts.

Plattig, K.-H. (1968). Über den electrischen Geschmack. *Z. Biol. (Munich)* **116**, 162.

Praetorius, H. M., Bodenstein, G., and Creutzfeldt, O. D. (1977). Adaptive segmentation of EEG records: A new approach to automatic EEG analysis. *Electroencephalogr. Clin. Neurophysiol.* **42**, 84.

Purpura, D. P. (1953). Nature of electrocortical potentials and synaptic organizations in cerebral and cerebellar cortex. *Int. Rev. Neurobiol.* **1**, 47.

Rapin, I., and Graziani, L. J. (1967). Auditory-evoked responses in normal, brain-damaged and deaf infants. *Neurology* **17**, 881.

Regan, D. (1970). Objective method of measuring the relative spectral luminosity curve in man. *J. Opt. Soc. Am.* **60**, 856.

Regan, D. (1972). "Evoked Potentials in Psychology, Sensory Physiology and Clinical Medicine." Chapman & Hall, London.

Regan, D. (1973). Rapid objective refraction using evoked brain potentials. *Invest. Ophthalmol.* **12**, 669.

Regan, D. (1977a). Speedy assessment of visual acuity in amblyopia by evoked potential method. *Ophthalmologia* **175**, 159.

Regan, D. (1977b). Steady-state evoked potentials. *J. Opt. Soc. Am.* **67**, 1475.

Regan, D., and Spekreijse, H. (1974). Evoked potential indications of color blindness. *Vision Res.* **14**, 89.

Salamy, A., Mendelson, T., Tooley, W. H., and Chaplin, E. R. (1980). Differential development of brainstem potentials in healthy and high-risk infants. *Science* **210**, 553.

Saunders, M. G. (1963). Amplitude probability density studies on alpha and alpha-like patterns. *Electroencephalogr. Clin. Neurophysiol.* **15**, 761.

Sclabassi, R. J., Risch, H. A., Hinman, C. L., Kroin, J. S., Enns, N. F., and Namerow, N. S. (1977). Complex pattern evoked somotosensory responses in the study of multiple sclerosis. *Proc. IEEE* **65**, 626.

Shagass, C. (1972). "Evoked Brain Potentials in Psychiatry." Plenum, New York.

Shipley, T. (1980). "Sensory Integration in Children." Thomas, Springfield, Illinois.

Sholl, D. A. (1956). "The Organization of the Cerebral Cortex." Methuen, London.

Starr, A., and Anchor, L. J. (1975). Auditory brain stem responses in neurological disease. *Arch. Neurol. (Chicago)* **32**, 761.

Stockard, J. J., Stockard, J. E., and Sharbrough, F. W. (1977). Detection and localization of occult lesions with brainstem auditory responses. *Mayo Clin. Proc.* **52**, 761.

Sutton, S., Braren, M., Zubin, J., and John, E. R. (1965). Evoked-potential correlates of stimulus uncertainty. *Science* **150**, 1187.

Truxal, C. (1983). Watching the brain at work. *IEEE Spectrum*, March, 52.

Tuchmann-Duplessis, H., Auroux, M., and Haegel, P. (1975). "Illustrated Human Embryology," Vol. III. Springer-Verlag, Berlin and New York; original French edition published by Masson S. A., Paris.

van der Gon, J. J. D., and Strackee, J. (1966). Some aspects of EEG frequency analysis, *IEEE Trans. Biomed. Eng.* **BME-13**, 120.

Ward, J. H. (1963). Hierarchical grouping to optimize an objective function. *Am. Stat. Assoc.* **58**, 236.

CHAPTER **4**

Chemical and Electrical Properties of Synapses

INTRODUCTION

This chapter contains a description of some of the physical processes occurring at the *synapses* of neurons, a term originally coined by Sherrington (1906) to denote the functional and structural sites within the nervous system at which transmissions between neurons and between neurons and muscles occur. Transmission between a neuronal axon and muscle is treated first since the neuromuscular junction constitutes the most extensively studied and the best understood synapse. In fact, an understanding of chemical transmission at this synapse has served as a guide in elucidating synaptic properties at other junctions. After developing the quantitative aspects of transmission at the neuromuscular junction, some properties of drug action at other synapses are discussed.

In contrast to other cells, neurons are characterized by one or more lengthy processes, or elongations. Histologists divide these neuronal processes into two classes, *axons* and *dendrites*, on the basis of their shape, size, and branching patterns. The dendritic patterns in different neurons are quite varied, although some discernible regularity in branching appears to exist

(Scheibel and Scheibel, 1970; Hillman, 1979). Figure 4.1 illustrates a few of the common neuronal morphologies although the number of different forms of neurons as classified by neurologists probably exceeds a hundred. The number of branches for a given neuronal type can vary, and research has shown that the complexity and the total length of the branches can be affected quite profoundly during neuronal development by nutritional and sensory deprivation (Coleman and Riesen, 1968; Volkmar and Greenough, 1972). This will be discussed in Chapter 5.

The cell body or *soma* is that part of the neuron containing the nucleus, and its size varies greatly, ranging from less than 10 μm to more than 100 μm in diameter. There is, corresponding to this variation in soma size, a similar variation in the size of the cell's nucleus. Neurons constituting the central nervous system (CNS) generally have smaller cell bodies than those of neurons constituting the peripheral nervous system, although pyramidal and cerebellar Purkinje cells can be quite large. In some invertebrates the neurons can achieve macro dimensions with neuronal somas 30–50 times larger than those typically occurring in vertebrates. Because electrophysiological measurements are easier to perform on larger cells, neurons of the *Loligo* (squid), *Aplysia* (sea slug), and other invertebrates have been extensively studied.

In the cytoplasm of the neuron there are a large number of organelles. These are discernible small structural units responsible for the metabolic requirements of the cell. Also within the cytoplasm surrounding the nucleus is a complex system of membranes called the *endoplasmic reticulum*, which in some regions is organized into aggregates of pancakelike folds called *cisternae*, or *Nissl bodies*. These sites are where the synthesis of structural and secretory proteins are thought to occur. Most of the cytoplasm of the neuron is formed in the body, or soma, of the cell and is transported to the other parts. The soma is the nutritional, or *trophic*, center of the neuron.

Generally, only a single axon originates from the soma, although this axon may branch into a number of smaller axons before terminating on one or more cells. An axon may give off a relatively large branch, i.e., a large-diameter axon, which is referred to as a *collateral*. Axons vary in length from microns to meters. The region where the axon joins the soma is called the *axon hillock*, or the initial segment of the axon or axon neck. This specialized membrane segment in some types of neurons will transduce a change in the soma *transmembrane potential* into an active response, i.e., an action potential in the axon. With the light microscope the axon hillock appears devoid of structure. However, in electron micrographs specialized structures called *filaments, tubules,* and *polyribosomes* can be seen. It is thought that the microstructural aspects of the axon neck impart the unidirectional rectifying aspect to nerve conduction; it transmits the potential signal away

INTRODUCTION 87

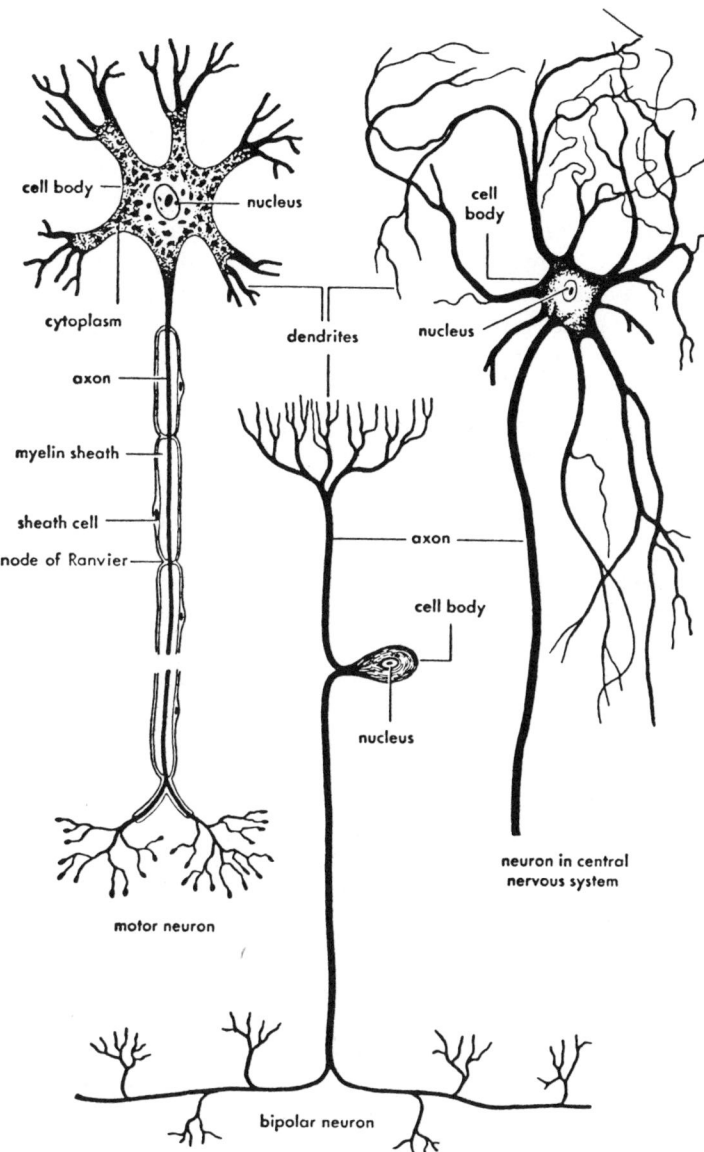

FIG. 4.1 Three common neuron cell types. The one on the left shows the myelin sheath formed from Schwann (sheath) cell membranes, which has been rolled to several layers thick around the axon. [From Griffin (1962).]

from the cell body, even though an action potential initiated by an experimenter at the *lower* threshold usually does invade the soma as well as traveling down the axon. An action potential (Vol. I, Chapter 3) created at the axon hillock travels undiminished in amplitude to the axon terminal ends with velocities of up to 100 m/sec. It is important to emphasize, however, that not all neurons possess axons. Indeed many cells, probably 50% of the CNS neurons, are without axonal processes. In this case neuronal communication is performed by *dendrodendritic* synapses, synapses between dendrites.

In contrast to axons, there are generally many dendrites arising from the cell body, and in many, but not all cases, each dendrite may branch many times, forming a complicated treelike structure, known as the dendritic tree. A discussion of the electrical and integrative properties of dendrites is given in Chapter 5.

The *information content* of some types of neurons are represented by the state of *polarization* of their somas. By the state of polarization we mean both the spatial and temporal characteristics of the soma's transmembrane potential. If the soma's membrane is depolarized sufficiently, then this depolarization is converted at the axon hillock into an action potential, which then travels undiminished to the axon terminals. Transmission from one cell to another occurs at the synapse. The past history of the synapse, the state of depolarization of the terminal, and the intracellular and extracellular environment of the synapse influence the operation of synaptic transmission. In contrast to nerve impulses, which are subject to little variability, the transmission of impulses at synapses is a labile process, and for this reason, the morphology, chemical constitution, and function of synapses are areas of intense investigation by neuroanatomists, physiologists, biochemists, and pharmacologists. Despite intensive research, current understanding of the regulation and control of synaptic transmission is limited.

KINDS OF SYNAPTIC CONNECTIONS

Most morphologists accept as *synapse junctions* those regions of apposition between two neurons having only two features: (1) a specialized form of intercellular contact between presynaptic and postsynaptic processes; and (2) in the case of chemical synapses, the presence of synaptic vesicles, which are spherical organelles 300–500 Å in diameter in the presynaptic terminal. Nevertheless, there are specialized synaptic junctions where vesicles do not occur. In general, vesicles are not seen in the postsynaptic cytoplasm, although in chemical synapses, as seen in electron micrographs, there is attached underneath the postsynaptic membrane a disk-shaped structure called the *postsynaptic density*. The postsynaptic density is composed mainly

KINDS OF SYNAPTIC CONNECTIONS 89

of musclelike proteins and is presumably involved in mediating the postsynaptic events of synaptic transmission. Synapses have many different types of morphologies. In some cases the axon ends with swollen terminals called *knobs*, and in some cases the dendrites develop specialized protruding structures called *spines*. The gross features of a few synapses are illustrated in Fig. 4.2. The actual number and spatial arrangement of synapses formed by a single axon with a cell body can be quite complex. Figure 4.3 illustrates this for a frog sympathetic ganglion neuron. (Ganglia are small ensembles of cell bodies in the peripheral nervous system lying outside the spinal cord.) Such neurons are known to be *unipolar*, that is, they lack dendrites and have only a single axon. Each neuron receives the termination of one, possibly two, preganglionic nerve fibers. Because their neuronal inputs and outputs are simple and because of their in vitro physiological viability, sympathetic ganglion cells have been the object of intensive research. As Fig. 4.3 illustrates, the preganglionic fiber is spirally wrapped around the axon hillock of the

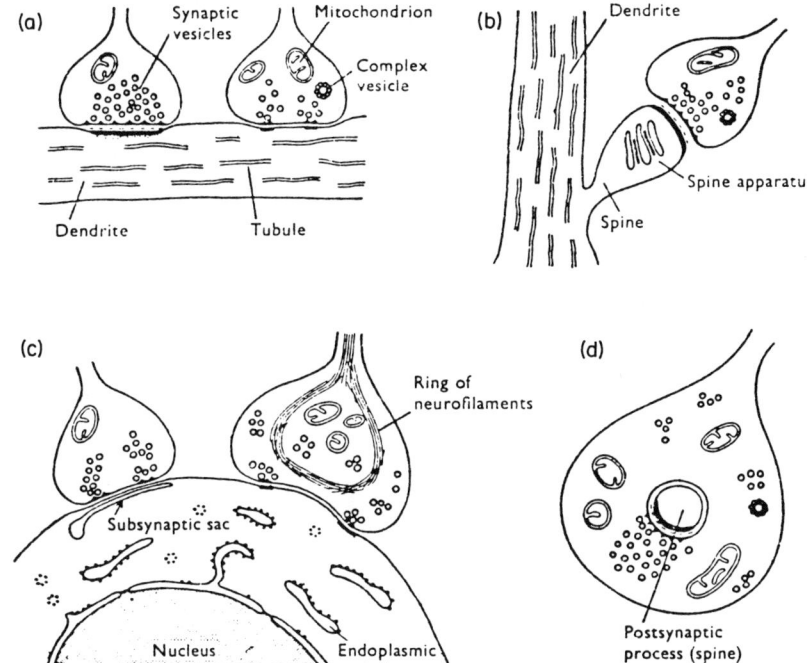

FIG. 4.2 A few synaptic types: (a) two types of synapse between axon terminals and a dendrite; (b) synapse between an axon terminal and a dendritic spine; (c) two types of synapse between axon terminals and a neuron soma; (d) a synapse whose postsynaptic spine is invaginated into the presynaptic terminal. [From Whittaker and Gray (1962).]

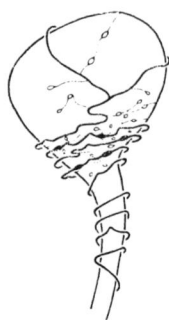

FIG. 4.3 The spiral ending of a nerve fiber upon a sympathetic ganglion neuron in the frog. Note the presence of many boutons (presynaptic knobs) near the origin (hillock) of the axon. [After Taxi (1965).]

neuron. The small circles are the axon terminals' endings, called presynaptic knobs, or *boutons*. Approximately 10% of the large cell body is covered by these boutons, as estimated from electron micrographs (see Fig. 2.19).

In the cerebral cortex approximately 2% of the synapses are on cell bodies and large dendritic trunks, approximately 50% are located on dendritic spines, and virtually all the rest terminate on fine dendritic processes. Although there are numerous types of synapses, we shall find it convenient to divide them into two classes, depending on the specific mechanism employed for the transmission of the nerve impulse from one neuron to another. These two classes are (a) chemical synapses and (b) electrical synapses.

CHEMICAL SYNAPSES

The chemical synapse is characterized by a narrow gap of 200–300 Å, termed the *synaptic cleft*, which separates the two cells. The presynaptic terminal consists not only of synaptic vesicles, both coated and uncoated (see p. 125), but it also contains an elaborate, intricate network of energy-generating *mitochondria*, voidlike structures called *vacuoles*, and elongated tube structures, called *microtubules* and *neurofilaments*. The names of the latter two structures arise because of their appearance in electron micrographs; microtubules are 200–300 Å in diameter, whereas neurofilaments are about 100 Å in diameter. They are thought to be composed of protein and to be involved in intraneural transport, the shaping of cells, and cell division. Electron microscopy (EM) has been useful not only in identifying structural aspects of synaptic terminals on the basis of their electron density, but has also provided morphological verification of *synaptosomes*. These vesicles are resealed single presynaptic terminals of neuronal tissue and their constituent subcellular organelles (Gray and Whittaker, 1962). The ability of the biochemists to make synaptosomes from different regions of the brain that retain their functional integrity has proven helpful in elucidating biochemical pathways. Although with EM it is theoretically possible to resolve objects of less than 1 Å diameter, such as single atoms, or bonds less than 1 Å in length, such as chemical bonds in organic molecules, the design limitations and lack of scattering power by atoms of low atomic number limit resolution to about 10 Å with a good instrument and to samples that contain atoms that are heavy enough to scatter electrons strongly from the beam.

When the nerve impulse arrives at the axon terminal the local depolarization of the presynaptic membrane increases the rate of secretion of a special chemical, the *neurotransmitter*, from the presynaptic membrane. This rate of secretion is dependent on the local calcium-ion concentration, for it is known that if calcium ions are absent from the extracellular medium, the chemical synapse ceases to transmit impulses. This apparent requirement for calcium in the extracellular fluid has been established for all chemical synapses tested. This is also true for other secretory processes, such as the liberation of hormones; however, this specificity with respect to calcium is not absolute, since in vitro studies of the frog neuromuscular junction and squid synapse have shown that strontium can replace external Ca^{2+} for initiation of transmitter release (Miledi, 1966). In contrast to strontium it is well known that magnesium ions antagonize the action of calcium and reduce the secretion of transmitters and hormones.

There exists a wide variety of neurotransmitters, ranging from the classic skeletal neuromuscular transmitter acetylcholine (abbreviated ACh) to amino acids, serotonin, catecholamines, and small polypeptides. The collective term *catecholamine* refers to all organic compounds that contain a catechol nucleus, i.e., a benzene ring with two adjacent hydroxyl substituents, and an amine group. Figure 4.4 illustrates the catechol and catecholamine structure. When the term catecholamine is applied to neurotransmitters, one usually means dopamine or its metabolic products norepinephrine and epinephrine. Table 4.1 lists a few of the better-known transmitters, their structural formula, and some of their presumed locations in the nervous

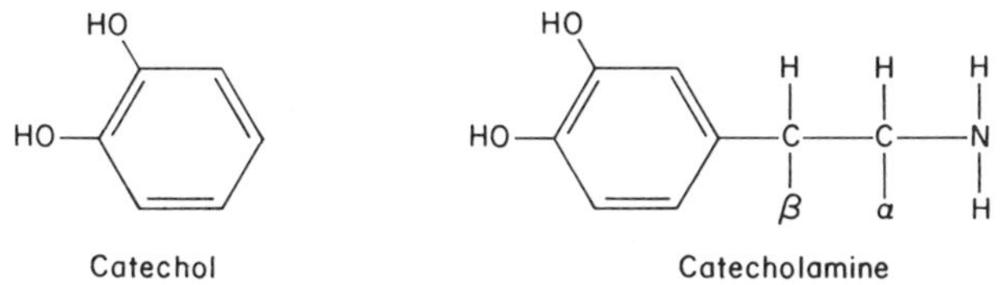

FIG. 4.4 The catechol and catecholamine molecular structure. Dopamine: $\alpha = \beta = H$; norepinephrine: $\alpha = H$, $\beta = OH$; and epinephrine: $\alpha = H$, $\beta = OH$, and NH_2 is replaced by $N(CH_3)H$. (See Fig. 4.39.)

TABLE 4.1

Some Neurotransmitters, Their Chemical Structure, and Some of Their Locations

Compound	Structure	Example Locations[a]
Acetylcholine (ACh)	$H-\underset{\underset{H}{\mid}}{\overset{\overset{H}{\mid}}{C}}-\overset{\overset{O}{\|}}{C}-O-\underset{\underset{H}{\mid}}{\overset{\overset{H}{\mid}}{C}}-\underset{\underset{H}{\mid}}{\overset{\overset{H}{\mid}}{C}}-\underset{\underset{CH_3}{\mid}}{\overset{\overset{CH_3}{\mid}}{N^+}}-CH_3$	neuromuscular junctions, autonomic ganglia
Norepinephrine (NE)	HO—(ring)—$\underset{\underset{H}{\mid}}{\overset{\overset{OH}{\mid}}{C}}-\underset{\underset{H}{\mid}}{\overset{\overset{H}{\mid}}{C}}-NH_2$; ring has OH	sympathetic ganglia, hypothalamus, spleen, heart, lungs
Dopamine (DA)	HO—(ring)—$\underset{\underset{H}{\mid}}{\overset{\overset{H}{\mid}}{C}}-\underset{\underset{H}{\mid}}{\overset{\overset{H}{\mid}}{C}}-NH_2$; ring has HO	neostriatum nucleus (in the brain)
Serotonin (5-hydroxy-tryptamine)	HO—(indole ring)—$\underset{\underset{H}{\mid}}{\overset{\overset{H}{\mid}}{C}}-\underset{\underset{H}{\mid}}{\overset{\overset{H}{\mid}}{C}}-NH_2$	pineal gland, cells of the gastrointestinal tract, platelets
Histamine	(imidazole ring)—$\underset{\underset{H}{\mid}}{\overset{\overset{H}{\mid}}{C}}-\underset{\underset{H}{\mid}}{\overset{\overset{H}{\mid}}{C}}-NH_2$	highest concentration found in certain nuclei of the hypothalamus, sympathetic postganglionic fibers
γ-Aminobutyric acid (GABA)	$H_2N-\underset{\underset{H}{\mid}}{\overset{\overset{H}{\mid}}{C}}-\underset{\underset{H}{\mid}}{\overset{\overset{H}{\mid}}{C}}-\underset{\underset{H}{\mid}}{\overset{\overset{H}{\mid}}{C}}-COOH$	substantia nigra[b] hypothalamus, *not* present in peripheral nerve tissue
Substance P	polypeptide mol. wgt. ~1400 daltons	digestive tract, peripheral nerves

[a] Neurons that are involved in controlling regulator internal processes of the body and that are not under voluntary control are said to belong to the autonomic nervous system. The autonomic nervous system is divided into two neuronal classes: *sympathetic* and *parasympathetic*. Sympathetic neurons have axons that originate from nuclei in the spinal cord whereas parasympathetic neuron fibers are derived from the nuclei of some cranial nerves. The term *ganglia* refers to small ensembles of neuron cell bodies that are located outside of the brain and the spinal cord.

[b] See p. 139.

system. In many cases their specific sites of action are still unknown, although acetylcholine has been shown to be the transmitter in autonomic ganglia, (see footnote a to Table 4.1) and in motoneurons' collateral–Renshaw-cell synapses (see p. 135) in the spinal cord. Dopamine is thought to be a transmitter in the corpus striatum, and evidence indicates that γ-aminobutyric acid, commonly abbreviated as GABA, is an inhibitory transmitter in many regions of the central nervous system, particularly in the cerebellum.

A major area of research concerns itself exclusively with the identification of chemicals that can serve as transmitters, their biochemistry, and their sites of action with the nervous system. In order for a chemical to be a neurotransmitter at a synapse some specific criteria must be satisfied. The two most important criteria for neurotransmitter identification were initially enunciated by Loewi (1921): (1) the substance must be released in appropriate amounts when the presynaptic nerve is stimulated; and (2) the application of the substance to the postsynaptic sites must mimic the effect of presynaptic stimulation. Not all transmitters listed in Table 4.1 rigorously satisfy these criteria. Furthermore, that a chemical has been proven to be a neurotransmitter at some synaptic junctions does not imply that it is a neurotransmitter at other locations even if it is present.

After the neurotransmitter is released from the presynaptic terminal, it then diffuses across the narrow gap in a manner describable by standard diffusion kinetics, i.e., Fick's law of diffusion (Vol. I, p. 61). For a synaptic cleft 200 Å wide, the acetylcholine molecule, which has a diffusion coefficient D within the synaptic cleft of $\sim 7.6 \times 10^{-6}$ cm^2 s^{-1}, would reach the *subsynaptic* (a synonym for postsynaptic) membrane, assuming one-dimensional random motion, in time

$$\tau = \text{(synaptic cleft dimension)}^2/2D \approx 0.26 \ \mu\text{sec} \quad (4.1)$$

The derivation of Eq. (4.1) is given in Appendix I.

When the number of neurotransmitter molecules considered is tens of thousands and the effects of the dimensions and shape of the synaptic junctional regions are included, calculations have shown that a fairly uniform distribution of the transmitter over the postsynaptic membrane requires about 50 μsec (Eccles and Jaeger, 1958). This calculation shows that a random process of diffusion is sufficiently fast to be consistent with experimental evidence since experiments indicate that maximum transmitter action on the postsynaptic membrane is attained approximately 1 msec after transmitter release.

Secreted molecules, which are not degraded by enzymatic reactions or reabsorbed by the presynapse, upon reaching the subsynaptic membrane, interact with receptor molecules. The interaction of the transmitter with its receptor leads to changes in the local permeability and therefore changes in the conductance (the inverse of the electrical resistance) of the postsynaptic membrane to selective ions. At the junction between a motoneuron and the muscle, the coupling between the receptor and transmitter results in a conformational change of the receptor–channel complex, which produces an opening in the channel, thereby permitting a sudden and massive flow of sodium and potassium ions through the membrane. This kinetic reaction of the transmitter (T) with the receptor (R) at the neuromuscular junction can be represented by the following scheme:

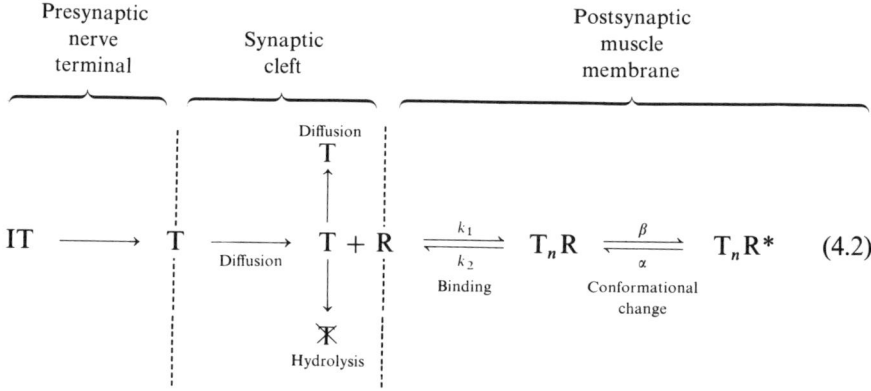

where IT is the transmitter still inside the presynaptic terminal, $T_n R$ is the initial transmitter–receptor complex, and $T_n R^*$ denotes the active conformation. The subscript n is used to denote the possibility that two or more transmitters may be required to be bound to a receptor molecule to transform a closed channel into an open channel. When the receptor–transmitter complex is in the conformational state $T_n R^*$, the channel associated with that receptor molecule is open. These transient permeability changes result in corresponding transient current and voltage changes. In some cases the changes produce a net depolarization, in which case the response is called an *excitatory postsynaptic potential* (EPSP). In the opposite case, where a hyperpolarization results, the response is termed an *inhibitory postsynaptic potential* (IPSP). Both EPSPs and IPSPs can be recorded by inserting a microelectrode into the postsynaptic cell and stimulating the presynaptic nerve fibers. Typical EPSP and IPSP responses are illustrated in Fig. 4.5. In both cases the response duration is within the order of a few milliseconds. In some neurons, such as ganglion cells of the bullfrog sympathetic nervous system, there are, in addition to these *fast* voltage changes, slow potential components that last for seconds or longer. Slow responses are also found

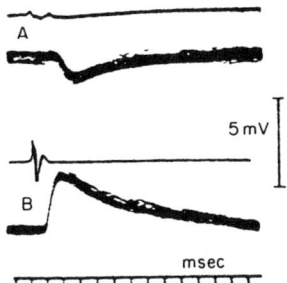

FIG. 4.5 The IPSP(A) and EPSP(B) recorded in a biceps-semitendinosus motoneuron of the cat stimulated by the appropriate afferent volleys. Note that the time course of the IPSP is almost identical to that of the EPSP and is hyperpolarizing (upward direction is positive). [From Coombs et al. (1955a).]

centrally and are thought to be a very important mode of innervation in the CNS.

An important property of chemical synapses is their unidirectional behavior, which allows for the passage of information from one neuron to another, but not in reverse. Until rather recently it was thought that each neuron was able to secrete but one type of neurotransmitter, a statement of which is commonly called *Dale's principle* (Dale, 1935). There is, however, a growing body of data suggesting that, while many nerves have but a single type of transmitter, others secrete more than one type of molecule, where one type of molecule may serve more as a modulator than as a transmitter. For example, the giant cerebral neuron in *Helix* (snail) releases both acetylcholine and serotonin (Cottrell, 1976). Although a synapse may have multiple transmitters, it is nevertheless accepted that a chemical synapse cannot be both inhibitory at one time and excitatory at other times, provided both the inhibitory and excitatory responses refer to the same time scale. An example clarifies this statement. There are neurons that from stimulation by the same input fibers elicit both inhibitory and excitatory responses, although possibly from different synapses on the soma. The well-studied sympathetic ganglion neurons in bullfrogs display inhibitory and excitatory postsynaptic potentials when certain input axons are activated. These cells have no dendrites and possess a single axon, so nature has apparently simplified the inhibitory and excitatory processes. Activation of the input axon that synapses directly on the soma elicits two responses simultaneously: a fast excitatory potential having a duration the order of a few tens of milliseconds, and a much slower inhibitory potential lasting hundreds of milliseconds. These quite different responses originate from the activation of two types of soma receptors by the same neurotransmitter, ACh.

ELECTRICAL SYNAPSES

At electrical synapses no chemical transmitters are involved, and therefore inhibition is not expected. Transmission results from the passive spreading of the presynaptic depolarization, called *electrotonic spreading*, usually along a low-resistance pathway connecting the two cells. However, electrical coupling between neurons can be either *conductive* or *inductive*. Conductive electrical coupling occurs when there is structural contact between the neurons, whereas inductive coupling involves modifying neighboring neurons by the ion currents in the extracellular medium from a conducting neuron in close apposition (Marrazzi and Lorente de Nó, 1944). Morphological studies of electrical conductive synapses show evidence of small intercellular channels connecting the cytoplasm of the coupled cells. Such close appositions of membranes are called *gap junctions*, a term initially introduced in electron microscopy to label such structures. There are two conclusive demonstrations of structural connections between the electrically coupled cells: (1) They are seen by electron microscopy in which heavy elements such as lanthanum are used to enhance contrast. (2) When a dye (e.g., procion yellow) that does not penetrate the plasma membrane is injected into one cell, it appears in the electrically coupled adjacent cell. From a variety of staining and freeze fracture studied, a proposed structure of the gap junction can be constructed. This is illustrated in Fig. 4.6.

Basically two types of electrical conductive synapses can be distinguished: (1) those in which transmission occurs across the synapse in only one direction (in this sense they are similar to chemical synapses); and (2) those that

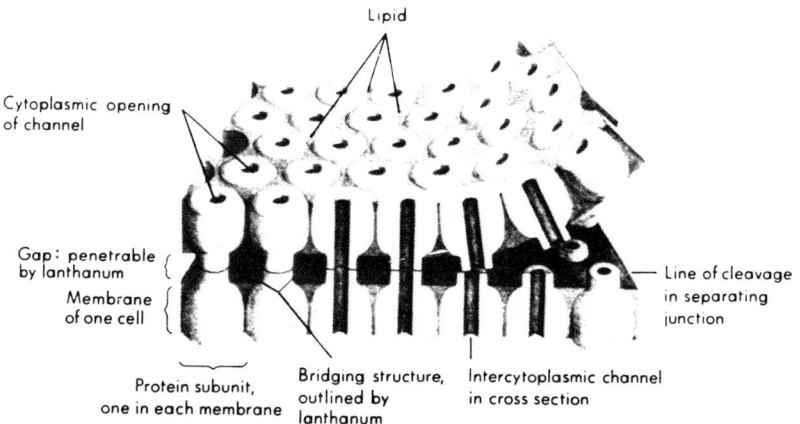

FIG. 4.6 Diagrammatic representation of a small area of a gap junction. At right the junction is split and the upper membrane is peeled back to reveal the external aspects of the lower membrane. [N. B. Gilula, unpublished diagram; see Bennett (1977).]

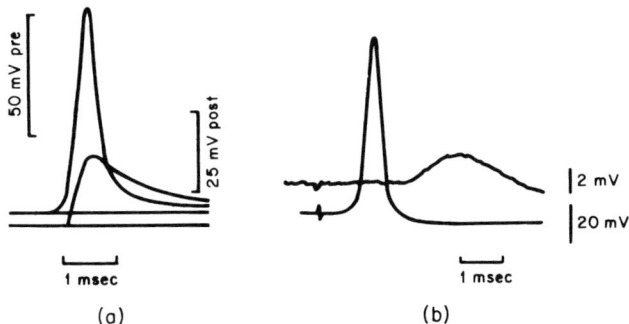

FIG. 4.7 Comparison of impulse transmission of electrical and chemical synapses. (a) Impulse transmission at the electrical synapse of the crayfish giant motor synapse. Potential changes were recorded with microelectrodes inserted into the presynaptic and postsynaptic fibers. The sharp trace denotes the presynaptic fiber potential. Smaller trace is the postsynaptic response. Note its short latency. (See Fig. 4.8 for an attempt to reverse the direction of the impulse in the same fiber.) [From Furshpan and Potter (1959).] (b) Impulse transmission at the chemical synapse of the squid giant synapse. Lower trace: Presynaptic potential. Upper trace: Postsynaptic potential. Note the latency of approximately 2 msec of the post potential due to time for diffusion of the neurotransmitter. [Reproduced from Takeuchi and Takeuchi, *J. Gen. Physiol.* **45**, 1181 (1962) by copyright permission of the Rockefeller University Press.]

support transmission in both directions. Figures 4.7a and 4.8 illustrate the unidirectional, rectifying nature of transmission in the crayfish giant motor synapse, where transmission is known to be electrical. In Fig. 4.7a the stimulus, the sharp peak, is applied to the presynaptic fiber, and the response, the smaller curve, is of reasonable magnitude. Figure 4.8 illustrates the reverse experiment in which the impulse, the sharp peak, is applied to the postsynaptic fiber (called *antidromic* stimulation). The response in the presynaptic fiber, lower curve, is negligible. For most electrical synapses, however, impulses are transmitted in both directions, and the membrane does not exhibit rectification. A defining characteristic of most electrotonic junctions is that the junctional resistance is constant, i.e., the current–voltage characteristics are linear. For chemical synapses, the current–voltage response in the postsynaptic membrane is highly nonlinear. Generally, the resistivity of the junctional membrane is very low, on the order of a few ohm centimeters squared. A striking aspect of electrical transmission is the brevity of its synaptic delay. As Fig. 4.7 illustrates, the synaptic delay in the giant motor synapse is approximately 0.1 msec. However, for the squid giant synapse (see Fig. 4.7b), which is chemically mediated instead of electrically, the delay as measured between the peak of the presynaptic spike and the start of the postsynaptic response is about 2 msec. For this reason, electrical transmission is clearly better where speed is important, for example, in those neurons involved in escape mechanisms, and where highly synchronized

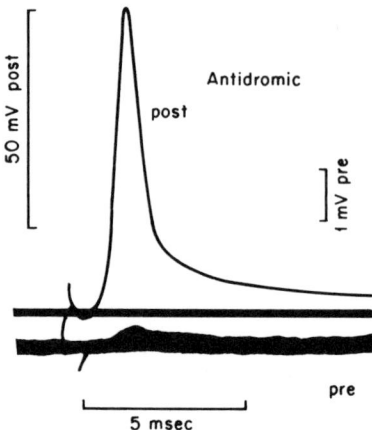

FIG. 4.8 The rectifying properties of the electrical synapse of the giant motor synapse of the crayfish. The antidromic stimulation of postsynaptic fibers (sharp peak) does not evoke a response in the presynaptic cell. [From Furshpan and Potter (1959).]

activity is needed for the survival of the organism. Electrical synapses are relatively common in both invertebrates and nonmammalian vertebrates. Although electrical synapses do occur in mammals, their importance in the operation of the mammalian central nervous system is unclear. A large presynaptic area compared to the postsynaptic area appears to be necessary for effective electrical transmission, and this may be a serious limitation when large numbers of such elements are necessary, such as occurs in the CNS of mammals. There is little doubt that the vast majority of synapses in the central nervous system of mammals are chemically mediated, although electrotonic transmission does occur at some synapses in the sensory cortex and olfactory bulb and has been directly demonstrated between pyramidal cells in the rat hippocampus (MacVicar and Dudek, 1981).

PROPERTIES OF POSTSYNAPTIC POTENTIALS AND CURRENTS AT CHEMICAL SYNAPSES

The following discussion will be restricted to chemical synapses. Studies have revealed that to a first approximation both the magnitude and time course of the postsynaptic potential (PSP) are determined by the neurotransmitter concentration in the *immediate* vicinity of the postsynaptic membrane. A more correct statement would be that PSPs are proportional to the number of transmitters bound to receptors, although this way of expressing the concept need not concern us here. A response of this nature is called a *graded response*. This is illustrated graphically in Fig. 4.9, where a

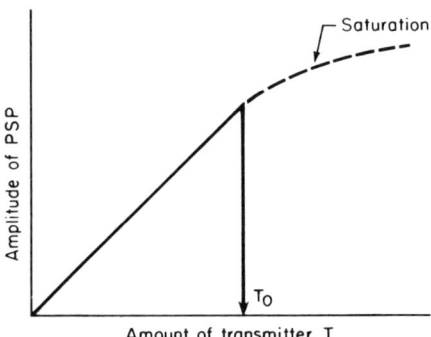

FIG. 4.9 Relation between the postsynaptic potential (PSP) amplitude and amount of transmitter released from the axon terminal.

plot of the peak of the PSP amplitude as a function of the transmitter released from the presynaptic membrane is shown. Up to a maximal concentration T_0, the relation between the PSP amplitude and the transmitter concentration to a first approximation is linear. A second property of PSPs is referred to as *temporal summation*. This is best explained by noting that a graded response implies that within the linear region the PSP caused by a second impulse will simply add to what remains of the first PSP. If $T_1(t)$ and $T_2(t)$ are the concentrations of the transmitter at time t corresponding to inputs 1 and 2, with resulting postsynaptic potentials PSP_1 and PSP_2, then the net postsynaptic potential PSP when both impulses occur obeys the relation

$$PSP(t) = PSP_1(t) + PSP_2(t) \tag{4.3}$$

provided $T_1(t) + T_2(t) < T_0$. Figure 4.10 illustrates this temporal summation property. The peak amplitude A of the composite potential increases as the time delay Δt between the two inputs decreases. An example illustrating the temporal summation of excitatory potentials in a motoneuron is shown in Fig. 4.11. An afferent volley is unable to generate an action potential either alone or when separate from a preceding volley by an interval of 5 msec or more (Fig. 4.11a). When the time interval between two afferent volleys is sufficiently short, as illustrated in Fig. 4.11b, an action potential is initiated. Temporal summation also holds when there are more than two inputs. In the special case when an axon feeding the synaptic terminal is firing at a rate f, the PSPs will sum into a steady-state depolarization with an amplitude proportional to the frequency at which the nerve impulses arrive at the terminal. This is an important property since for many neurons the magnitude of the state of constant depolarization is coded in the nerve impulse frequency,

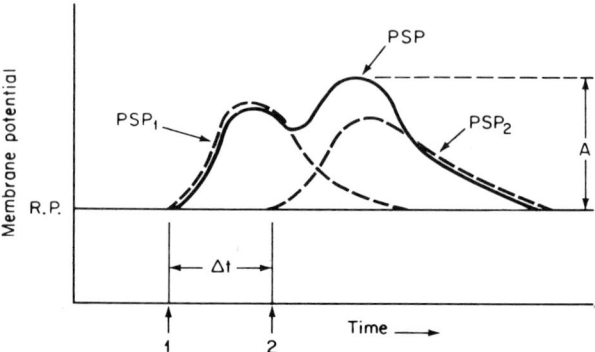

FIG. 4.10 Schematic of temporal summation of postsynaptic potentials for synaptic inputs at times 1 and 2. Dashed lines denote PSPs if only a single neuron is activated either at time t or $t + \Delta t$. The composite PSP is indicated by a solid line and is what is actually seen on the oscilloscope when both neurons are activated. Dashed and solid lines slightly separated vertically for clarity. R.P. denotes resting membrane potential.

a property of the all-or-none behavior of axons (see Vol. I, Chapter 3). It is evident therefore that the temporal summation property of PSPs provides a scheme for decoding the nerve impulse frequency into a membrane depolarization of the soma.

The above discussion was limited to excitatory potentials; however, inhibitory postsynaptic potentials (IPSPs) share many of the same properties as EPSPs. Thus, IPSPs show temporal summation, and if the IPSPs arrive at the same site and at a constant rate, they also sum to give an average hyperpolarization proportional to the frequency of the arriving nerve impulses. Although EPSPs of sufficient magnitude can generate a nerve

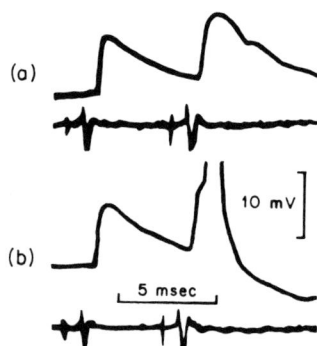

FIG. 4.11 (a) The temporal summation of two EPSPs as in the schematic of Fig. 4.10. (b) The second EPSP followed soon enough that its addition to the first achieved a threshold depolarization, initiating an action potential. [From Brock et al. (1952).]

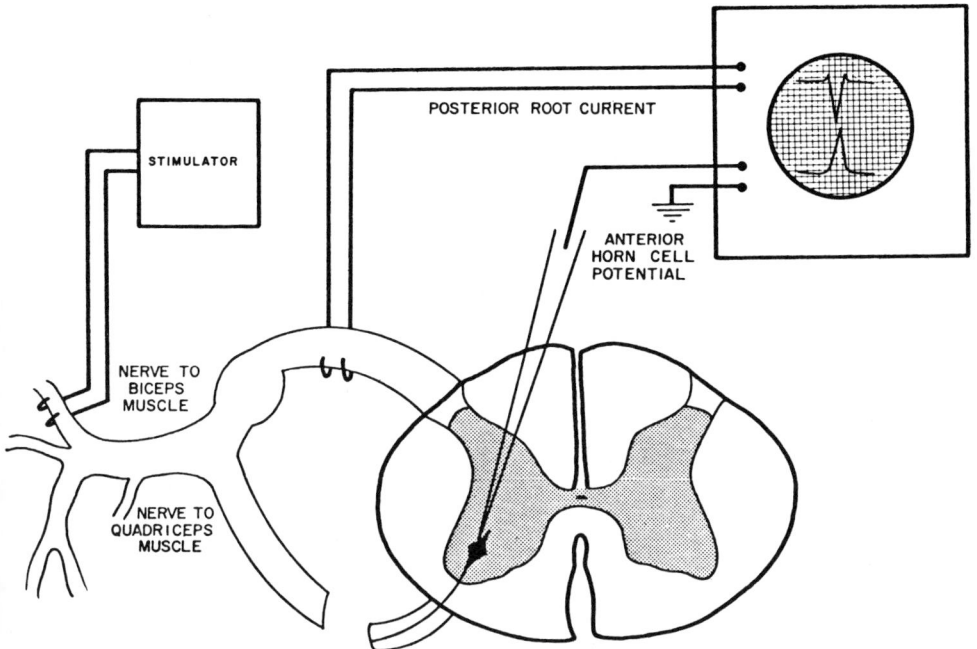

FIG. 4.12 The experimental setup for recording from anterior horn cells in the spinal cord. [From Curtis (1972).]

impulse, an IPSP, because it corresponds to a membrane more hyperpolarized, makes it more difficult for EPSP inputs to produce an action potential. In the case where a neuron has both excitatory and inhibitory synaptic inputs, the EPSPs and IPSPs sum algebraically; the inhibitory potentials are subtracted from the excitatory potentials. This summation property of an IPSP and EPSP has been recorded from an anterior horn cell. The recording setup is shown in Fig. 4.12. Stimulation of the nerve to the biceps initiates an action potential (Fig. 4.13a), whereas when the nerve to the quadriceps is stimulated an IPSP is evoked (Fig. 4.13b). If both nerves are stimulated, then, depending on their temporal separation, either an action potential is still evoked (Fig. 4.13c) or it is suppressed (Fig. 4.13d). Figure 4.13 illustrates not only the algebraic properties of IPSPs and EPSPs, but also their temporal summation properties.

The above discussion is highly simplified since it was assumed that the PSPs are strictly additive. This is only an approximation, however, because IPSPs (and EPSPs) are greatly modified by even relatively small changes in membrane potential, being increased (decreased) in amplitude by membrane depolarization and diminished (increased) by hyperpolarization. As Fig. 4.14a

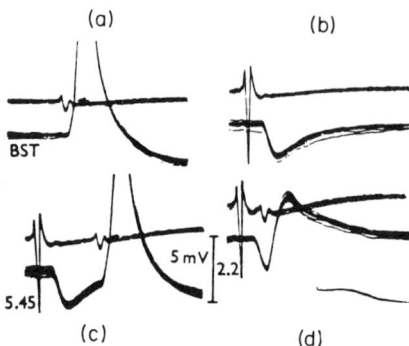

FIG. 4.13 Summation of an IPSP and an EPSP. Lower trace of each pair is an intracellular recording from a biceps anterior horn cell. The upper trace is the action current in the posterior root. (a) The nerve to the biceps was stimulated and an action potential was evoked. (This is off scale in the figure.) (b) The nerve to quadriceps was stimulated and an IPSP was evoked. (c) Both nerves to biceps and quadriceps were stimulated 45 msec apart and an action potential was evoked, whereas in (d) they were stimulated 2.2 msec apart and no action potential resulted, because the IPSP generated by the quadriceps stimulus inhibited the biceps anterior horn cell. [From Coombs *et al.* (1955b).]

illustrates, when the membrane is hyperpolarized sufficiently the IPSP will actually reverse direction and exhibit a depolarizing response rather than hyperpolarizing behavior. At some membrane potential, known as the *inhibitory equilibrium potential*, the amplitude of the IPSP is zero. This is clearly evident in Fig. 4.14b, where a plot of the amplitude of the IPSP as a function of membrane potential is shown. Similarly, the size of an EPSP also depends on the membrane potential, although its dependence is exactly opposite to that described above for the IPSP. That is, the EPSP peak amplitude increases for membranes hyperpolarized, and it decreases for membranes depolarized. Just as in the case of IPSPs, there exists a membrane potential for which the amplitude of the EPSP vanishes. This potential is called the *excitatory equilibrium potential*. However, the linear additivity of EPSPs, like that of IPSPs, is often a sufficiently accurate approximation.

Figure 4.14a illustrates this dependence of the IPSP amplitude on membrane potential for the disynaptic Ia cat motoneuron recorded by stimulating the afferent inhibitory nerve fibers (Ia group). A disynaptic motoneuron is a motoneuron with a pathway in which the incoming impulse crosses two synapses before termination. Figure 4.15 illustrates the definition of *monosynaptic*, *disynaptic*, and *polysynaptic* pathways. Since "polysynaptic" means more than a single synapse, the term can be used with reference to disynaptic pathways. The intervening neurons as shown in Fig. 4.15 are

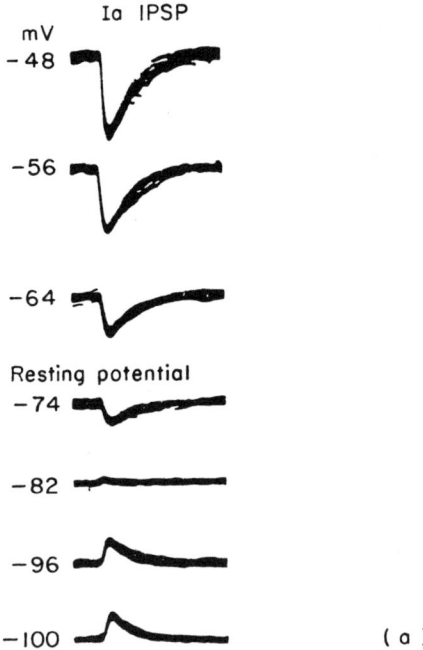
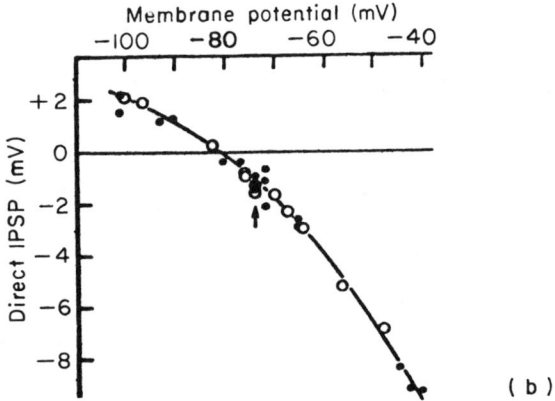

FIG. 4.14 (a) The dependence of the inhibitory postsynaptic potential (IPSPs) magnitude on membrane potential. These records are from a disynaptic Ia cat motoneuron cell. The resting potential is -74 mV. (b) Maximum amplitude of the IPSPs of the series shown in (a) plotted against the respective membrane potentials. Note the absence of a response when the membrane potential is the reversal potential E_s. Arrow indicates the resting potential. [From Coombs et al. (1955a).]

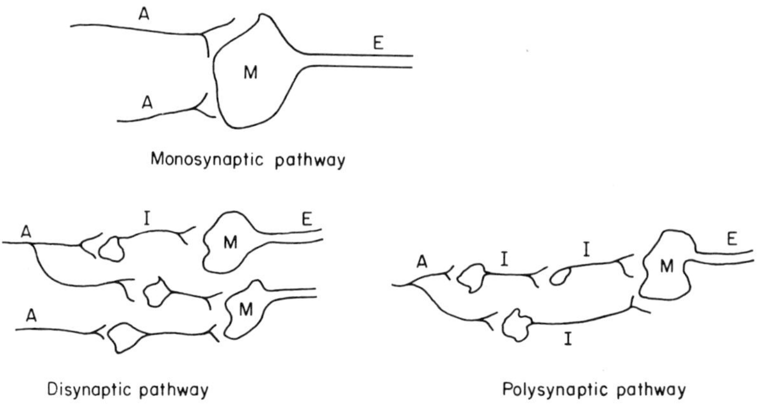

FIG. 4.15 Monosynaptic, disynaptic and polysynaptic pathways for motoneurons (M) which connect afferent (input) fibers (A) with efferent (output) fibers (E). I labels the interneurons.

called *interneurons*, and they typically exert an inhibitory effect on motoneurons. It is evident from Fig. 4.14a that since a single IPSP corresponds to a membrane potential change, the size of two IPSPs occurring simultaneously will not exactly equal the sum of each occurring separately.

The above discussion of the behavior of the PSP's amplitude on membrane potential is valid only for those synaptic potentials arising from ion conductance increases in the postsynaptic membrane. Not all postsynaptic membrane responses, however, result in ion conductance increases when neurotransmitters bind to membrane receptors. In some neurons the PSPs are due to conductance decreases. In these neurons the behavior of the IPSPs and EPSPs on membrane potential is just opposite to that discussed above. That is, EPSPs decrease for membranes hyperpolarized and increase for membranes depolarized, whereas IPSPs increase with hyperpolarization and decrease for membranes depolarized. Long-lasting synaptic potentials in sympathetic ganglion neurons of the American bullfrog, *Rana catesbeiana*, are thought to arise from ion conductance decreases.

At the neuromuscular junction the excitatory postsynaptic potential is usually called the *end-plate potential*. End-plate potentials have been extensively studied and the properties of the synaptic membrane have been shown to be characterized by a constant *equilibrium potential* (reversal potential) E_s. An example of an equilibrium potential for the disynaptic cat motoneuron is shown in Fig. 4.14a, third from bottom, for an inhibitory synapse. Because potentials above or below this value alter the sign of the response, it is sometimes called the *reversal potential* instead of equilibrium potential. Reversal potentials exist for both IPSPs and EPSPs. This potential

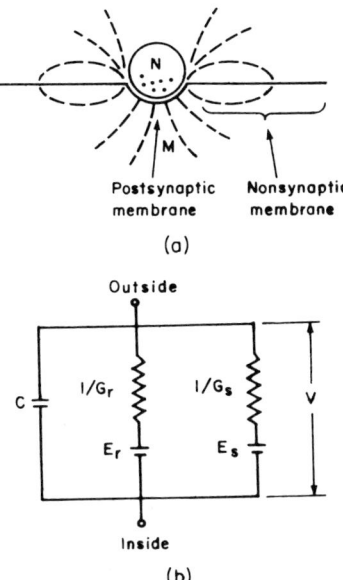

FIG. 4.16 (a) Schematic illustration of the lines of current flow during the action of the transmitter at the end plate. N denotes the nerve ending; M is the muscle fiber. (b) Equivalent circuit for end plate and the muscle fiber membrane. C is the capacitance of the nonsynaptic membrane, E_r and G_r are the reversal potential and membrane conductance, respectively, of the nonsynaptic membrane, whereas E_s and G_s denote the same quantities for the postsynaptic membrane. [From Takeuchi (1977).]

in series with a time-varying resistance $R_s = G_s^{-1}$, where G_s is the transmembrane conductance, forms one branch of a circuit, which represents the electrical behavior of the postsynaptic membrane. When the transmitter acts at the postsynaptic membrane a current, called the *end-plate current* (EPC), flows through the postsynaptic membrane, with a net flow of positive ions inward for EPSP and outward for IPSP. Completion of the current loop is provided by the flow of current across the nonsynaptic part of the membrane, as shown in Fig. 4.16a. The dotted lines symbolize the positive-ion current direction. The nonsynaptic membrane can be shown to be characterized by a voltage-independent capacitance C, an equilibrium potential E_r, and a conductance G_r. A simplified equivalent circuit of the end plate and the neighboring muscle fiber membrane is shown in Fig. 4.16b. It is possible to measure the end-plate current by employing voltage-clamp methods (see Vol. I, Chapter 3). In most applications of this technique two intracellular microelectrodes are employed, one for measuring the transmembrane voltage and the other for supplying current flow through the muscle membrane. By a properly constructed feedback circuit the transmembrane

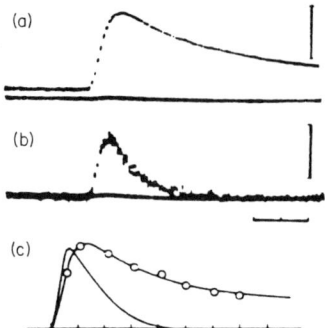

FIG. 4.17 (a) End-plate potential (EPP) recorded intracellularly without feedback. (b) End-plate current (EPC) recorded under voltage-clamped conditions at its resting potential from the same end plate. Decay is exponential. (c) Superimposed tracings of EPP and EPC. Circles indicate potential reconstructed from the observed end-plate current. Voltage scale, 5 mV; current scale, 1×10^{-9} A; time scale 1 msec. Upward direction denotes inward flow of positive current. [From Takeuchi and Takeuchi (1959).]

potential is held constant in the presence of the combined effects of brief electrical changes at the end plate because of nerve fiber stimulation and transmembrane current supplied by the current electrode. With a constant membrane potential, the capacitance current, which is proportional to dV/dt, is eliminated. Thus the feedback current, that is, the current supplied by the current electrode, is a direct measure of the EPC. Figure 4.17 illustrates the end-plate potential and end-plate current measured from a curarized (see pp. 4, 133) frog end plate when the sciatic nerve is stimulated. This figure shows that the EPC has a relatively brief duration, reaching a peak in approximately 0.77 msec and having a total duration of about 4 msec. The superimposed tracings of EPP and EPC are shown in Fig. 4.17c. This figure also gives the reconstructed potential changes (indicated by circles) using the measured end-plate current and Hodgkin and Rushton (1946) equation (Eq. 5.10, p. 159), which describes the passive electrotonic spreading of the current along and across the nonsynaptic membrane. The declining phase of the EPP is a measure of this passive electrotonic spreading. For the voltage-clamped synapse, corresponding to a fixed value of the membrane potential (V), the total postsynaptic current (I_s) can be written

$$I_s = G_s(V - E_s) \qquad (4.4)$$

where G_s is the postsynaptic membrane conductance. From a family of membrane currents, I_s-versus-time curves at different fixed membrane potentials, referred to as holding potentials, the time dependence of the end-plate conductance G_s, defined by Eq. (4.4), can be inferred. Figure 4.18

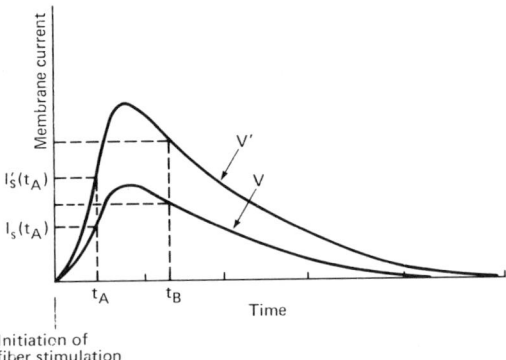

FIG. 4.18 Schematic illustration of endplate current for two different clamped membrane potentials, V and V'. $I_s(t_A)$ and $I'_s(t_A)$ denote the membrane currents at time t_A for potentials V and V', respectively.

illustrates two typical EPCs for holding potentials V and V', and as shown in the figure, the current through the membrane at time t_A is $I_s(t_A)$ and $I'_s(t_A)$, respectively, for these two potentials. If many different membrane holding potentials are examined, then for this fixed time t_A, a curve such as that labeled A in Fig. 4.19 can be constructed where $I_s(t_A)$ is plotted on the ordinate and the membrane potential is plotted along the abscissa. It follows from the relation

$$I_s(t_A) = G_s(t_A)(V - E_s) \tag{4.5}$$

that the slope of $I_s(t_A)$ versus V is the conductance G_s at time t_A. A similar construction can be performed at each instant after initiation of the EPC. Four curves constructed in the manner described above are shown in Fig. 4.19. In this way the complete time dependence of the end-plate conductance can be determined.

It is important to note that the I_s-versus-V curves extrapolated to zero current all intersect at a point on the abscissa. This result shows that the reversal potential E_s is independent of time, and, in addition, the intercept value provides an estimate for E_s. For the case illustrated in Fig. 4.19, the value of the reversal potential is 10–20 mV negative to the outside Ringer solution.† Note that in Fig. 4.19 a different zero reference potential has been employed so it appears that the intercept with the abscissa is at -60 mV.

Formation of the active transmitter–receptor complex, indicated by $(T_n R)^*$ on p. 94, opens ion pathways across the synaptic membrane. Ions

---

† An aqueous solution of ions with the same concentrations as those in plasma.

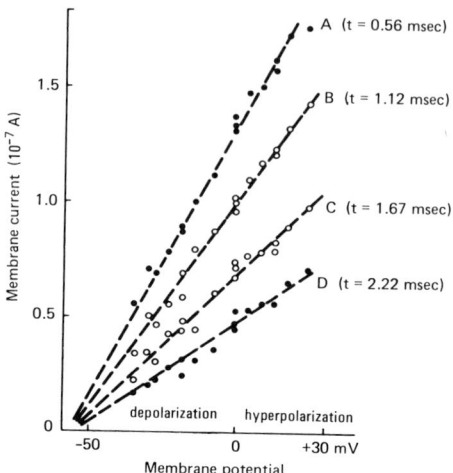

FIG. 4.19 Size of end-plate current at various times from the start of curent as a function of the displacement of the membrane potential from its resting value. A, 0.56 msec; B, 1.12 msec; C, 1.67 msec; and D, 2.22 msec. from the onset of end-plate current. Slope of $I-V$ curves gives the value of the conductance at the indicated times. [From Takeuchi and Takeuchi (1959).]

that can penetrate the membrane can be determined experimentally by measuring changes in the reversal potential (E_s) in various ionic environments. If the currents are generated by Na^+, K^+, and Cl^- only and these different ions are independent of each other, then the total synaptic current can be written

$$I_s = I_{Na} + I_K + I_{Cl} \tag{4.6}$$

where I_{Na}, I_K, and I_{Cl} are the sodium, potassium, and chloride ion currents, respectively. These currents are defined as

$$I_{Na} = \Delta G_{Na}(V - E_{Na}) \tag{4.7}$$

$$I_K = \Delta G_K(V - E_K) \tag{4.8}$$

$$I_{Cl} = \Delta G_{Cl}(V - E_{Cl}) \tag{4.9}$$

where the ΔGs are the associated conductance changes and E_{Na}, E_K, and E_{Cl} are the Nernst equilibrium potentials [Vol. I, Chapter 3 and Eq. (A.26)], i.e.,

$$E_{Na} = 61 \log_{10} \frac{[Na]_o}{[Na]_i} \text{ mV} \quad \text{(at 37°C)} \tag{4.10}$$

$$E_K = 61 \log_{10} \frac{[K]_o}{[K]_i} \text{ mV} \quad \text{(at 37°C)} \tag{4.11}$$

$$E_{Cl} = 61 \log_{10} \frac{[Cl]_i}{[Cl]_o} \text{ mV} \quad \text{(at 37°C)} \tag{4.12}$$

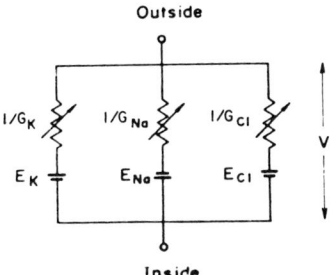

FIG. 4.20 Equivalent electrical circuit for the synaptic membrane. [From Takeuchi (1977).]

The quantities in brackets refer to concentrations, and the subscripts denote the inside (intracellular) and outside (extracellular) values (see Vol. I, p. 349). The equivalent electrical circuit for the synaptic membrane is shown in Fig. 4.20. However, as explained below, chloride conductance does not affect membrane current at the neuromuscular junction. At the reversal potential, $I_s = 0$ and, from Eq. (4.5), $E_s = V$. It therefore follows from Eqs. (4.6)–(4.9) that

$$E_s = \frac{(\Delta G_{Na})E_{Na} + (\Delta G_K)E_K + (\Delta G_{Cl})E_{Cl}}{\Delta G_{Na} + \Delta G_K + \Delta G_{Cl}} \quad (4.13)$$

For the specific case of the frog end-plate experiments, there is no change in E_s even with drastic changes in chloride concentration. However, the reversal potential is sensitive to changes in the Na$^+$ and K$^+$ concentrations. This dependence of E_s on external sodium and potassium is illustrated in Fig. 4.21. The solid lines are drawn according to Eq. (4.13) with $\Delta G_{Cl} = 0$ and $\Delta G_{Na} = 1.29 \, \Delta G_K$. This value for the ratio of the sodium and potassium conductances was determined by noting that at the reversal potential the end-plate current is zero,

$$I_s = 0 = \Delta G_{Na}(E_s - E_{Na}) + \Delta G_K(E_s - E_K) \quad (4.14)$$

and therefore

$$\Delta G_{Na}/\Delta G_K = -(E_s - E_K)/(E_s - E_{Na}) \quad (4.15)$$

For a reversal potential $E_s = -15$ mV, and with normal Ringer conditions for a frog Eqs. (4.10) and (4.11) give $E_K = -99$ mV and $E_{Na} = 50$ mV and thus $\Delta G_{Na}/\Delta G_K = 1.29$. Hence, the end-plate membrane is slightly more permeable to sodium ions than it is to potassium ions. This technique of varying the ion concentration is quite general and can be used to identify the ions involved in the generation of PSPs at other postsynaptic membranes. However, when more than two different ions are involved the analysis becomes difficult. It can be shown that since EPC is monophasic (the pulse

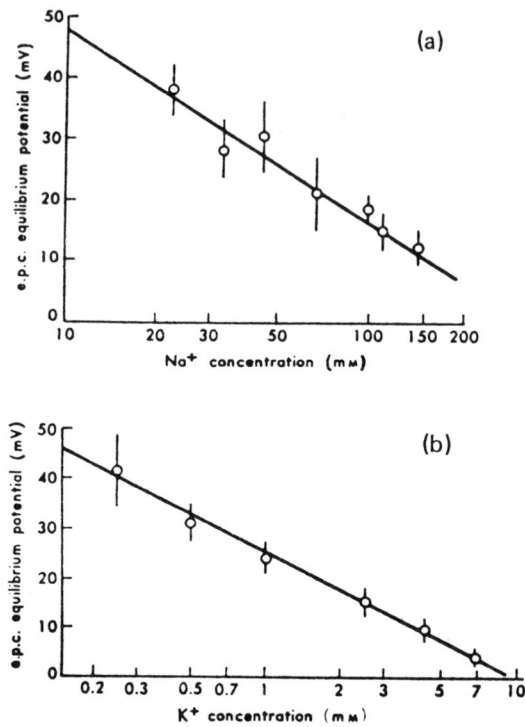

FIG. 4.21 The relation between the reversal potential of the end-plate current and the external ion concentrations, $[K]_o$ and $[Na]_o$. [From Takeuchi and Takeuchi (1960).]

exhibits only one direction of polarity) at all membrane potentials and $\Delta G_{Na}(t)/\Delta G_K(t)$ is independent of time when V equals the reversal potential, the time dependences of ΔG_{Na} and ΔG_K are the same for all membrane potentials. This result is in sharp contrast to the permeability changes of Na^+ and K^+ ions that accompany the propagation of an action potential (see Vol. I, Fig. 3.6). In the case of the axon action potential the conductance changes associated with these ions are distinctly different.

PRESYNAPTIC PROPERTIES OF CHEMICAL SYNAPSES

The frog neuromuscular junction has played a central role in forming our understanding of presynaptic mechanisms. This is because the neuromuscular synapses are relatively large, viable, and therefore easily accessible experimentally. Processes operative at this junction are thought to occur at other chemically mediated synapses both in the central and peripheral

PRESYNAPTIC PROPERTIES OF CHEMICAL SYNAPSES

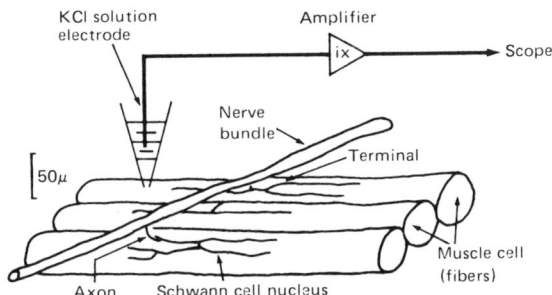

FIG. 4.22 Illustration of the branching of a motor nerve after entering the muscle. Very fine branches containing only a single axon leave the nerve bundle and form a junction on a muscle fiber, as shown in Fig. 4.23. Penetration of a muscle fiber with a microelectrode as illustrated records the resting potential of muscle fiber with respect to the outside bath; generally the resting potential is about -90 mV. [Modified from Steinbach and Stevens (1976).]

nervous systems. The first experimental evidence that the quantity of neurotransmitter released per nerve impulse occurs in small packages, or quanta, was provided by careful electrophysiological studies of the neuromuscular synapse (Fatt and Katz, 1952). Similar behavior has subsequently been observed at other synapses.

Before discussing the details of the studies by Katz and co-workers, it is helpful to examine the structural aspects of the neuromuscular junction. Figure 4.22 illustrates how the motor nerve bundle, composed of several axons, innervates (supplies with nerves) the muscle. Eventually each single axon of the motor nerve makes individual contact with a muscle fiber. The finer details of the junction formed between the axon and the surface of the muscle fiber is shown in Fig. 4.23. The region of the muscle fiber facing the nerve terminal and consisting of a series of folds is called the *end plate*. As revealed by electron-microscope studies, a characteristic feature of the presynaptic terminal is the clustering of synaptic vesicles around certain preferred sites, called *releasing* sites or "hot spots." These sites usually face the postsynaptic fold, denoted by the letter S in Fig. 4.23. The vesicles contain the neurotransmitter acetylcholine (ACh).

As with all excitable tissue, if a microelectrode, which is a very fine glass pipette drawn out to form a tip of less than 0.5 μm outside diameter and filled with 3-M KCl, is inserted into muscle fiber, a potential difference of approximately -70 to -90 mV with respect to the ground electrode is recorded. This potential is called the resting potential of the muscle fiber, and its polarity is such that it is negative inside the fiber (see Vol. I, Chapter 3). An important property of the resting potential is its constancy along the muscle fiber. It is also constant with time provided no external stimulus is

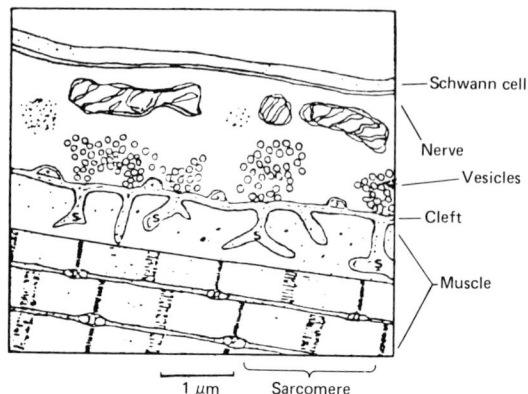

FIG. 4.23 Features of the neuromuscular junction as seen in transmission electron micrograph. The muscle fiber's sarcomeres and longitudinal filaments are evident as are the postsynaptic folds (S) in the muscle surface. The nerve ending is overlaid by a Schwann cell layer. The synaptic vesicles tend to accumulate around certain selective sites within the cell near the presynaptic membrane. These selective sites usually face a postsynaptic fold (S) on the muscle surface. [From Steinbach and Stevens (1976).]

applied, either in the form of adding external chemicals to the bath in which the preparation is immersed or by stimulating the motor nerve. Nevertheless, on closer examination of the resting state, Fatt and Katz (1952) found small subthreshold depolarizations in the muscle fiber near where the nerve makes contact with the muscle fiber. These small depolarizations, about 0.5 mV in maximum amplitude and having a total duration of approximately 20 msec, occur spontaneously, and are called *miniature end-plate potentials* (mepp). In the resting state mepps occur with the extremely low frequency of about one per second. If, however, the axon terminal is depolarized either by chemicals or by stimulation of the nerve fibers, then there is an associated increase in the frequency of mepps. When the microelectrode is inserted at some distance from the end plate of the muscle these random bursts of depolarization are not observed. These observations provide evidence for the spatial localization of receptors on the postsynaptic membrane. Figure 4.24a illustrates these mepps at the end plate and their absence in the same muscle fiber 2 mm away (Fig. 4.24b).

Each miniature end-plate potential is associated with the arrival at the postsynaptic membrane of a small amount of acetylcholine, termed a *quantum*. An estimate of the size of one quantum can be determined by performing *extracellular iontophoretic* experiments. In this technique a fine-tipped glass electrode containing the drug, in this case the organic salt acetylcholine chloride, is positioned close to the muscle fiber. Passage of an appropriate

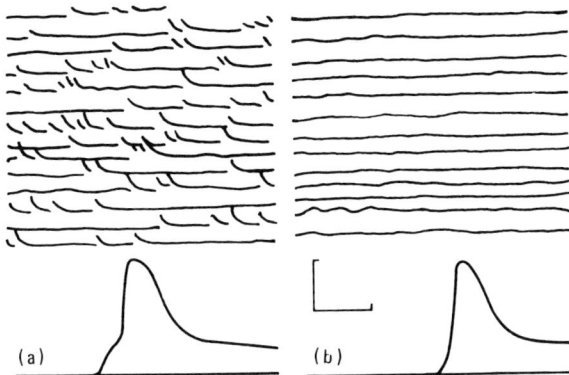

FIG. 4.24 Intracellular recording at frog neuromuscular junctions. (a) recorded at the end plate, (b) recorded at a distance 2 mm away, in the same muscle fiber. (a) shows the spontaneous activity (mepps) at the end plate. Lower records are action potentials recorded at same positions as upper data. [From Fatt and Katz (1952).]

current to ground (generally taken to be the bath in which the preparation is immersed) releases some of the acetylcholine from the electrode because ACh is an ion. A second electrode positioned close to the iontophoretic electrode and located *intracellularly* records the potentials in the postsynaptic membrane due to the ACh-induced channel opening. A schematic of the experimental setup for extracellular iontophoresis of a monosynaptic neuron (N) is shown in Fig. 4.25. Experiments performed in this manner

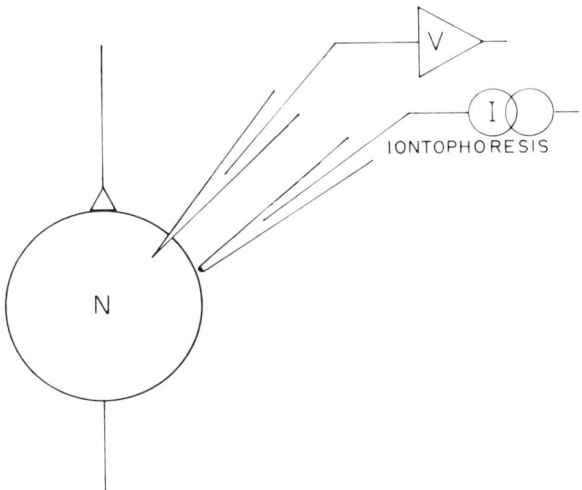

FIG. 4.25 Experimental setup illustrating extracellular iontophoresis technique for a neuron (N) lacking dendrites and containing a single axon.

have shown that the "induced" ACh potentials were capable of continuous reduction in size and duration by simply lowering the strength and duration of the current pulse applied to the iontophoretic electrode. In fact, it was possible to evoke ACh-induced potentials of much shorter duration and smaller amplitude than the naturally occurring mepps. The important conclusion from experiments such as these is that the mepp signals the arrival of a large multimolecular packet of ACh (the quantum) at the postsynaptic membrane. By adjusting the duration and magnitude of the iontophoretic current pulse so that the induced ACh potential equals a mepp, the size of a quantum can be estimated. These experiments (Fatt and Katz, 1952) provided an upper bound estimate of approximately 10^5 molecules per quantum. More recent estimates for the number of ACh molecules in one quantum are on the order of 10^4 molecules. Additional experimental evidence that the mepps at the neuromuscular junction owe their origin to acetylcholine is provided by examination of the effects of other chemicals on their amplitude. For example, both end-plate potentials and mepps amplitudes are suppressed by curare and enhanced by cholinesterase inhibitors to the same extent by the same doses.

Quanta are now identified with the synaptic vesicles, which are small spherical organelles observed on the presynaptic side of the nerve terminals in electron micrographs. The quantal hypothesis of transmitter release of del Castillo and Katz (1954) can be summarized as follows. Miniature endplate potentials are considered as arising from a leakage of acetylcholine from the nerve terminal, one quantum corresponding to one mepp. Therefore, a nerve impulse does not initiate a new process, but rather accelerates an existing one, intensifying for a very brief moment the terminal activity so that several hundred quanta are released within less than 1 msec. An alternative suggestion that the size of the acetylcholine quantum is altered by the nerve impulse has been shown experimentally by del Castillo and Katz (1954) and others to be untenable. The end-plate potential (EPP), which initiates a muscle twitch, is the summed response of many quantum components. The observation that *botulinum toxin*, one of the many chemicals responsible for food poisoning, abolished both EPPs and mepps by preventing the release of ACh from the presynaptic terminal offers additional support for a common origin of EPPs and mepps. As in the case of postsynaptic potentials, the mepps are not strictly additive, although their individual conductances are additive.

When the synaptic vesicle is identified as the storage unit of one ACh quantum, the physical picture that emerges is the following. The relatively large population of vesicles (N) contained inside the nerve terminal is considered to be mobile and in random motion. Because of this random motion, some of these vesicles undergo "collisions" with the axon terminal

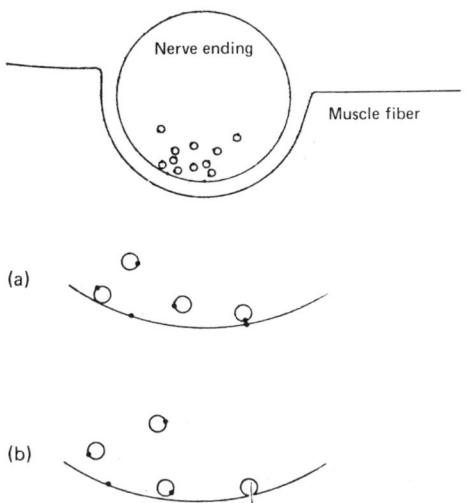

FIG. 4.26 Schematic diagram to illustrate the "random" theory of release of synaptic quanta. (a) Collision of synaptic vesicle with a releasing site. (b) Formation of passage channel for release of neurotransmitter. [From del Castillo and Katz (1957).]

membrane and release their ACh. Not all collisions result in transmitter release. Only when the synaptic vesicle strikes preferred terminal membrane locations, called "hot spots" or "releasing sites," indicated by solid dots in Fig. 4.26, does a fusion between the vesicle and axon membrane occur. These successful collisions produce a transitory channel connecting the interior of the vesicle with the cell's extracellular environment, resulting in the discharge of the contents of the vesicle into the synaptic cleft. This is illustrated in Fig. 4.26. The collision rate is assumed to be high, but successful collisions are statistically rare. The quantum hypothesis then has a statistical nature. The distribution function governing the number of quanta released by a given nerve impulse can be shown to obey a binomial distribution. Thus, if p denotes the probability that a quantum is released, then the probability P_k that k quanta ($k \leq N$) are released in a given trial is (see Appendix A and Vol. I, pp. 361–363)

$$P_k = \frac{N!}{(N-k)!k!} p^k (1-p)^{N-k} \qquad (4.16)$$

In cases where p is very small ($p \ll 1$), the distribution given by Eq. (4.16) may be approximated quite accurately by the Poisson distribution,

$$P_k = (m^k/k!) e^{-m} \qquad (4.17)$$

where m is the mean number of quanta released and equals pN. The derivation of Eq. (4.17) from Eq. (4.16) is given explicitly in Vol. I (pp. 361–363) (see also Appendix A). The advantage of the Poisson statistical distribution is that it requires only a knowledge of the average number m of packets released in order to predict the probability of releasing $0, 1, \ldots, k, \ldots$ packets. Unfortunately, knowing the value of m does not allow the values of p and N to be inferred.

Tests of the validity of Eq. (4.17) have been performed (Boyd and Martin, 1956) by recording the evoked end-plate potentials for fixed stimulus conditions at the neuromuscular junction in the presence of high magnesium concentration (~ 12.5 mM). Magnesium ions are added to the bath of Ringer solution since magnesium competes with calcium and is known to inhibit synaptic transmission at very high concentration. However, if proper concentrations are employed, it has the effect of suppressing the release of ACh sufficiently during nerve stimulation so that end-plate potentials are few (i.e., $p \ll 1$) and small in amplitude and therefore easily counted. Figure 4.27 illustrates a histogram distribution of the magnitudes of evoked EPPS from a cat's tenuissimus neuromuscular junction. In this example, out of 198

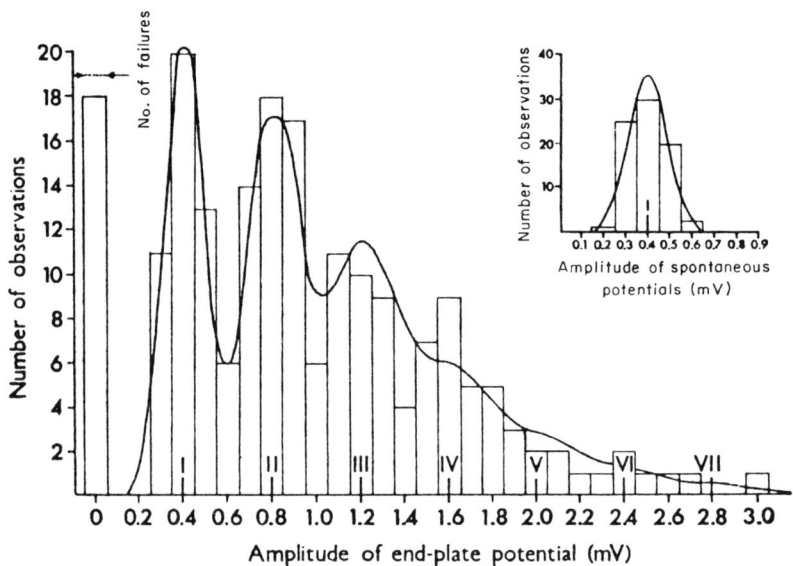

FIG. 4.27 Amplitude histogram of evoked endplate potentials (EPP) and of spontaneous miniature end-plate potentials (inset) recorded from a cat tenuissimus muscle end plate partially blocked in the presence of a high magnesium concentration. A Gaussian curve fitted to the mepp histogram was used to calculate the theoretical distribution of EPP amplitudes (solid line). [From Boyd and Martin (1956).]

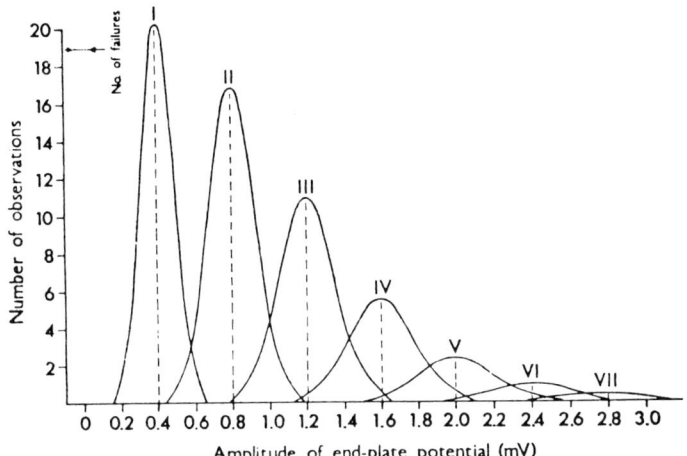

FIG. 4.28 The expected responses for 1, 2, ... (roman numerals in the figure) quanta calculated from the Poisson distribution (4.17). Expected responses are distributed normally about mean amplitudes equal to 1, 2, ... times mean amplitude of spontaneous potentials and with variances equal to integral multiples of the spontaneous potentials. [From Boyd and Martin (1956).]

identical nerve impulses 18 elicited no postsynaptic response, 6 EPPS had an amplitude of 0.6 mV, 9 occurred with an amplitude of 1.6 mV, etc. The peaks in the histogram distribution at amplitudes of 0.4, 0.8, ..., 2.8 mV correspond to quantum releases of 1, 2, ..., 7 packets (roman numerals in Fig. 4.27). Support for this assignment can be rationalized by the theoretical curves in Fig. 4.28, which were used in the construction of the theoretical amplitude distribution in Fig. 4.27. The solid line in Fig. 4.27 indicates the theoretical curve. The details of the analysis follow from the Poisson distribution. The fraction of failures, those inputs that do not evoke a response, obey the relation

$$P_0 = \frac{\text{number of failures}}{\text{total number of trials}} = e^{-m} = \frac{18}{198} \quad (4.18)$$

This gives $m = \ln(\frac{198}{18}) = 2.39$. A value for m can also be determined by dividing the mean amplitude of all responses including failures by the mean amplitude of the spontaneous potentials:

$$m = \frac{\text{mean size of EPP}}{\text{mean size of spontaneous mepp}} = \frac{0.933 \text{ mV}}{0.4 \text{ mV}} = 2.33 \quad (4.19)$$

This equation is merely a restatement of the quantum hypothesis and is valid only in the linear regime. For our purposes the slight differences in the value of m determined by Eqs. (4.18) and (4.19) will be neglected. With the value of m so determined, the number of events in which 1, 2, and 3 quanta are released simultaneously are, respectively,

$$n_1 = me^{-m}, \qquad n_2 = \frac{m^2}{2}e^{-m}, \qquad n_3 = \frac{m^3}{3!}e^{-m} \qquad (4.20)$$

Miniature end-plate potentials are not quite uniform in amplitude, and their amplitude distribution, shown in the inset in Fig. 4.27, can be fitted by a Gaussian distribution with an average value of 0.4 mV and a variance σ^2 (see p. 186). When k quanta are released simultaneously the expected amplitude distribution is characterized by a mean of $0.4k$ mV and a variance $\sigma_k^2 = k\sigma^2$, as indicated in Fig. 4.28. The latter relation follows since variances of summed events add linearly. If Poisson statistics govern quantum release, then the area under the distribution of k quanta released should equal n_k. This can be written mathematically as

$$n_k = \frac{m^k}{k!}e^{-m} = \int_{-\infty}^{\infty} A_k \exp\left\{-\frac{(E - 0.4k)^2}{2\sigma_k^2}\right\} dE \qquad (4.21)$$

where E is measured in millivolts. From Eq. (4.21) the normalization constants A_k are determined. A simple algebraic summation for $k = 1, 2, \ldots$ quanta released in a series of trials gives the expected number of observations at a fixed end-plate potential amplitude E. The theoretical curve $N(E)$ in Fig. 4.27 is therefore simply the sum

$$N(E) = \sum_{k=0}^{\infty} A_k \exp\left[-\frac{(E - 0.4k)^2}{2\sigma_k^2}\right] \qquad (4.22)$$

As can be seen in Fig. 4.27, the experimental results agreed quite closely with those expected theoretically, and it is this agreement that provides strong evidence for the statistical nature of transmitter release. Similar studies at many other synapses reveal fluctuations in transmitter release in accordance with Poisson statistics, although the average number of quanta released at different synapses for a single physiological response varies enormously. At the normal vertebrate neuromuscular junction m is on the order of 100–300, whereas at single terminals in sympathetic ganglia m is about 1–3. For inhibitory and excitatory synapses in the spinal cord, m is approximately unity. A general result is that synapses having larger numbers of integrative inputs have smaller m values.

An additional experimental test of the random succession of spontaneous transmitter release is provided by examining the distribution of time intervals between mepps. A characteristic property of a random process where each event is independent of preceding events is that the number of occurrences separated in time intervals between t and $t + \Delta t$ (see Fig. 4.29a) is distributed

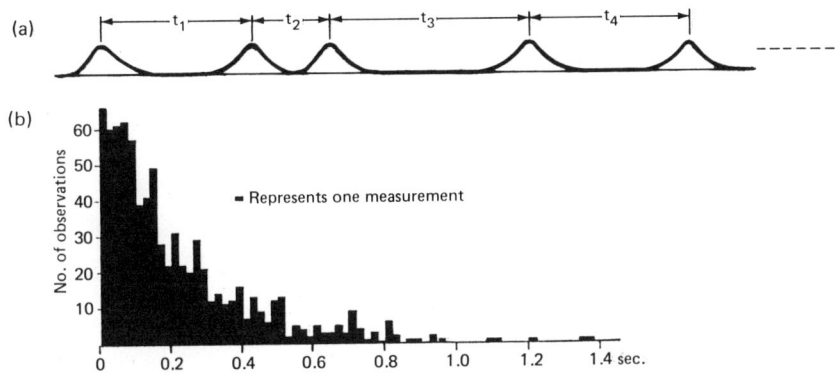

FIG. 4.29 (a) A series of schematic miniature end-plate potentials and the time intervals as measured between their peaks. (b) Distribution of time intervals between spontaneous releases for a series of 800 mepps. Observations were grouped in classes with $\Delta t = 20$ msec. Mean interval $T = 0.221$ sec. [From Fatt and Katz (1952).]

exponentially. This can be shown as follows. Suppose a mepp occurs at $t = 0$. Let $p(t)$ be the probability that no mepp occurs during the time interval 0 to t. The probability for the absence of mepps between 0 and $t + \delta t$, $p(t + \delta t)$, can be related to the probability $p(t)$ by using a standard theorem from probability theory, namely, that the *joint probability* of observing two distinct events is equal to the product of their respective probabilities if the two events are mutually independent. Hence $p(t + \delta t)$ equals $p(t)$ times the probability of no mepps during the interval t to $t + \delta t$. If δt is a small unit of time, then the probability of a mepp occurring in a time δt is $v\,\delta t$, where v is the average frequency of occurrence, i.e., $v = T^{-1}$, where T is the mean time separation between mepps. The probability of no mepp occurring during the time t to $t + \delta t$ is $1 - v\,\delta t = 1 - \delta t/T$. Thus

$$p(t + \delta t) = p(t)(1 - \delta t/T) \tag{4.23}$$

if δt is sufficiently small. In the limit where δt becomes infinitesimal, by definition

$$\frac{dp}{dt} = \left.\frac{p(t + \delta t) - p(t)}{\delta t}\right|_{\delta \tau \to 0} = -\frac{p(t)}{T} \tag{4.24}$$

The desired solution of this equation is $p(t) = e^{-t/T}$ since at $t = 0$ the probability must be unity. If N denotes the total number of mepps observed, then the number of occurrences n between t and $t + \Delta t$ is equal to N times the mean probability $\bar{p}$ over this time interval, i.e.,

$$n = N\bar{p} = \frac{N}{T} \int_{t}^{t + \Delta t} e^{-t/T}\,dt \tag{4.25}$$

By taking Δt sufficiently small compared to the average time interval T of separation between mepps (i.e., $\Delta t \ll T$), the integral in Eq. (4.25) can be approximated by $\Delta t \, e^{-t/T}$, and doing this gives

$$n = (N \, \Delta t/T) e^{-t/T} \qquad (4.26)$$

Such a dependence of n on time is illustrated in Fig. 4.29b.

The above discussion of transmitter release was restricted to mepps in the well-studied neuromuscular synapse. Experimental evidence indicates that quantum release of transmitters also occurs at other synapses. Whether there are nonquantum releases, i.e., a flow of single molecules, across the terminal membrane of a neuron constitutes a much disputed and active area of research. There is no doubt that at the neuromuscular junction there is nonquantum leakage of acetylcholine (see MacIntosh, 1980); however, the physiological importance of this process is unknown.

Although much is now known about ACh quantum leakage and its response, the formation of mepps, the physiological reason for this process is not known. Some suggestions on this fall into two categories: (1) the trivial explanation, namely, that there is a spurious leakage of ACh with no physiological corollary; and (2) that the leakage is essential for some physiological regulatory process. In this latter category there are two possibilities: (a) the leakage is an overflow from a metabolic process in the presynaptic terminal; or (b) the arrival of the ACh or other factors released during exocytosis, e.g., ATP or peptides, at the postsynaptic membrane may serve as a regulating factor in the formation of or the density of the receptors or both. In the latter case the resting quantum release rate would have a direct relation to the average lifetime of a postsynaptic receptor, estimated to be about 200 hr for acetylcholine receptors at postsynaptic membrane junctions (sometimes called junctional ACh receptors) in rat skeletal muscle and 30 hr for extrajunctional ACh receptors, those receptors not facing synaptic terminals (Steinbach et al., 1979).

There is good biological evidence that transmitter substances are contained in the synaptic vesicles. For example, when subcellular particles from cholinergic (see p. 131) terminals are separated according to their densities by using ultracentrifugation techniques, the vesicle population occurs at the same location as does the ACh. In spite of this, a close examination of the cytoplasm in the presynaptic terminals reveals the presence of acetylcholine not contained within vesicles.

A major difficulty with the Poisson model is its inability to give independent meaning and values to N and p. Presumably, N is the total number of vesicles capable of responding to an input stimulus since experiments at invertebrate neuromuscular junctions and some central vertebrate synapses have shown that N is much less than the total number of available vesicles, whereas p is thought to represent the average probability for the release of one unit, although there are no tests of this assignment. At synapses where the binomial equation (4.16) provides a more adequate description of the observed postsynaptic amplitude distribution, values of N and p can be inferred. In such cases the N and p parameters can be correlated with identifiable structural components of the presynaptic neurons. A synapse that obeys binomial statistics and where careful studies have been performed is the goldfish Mauthner (M) cell (Korn *et al.*, 1981), and in this case N was found to be equal to the number of presynaptic boutons (see p. 35). Unfortunately, it is unlikely that this structural correlation with N is true for all synapses, and it is more likely that there are structural differences as there are functional differences.

EXOCYTOSIS

The transfer of substances from a cell's interior into the external medium is called *secretion*. Secretion can occur in many forms. For example, some chemicals pass unhindered through the lipid bilayer by dissolving in the fatty-acid constituents of the plasma membrane; others pass through prescribed locations, namely, pores, whereas other substances are prepackaged into small intracellular vesicles bound by membranes. If the vesicles release their *entire* contents, including all diffusible ingredients in addition to the chemical transmitter, into the extracellular medium following *fusion* or coalescence of the vesicle membrane with the inner surface membrane of the cell, then the secretion process is called *exocytosis*. This process is illustrated in Fig. 4.30. That the exocytotic process occurs at only certain

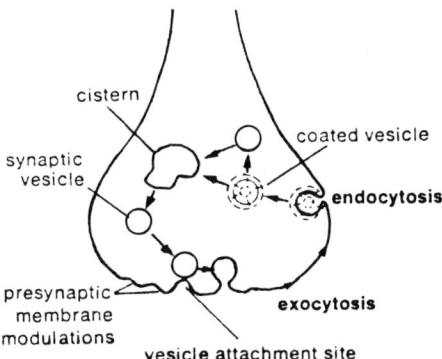

FIG. 4.30 Schematic of the exocytosis mechanism of transmitter release and the recycling of the synaptic vesicle membrane. The endocytosis (uptake) and exocytosis (release) processes are assumed to take place at different sites. [From Jones (1981).]

presynaptic membrane sites, called vesicle attachment sites (also called "hot spots" or releasing sites), is suggested by a variety of electron-microscopy studies such as those by Heuser and Reese (1979). Current understanding of how the vesicle membrane and the vesicle attachment sites fuse and form a channel for the release of the transmitter substance is almost nonexistent.

Exocytosis is not only involved in the release of neurotransmitters from nerve terminals, but also occurs in the release of numerous hormones and in the liberation of digestive enzymes into the gut. That exocytosis is the principal mechanism of evoked transmitter release is supported by several observations. (1) The appearance of Ω-shaped profiles in thin-section micrographs in freeze-etched fractions through the synaptic cleft. This is shown in Fig. 4.31. To capture the exocytotic process in this particular case the frog cutaneous pectoralis muscle was exposed to 4-aminopyridine, a chemical known to augment the amount of transmitter released in response to each nerve stimulus. (2) Measurement of the presynaptic membrane area after prolonged high-frequency electrical stimulation of the axon shows that it is larger than before activation. The area increase in the presynaptic membrane matches the decrease in total surface area of vesicles membranes. In addition exocytosis is consistent with the release by nerve stimulation of other chemicals present in vesicle interior. Even though this evidence appears

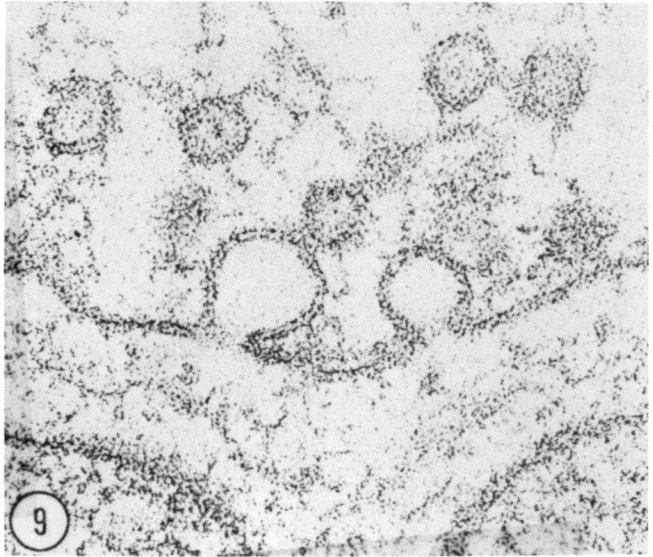

FIG. 4.31 High magnification of an extremely thin section through one active zone of a nerve that was quick-frozen. The synaptic vesicle fusion with the plasma membrane as the neurotransmitter is discharged is evident by the Ω shapes. [From Heuser (1976).]

convincing for the exocytosis mechanism of transmitter release, a critical analysis of the experimental data still leaves some doubt (Zimmermann, 1979).

Cells, in general, do not secrete continuously, but rather in response to a stimulus. For example, the elevation of the blood-sugar level initiates the release of the hormone insulin from the β cells of the pancreas. In the case of transmitter release the nerve impulse depolarization triggers the influx of calcium ions through the presynaptic membrane, which then initiates the transmitter release. The time lapse from calcium ion entry into a nerve to the time of transmitter release can be estimated from electrophysiological measurements and is found to be only 200 μsec. Such a short time interval places severe restrictions on the amount of vesicle movement. An estimate of this displacement, if the vesicles' motion is considered random, can be made as follows (Parsegian, 1977). As shown in Appendix I the mean-square displacement $(\Delta x)^2$ that occurs during a time t for one-dimensional random motion obeys the relation

$$(\Delta x)^2 = 2Dt \qquad (4.27)$$

where D is the diffusion coefficient. An estimate of the vesicle diffusion coefficient in the cell cytoplasm can be calculated from the Stokes–Einstein relation for the hydrodynamic motion of a sphere through a fluid:

$$D = kT/6\pi a\eta \qquad (4.28)$$

where k is Boltzmann's constant, T the absolute temperature, a the vesicle radius, and η the viscosity of the intracellular medium. With $kT = 4 \times 10^{-14}$ erg (its body-temperature value), $a = 250$ Å, $\eta = 10^{-2}$ poise (the value for water), Eq. (4.28) gives a vesicle diffusion constant of 8.5×10^{-8} cm^2 sec^{-1}, a value approximately one order of magnitude smaller than the diffusion constant of calcium ions in squid axons, which is estimated to be 6×10^{-7} cm^2 sec^{-1}. Inserting the value of $D \approx 8.5 \times 10^{-8}$ cm^2 sec^{-1} and $t \sim 200$ μsec into Eq. (4.27) gives a vesicle displacement about equal to its diameter, ~ 580 Å, whereas a Ca^{2+} ion in an axon may travel as much as 1500 Å during the equivalent releasing time. These calculations thus imply that only those vesicles in the neighborhood of the inside of the presynaptic membrane can release their contents through the wall.

What the role of calcium is and how it performs its necessary function is poorly understood. Some clues are evident from the following observations. Experiments have shown that in a typical presynaptic evoked release at the squid giant synapse about 1.7×10^9 Ca^{2+} ions flow into the presynaptic terminal and cause the release of the contents of $\sim 5 \times 10^3$ vesicles. This corresponds to about 3.4×10^5 Ca^{2+} ions per vesicle. If the diffusion rate of calcium ions through the presynaptic membrane and in the cytoplasm are taken to be the same, $\sim 6 \times 10^{-7}$ cm^2 sec^{-1}, then it is possible to estimate the concentration of Ca^{2+} ions within the terminal. This can be done as follows. The concentration of calcium is by definition equal to the number of Ca^{2+} ions within the cell divided by the volume they occupy. This volume is approximately equal to the maximum distance a Ca^{2+} ion can traverse in 200 μsec [from Eq. (4.27) this distance is $\sim 2 \times 10^{-5}$ cm] times a typical presynaptic terminal area (estimated to be 1.5×10^{-3} cm^2). Thus, the concentration of Ca^{2+} ions in the presynaptic terminal during evoked release is $\sim 1.7 \times 10^9$ Ca^{2+} ions/$(2 \times 10^{-5}$ cm $\times 1.5 \times 10^{-3}$ cm$^2) \approx 0.1$ mM. This is an enormous concentration compared to the free cytoplasmic calcium concentration, which is about $10^{-7} M$ in nearly all cells. It has been suggested that calcium screens the electrostatic repulsive forces between the vesicle and the presynaptic membrane and in addition activates a Ca^{2+} enzyme system for the removal of the hydration layers between the vesicle and plasma membrane. This sharp rise in calcium-ion concentration in the presynaptic terminal may have additional functions other than screening of electrostatic repulsive forces such as specific binding to macromolecules, which may assist in the movement of synaptic vesicles in the cytoplasm and their fusion with the presynaptic membrane. This is evident since it has been shown that for membrane separations less than 30 Å, the repulsive forces are only weakly dependent on membrane charge (Cowley *et al.*, 1977). The qualitative behavior of the repulsive forces as a function of membrane separation is shown in Fig. 4.32.

Just as how vesicles discharge their contents is not completely understood, their reformation is also not known. One hypothesis views vesicle discharge and recovery as spatially separate, one-way interactions with the presynaptic

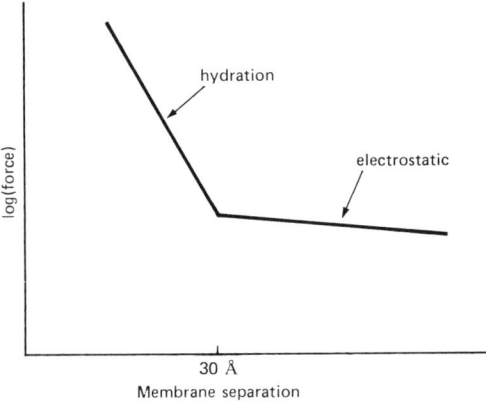

FIG. 4.32 Qualitative behavior of repulsive force between a vesicle and the presynaptic membrane. [After Parsegian (1977).]

plasma membrane, as illustrated in Fig. 4.30. Recovery of the vesicle membrane from the plasma membrane is by means of *endocytosis*, a term meaning the internal budding of a patch of plasma membrane, as shown in Fig. 4.30. Other investigators, however, have suggested that reclaiming and releasing sites are very close together, possibly at the same site. In either case a cytoplasmic material of unknown composition promotes the endocytotic process and, in addition, forms a *coat* on the newly formed vesicles. These *coated vesicles* either aggregate, forming irregular compartments called *cisternae*, or directly form mature synaptic vesicles. How and at what stage the vesicles are filled with the neurotransmitter are unclear. From this discussion it should be evident that the terminals of axons are quite complex little factories involved in a host of continuous processes vital to the functioning of neuronal transmission.

A MOLECULAR MODEL OF POSTSYNAPTIC RESPONSES

Having discussed the quantitative aspects of transmitter release we now examine the molecular aspects of postsynaptic currents and potentials. The theory to be described was initially developed for the analysis of the acetylcholine-activated channel at the neuromuscular junction, although the general aspects of the model are thought to be applicable to other synapses.

The essential features of the model are as follows. Those neurotransmitters (T) that are not destroyed by enzymatic processes (e.g., acetylcholinesterase in the case of cholinergic synapses) or lost from the synaptic cleft by passive diffusion bind to the acetylcholine-channel complex receptors (R). The receptor R is an intrinsic membrane protein with a molecular weight of about 250,000 daltons.[†] Initially, the bound transmitter–receptor complex (TR) has a molecular arrangement such that the receptor-associated ion channel is closed. The TR complex then undergoes a reorientation in space referred to as a conformational transformation. In this new configuration the channels open and ions are transported between the cytoplasm and extracellular medium. Since each possible spatial arrangement of atoms in a macromolecule has associated with it a particular energy, it is possible to envision the transmitter–receptor complex as having several principal local free-energy minima of conformation. Fortunately, most of the experimental data on end-plate currents can be explained by the two-state (open and closed) model. Figure 4.33 illustrates a plot of the free energy (F) (see Vol. I, p. 344) of a macromolecule as a function of different molecular conformations. The open and closed states are depicted as local free-energy minima.

[†] 1 dalton = 1.0000 on the atomic mass scale; very nearly equal to the mass of a hydrogen atom. The term is used interchangeably with the term molecular weight.

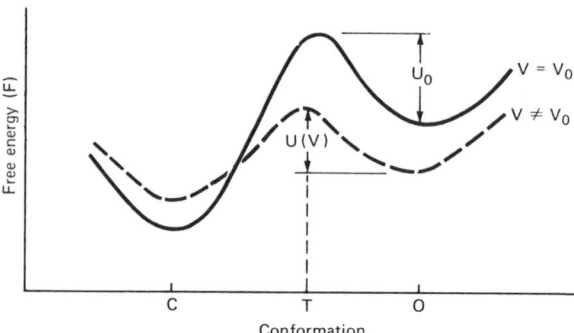

FIG. 4.33 The free energy associated with different macromolecular conformations. Dashed line corresponds to a fixed transmembrane potential V. Solid line refers to free energy of conformation when the transmembrane potential equals its resting state value V_0. The open (o) and closed (c) channel states are illustrated as two local minimum free-energy states, and T represents the transition barrier.

The above mechanism of action corresponds to the following kinetic scheme based on Eq. (4.2)

$$nT + R \underset{k_2}{\overset{k_1}{\rightleftarrows}} (T_n R) \underset{\alpha}{\overset{\beta}{\rightleftarrows}} (T_n R)^* \tag{4.2'}$$

where n denotes the number of transmitter molecules required to open a channel (n is 2 for the frog neuromuscular end plate, i.e., two transmitter molecules are required for each receptor) and the asterisk signifies an open channel; k_1, k_2, α, and β are the rate constants for the indicated processes; k_1 and k_2 are binding and dissociation rates and α and β are configurational change rates. A theoretical expression for the end-plate conductance $G_s(t)$, which appears in Eq. (4.4), can be determined in the following manner. Although decay of EPCs (p. 125) could reflect diffusion of ACh out of the synaptic cleft or its hydrolysis, experiments indicate that the time course of the decay of EPCs is not determined by the cleft concentration of acetylcholine (Stevens, 1974). The neurotransmitter cleft concentration can become vanishingly small in a time short compared with the decay of the end-plate current. Evidence for this comes from experiments that show that the decay of the end-plate current in both the presence and absence of anticholinesterases is very sensitive to membrane potential. These findings are difficult to rationalize if the EPC decay rate was determined by cleft concentration. Furthermore, since the end-plate currents decay with a single time constant, i.e., they exhibit exponential dependence on time, as illustrated in Fig. 4.17, it follows that only one of the backward reaction rates, k_2 or α in the kinetic scheme on p. 94, controls the rate. Experimental evidence is consistent with the conformational model in which the rate-limiting step is determined by α, the closing rate of the channels. In this model the binding and dissociation rates k_1 and k_2 of the transmitter to the receptor are assumed to be fast reactions compared with the reorientation time of the complex $(T_n R)$. This latter statement implies that an equilibrium constant K (Vol. I, p. 351) is sufficient for characterization of the intermediate state

$$K \equiv \frac{k_2}{k_1} = \frac{[R][T]^n}{[T_n R]} \tag{4.29}$$

Here $[R]$ denotes the number of unreacted receptors, $[T]$ is the cleft transmitter concentration, and $[T_n R]$ is the number of transmitter–receptor complexes in the closed configuration. It is evident from the kinetic scheme that the kinetic equation governing the time evolution of the open state is simply (Vol. I, p. 350)

$$\frac{d[T_n R]^*}{dt} = -\alpha[T_n R]^* + \beta[T_n R] \tag{4.30}$$

where the brackets denote the number of transmitter–receptor complexes. The asterisk signifies open channels. Unfortunately, this equation cannot be solved in general, since the time relation between $[T_n R]^*$ and $[T_n R]$ is not known. However, if the number of occupied receptors is assumed to be small compared with the total number of receptors N, then by setting $N = [R]$, Eq. (4.30) is amenable to a simple solution. It follows from Eq. (4.29) that in this approximation $[T_n R] = N[T]^n/K$, which on substituting into Eq. (4.30), gives

$$\frac{d[T_n R]^*}{dt} = -\alpha[T_n R]^* + \frac{\beta[T]^n N}{K} \tag{4.31}$$

If all open channels are characterized by a time-independent conductance γ, then the total conductance $G_s(t)$ as a function of time is simply

$$G_s(t) = \gamma [T_n R]^* \tag{4.32}$$

and from Eq. (4.31) it follows that $G_s(t)$ satisfies the equation

$$\frac{dG_s(t)}{dt} = -\alpha G_s(t) + \frac{\gamma \beta N}{K} [T]^n \tag{4.33}$$

The solution of this inhomogeneous linear differential equation can be obtained by the substitution $G_s(t) = e^{-\alpha t} X(t)$; thus

$$\frac{dX(t')}{dt'} = \frac{N\gamma\beta}{K} [T(t')]^n e^{\alpha t'} \tag{4.34}$$

The integral of Eq. (4.34) may be substituted for the $X(t)$ term in the above G_s expression:

$$G_s(t) = e^{-\alpha t} X(t)$$

$$= e^{-\alpha t} \frac{\beta N \gamma}{K} \int_0^t [T(t')]^n e^{\alpha t'} \, dt' \qquad (4.35)$$

Because of our lack of knowledge concerning the time dependence of the transmitter concentration, the integral in Eq. (4.35) cannot be explicitly calculated. Fortunately, as mentioned earlier, experiments indicate that the decline of the transmitter concentration in the cleft is rapid compared with the time duration of the end-plate current. Hence, there is a time t_0 at which $[T(t)] \cong 0$ for $t > t_0$, and in this time domain the synaptic conductance decays exponentially, i.e.,

$$G_s(t) = e^{-\alpha t} \frac{\beta N \gamma}{K} \int_0^{t_0} [T(t')]^n e^{\alpha t'} \, dt'$$

$$= (\text{const}) e^{-\alpha t} \qquad (4.36)$$

because the integral, whatever its value, is a constant. Hence, in the conformational model for transmitter–receptor interaction, the rate of decay of the end-plate current is identical to the channel closing rate. The behavior of drugs that modify the end-plate current decay either by prolonging or shortening the response can be interpreted as an alteration in how rapidly the open-channel state relaxes toward the closed-channel configuration or as a slowing or enhancement of presynaptic transmitter release.

Additional insight into the mechanism responsible for the decaying phase of the end-plate currents can be obtained from the following theoretical considerations. According to chemical rate theory the rate of a particular reaction is equal to the product of a characteristic encounter frequency of the reactant with the energy barrier that separates the initial state from its final state times a probability factor P for surmounting the energy barrier. The probability factor P is given by Boltzmann statistics as

$$P = e^{-U(V)/kT} \qquad (4.37)$$

where $U(V)$ is the energy barrier between the open and transition state and its value depends explicitly on the potential difference V across the membrane. This energy difference between the open and transition state is shown in Fig. 4.33. Thus, the channel closing rate can be expressed as

$$\alpha(V) = \nu e^{-U(V)/kT} \qquad (4.38)$$

A MOLECULAR MODEL OF POSTSYNAPTIC RESPONSES

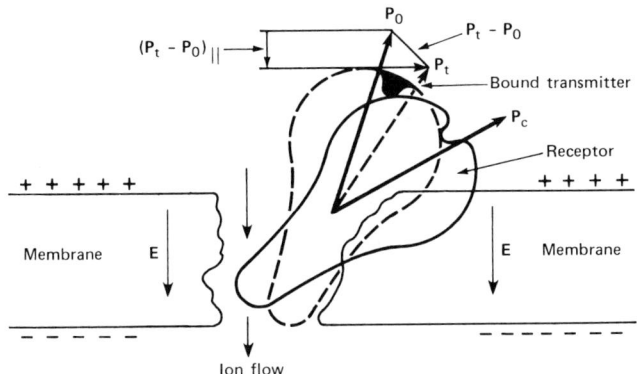

FIG. 4.34 Schematic illustration of the opening of a channel due to a conformational change of the receptor complex. Racket-shaped object denotes receptor complex. Solid line: closed configuration, passage of ions through channel is prohibited. Dashed line: open configuration, transport of ions is allowed. Other symbols denoting dipole moments of open (P_o), closed (P_c), and transition (P_t) states are discussed in the text.

The frequency of encountering the barrier, denoted by v, is generally taken to be an effective vibrational frequency of the molecular system and is typically on the order of 10^{11}–10^{13} sec^{-1} (see Vol. II, p. 209).

The dependence of the transitional energy difference $U(V)$ on transmembrane potential can be rationalized in the following manner. Each conformational state has associated with it an electric dipole moment, which characterizes the spatial arrangement of the charges on the protein. Neither the magnitudes nor the directions of the intrinsic electric dipole moment of the transmitter–receptor complex for the open and closed states are expected to be the same. A schematic illustration of the opening and closing of a channel by a conformational change in the receptor complex is illustrated in Fig. 4.34. From elementary physics it is known that an electric dipole moment **p** in an external electric field **E** has associated with it an interaction energy

$$W = -\mathbf{p} \cdot \mathbf{E} = -pE \cos \theta \tag{4.39}$$

where θ is the angle between the vectors **p** and **E**. In the present case the electric field arises from the potential difference V across the membrane. Assuming a linear behavior between E and V (see Vol. I, p. 67) gives

$$E = V/d \tag{4.40}$$

where d is the thickness of the membrane. If U_o denotes the energy difference between the open state and transition state in the absence of an electric field (see Fig. 4.33) and if the only energy difference due to the electric field arises

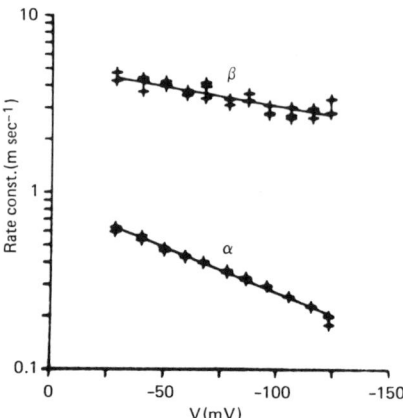

FIG. 4.35 Semilog plots of α and β determined from voltage-clamped frog sartorius nerve-muscle preparation at temperature of 20°C. [From Stevens (1974). Copyright 1974 Raven Press, New York.]

from the dipolar interaction energies, then it follows that the energy barrier between the open and transition state when the transmembrane potential is V is given by

$$U(V) = U_o - (W_{open} - W_{transition}) = U_o - (\mathbf{p}_t - \mathbf{p}_o)_\| V/d$$
$$= U_o - (\Delta p)_\| V/d \qquad (4.41)$$

Here $\mathbf{p}_t$ and $\mathbf{p}_o$ are the electric dipole moments of the transition state and open state of the receptor complex, respectively. The net change in the electric dipole moment vector between these two configurations in the direction of the membrane electric field has been denoted by $\Delta p_\|$. Inserting this expression for $U(V)$ into Eq. (4.38) gives a channel closing rate α exponentially dependent on the membrane potential, i.e.,

$$\alpha(V) = v \exp(-U_o/kT) \exp(V\Delta p_\|/dkT) \qquad (4.42)$$

Figure 4.35 illustrates this exponential behavior of α on transmembrane potential. A method of measuring the rate constants α and β is given in Chapter 6. From the slope of log $\alpha(V)$ versus V an estimate of the change in the electric dipole moment can be inferred. For the data shown in Fig. 4.35, the slope $\Delta p_\|/dkT$ is 0.008 mV^{-1}. Assuming an effective membrane thickness of 50 Å, taking $kT = 4.04 \times 10^{-21}$ J at 20°C, a value of $\Delta p_\| = 48$ D is calculated.[†] Proteins generally have intrinsic dipole moments the order of 200–500 D (McClell, 1963); hence the inferred value of $\Delta p_\|$ associated with

[†] One debye = 1 D = 10^{-18} esu cm = 3.34×10^{-30} C m.

conformational change of the receptor complex to its closed state configuration corresponds to approximately a 10–20% alteration of its value along the field. Using arguments similar to those given for α, the opening rate β is expected and indeed is found experimentally to depend exponentially on membrane voltage, as illustrated in Fig. 4.35. In this case a fit to the experimental curve gives a dipole moment difference of 19.2 D between the transition and closed state.

The equations developed above have been found to provide a good quantitative description of end-plate currents for most experimental conditions. The model presented is applicable to postsynaptic currents other than those generated at the neuromuscular junction, although the physical parameters characterizing the responses are expected to be different. However, the two-state model fails when the end plate is subjected to prolonged contact with cholinergic *agonists*. An agonist is defined as a chemical that binds to the receptor as does the naturally occurring transmitter and which generates a similar postsynaptic response. A constant application of an agonist at the subsynaptic membrane eventually elicits a reduced response. This reduction in sensitivity of the postsynaptic membrane is termed *desensitization*, and this effect cannot be explained using the two-state model discussed above.

DRUG ACTION AT SYNAPTIC RECEPTORS

The growth in our understanding of synaptic transmission has greatly clarified pharmacology. Knowledge of how a particular chemical acts, and where, has found use not only in clinical medicine but also in the elucidation of physiological mechanisms. Each step involved in neurotransmission represents a potential point for drug action. Table 4.2 lists a few types of drug actions at peripheral *cholinergic* and *adrenergic* synapses. Neurons that release acetylcholine from their terminals are known as cholinergic. In addition to the classical neuromuscular junction discussed in this chapter, cholinergic nerve endings occur in various regions of the brain, e.g., cerebral cortex, hippocampus, striatum, and hypothalamus, and in the peripheral nervous system, e.g., ganglion cells and parasympathetic excitation nerve fibers to smooth muscle. Adrenergic synapses are those synaptic junctions that release the transmitter noradrenaline (norepinephrine). In the peripheral nervous system adrenergic synapses occur at motor innervation of vascular smooth muscle where their effects are excitatory and at intestinal smooth muscle where norepinephrine has primarily inhibitory behavior. Norepinephrine has also been identified as a neurotransmitter in several regions of the CNS, e.g., the cerebellum, the olfactory bulb, the caudate nucleus, and the spinal cord.

TABLE 4.2

Types of Action of Representative Drugs at Peripheral Cholinergic and Adrenergic Synapses

Mechanism of Action	System	Drug
Interference with synthesis of transmitter	Cholinergic Adrenergic	Hemicholinium (HC-3) α-Methyl-p-tyrosine
Prevention of release of transmitter	Cholinergic Adrenergic	Botulinus toxin Bretylium

Blockage of endogenous transmitter at postsynaptic receptors	Cholinergic
	  Muscarine
	  Nicotine
	
	Adrenergic
	  Alpha
	  Beta
	  Beta$_1$
		
		Atropine
		d-tubocurarine
		hexamethonium
		
		Phenoxybenzamine
		Propranolol
		Practolol

Inhibition of enzymatic breakdown of transmitter	Cholinergic Adrenergic	Anticholinesterase (e.g., physostigmine) MAO inhibitors (e.g., pargyline, nialamide, tranylcypromine)

Drugs can act either presynaptically or postsynaptically. The poison *botulinum toxin*, produced by the bacterium *clostridium botulinum*, was a serious problem in the food industry in the 19th century, and occasionally still is. It prevents the release of acetylcholine from nerve terminals at all cholinergic fibers. The detailed mechanism of botulinum toxin still remains to be explored. The synthetic compound hemicholinium (HC-3) also blocks neuromuscular and ganglionic transmission; however, in this case the mechanism is known to be by inhibition of the transport across the presynaptic membrane of choline into the cell from the synaptic cleft. This compound is needed for the manufacture of acetylcholine and is a normal constituent in the extracellular fluid. Because HC-3 inhibits choline uptake, it has been used as a tool for the study of acetylcholine turnover and storage in cholinergic nerve terminals.

There are several ways in which drugs acting postsynaptically promote the blockage of the normal action of a transmitter. The cholinergic receptor drugs produce either one of two effects: (1) the same biologic response as that of ACh, in which case the drug is called an agonist [e.g., the compound *methacholine* when applied to ganglionic postsynaptic membranes produces a PSP similar to the endogenous (normal) response]; or (2) it interacts with the receptor and produces no membrane permeability changes. Drugs that interfere with or prevent the formation of an agonist–receptor complex are known as *antagonists*. Antagonists are classified as being either *competitive* or *noncompetitive*. An antagonist is competitive if it combines reversibly with the same binding sites as the endogenous chemical and can be displaced from these sites by an excess of the agonists. *Curare*, a poison used by various South American aborigines for killing wild animals, is a classic example of a competitive antagonist, or blocking agent at the neuromuscular junction. An antagonist is *noncompetitive* when its effects cannot be overcome by increasing concentrations of the agonist. A noncompetitive antagonist is *reversible* if the removal of the antagonist restores the system to its original state. *Irreversible noncompetitive antagonists* lead to the destruction of the receptor.

Receptors in excitable cells are key elements in the mediation of intercellular communication in the nervous system. The active site of a receptor is highly specific for its own transmitter, and even slight chemical alterations of the transmitter can prevent the molecule from activating the receptor. In fact, even stereochemical differences are crucial. For example, L-epinephrine and D-epinephrine are identical molecules with respect to the number and kinds of atoms, but they are isomeric in structure. They differ only in the spatial arrangement of the groups attached to the asymmetric carbon atom (see Fig. 4.39). The two structures cannot be superimposed. Because of this subtle difference, it is found that L-epinephrine is approximately ten times more potent in raising blood pressure than is D-epinephrine. It is this high degree of specificity in structure–function relation that allows for a classification of receptors according to how they interact with particular drugs. For example, the effects of the alkaloids *muscarine* and *nicotine* at cholinergic junctions provide the basis for the differentiation of the ACh receptor into two classes. ACh receptors that bind nicotine are called *nicotinic* receptors, whereas those that interact with muscarine are called *muscarinic*. An example of the first is the receptor at the skeletal neuromuscular end plate, which is nicotinic. The best known competitive antagonist for ACh muscarinic receptors is *atropine*. The drug (eye drops) used by ophthalmologists is a mixture of atropine (or homatropine) and cocaine. Atropine prevents the action of the normal transmitter, acetylcholine, in the parasympathetic nerves that control the *sphincter pupillae*. Since the sphincter normally balances the *dilator pupillae* for the adjustment of the iris to varying light intensities, its paralysis causes the iris to remain dilated. Some neurons, such as sympathetic ganglion cells, have both nicotinic and muscarinic acetylcholine receptors, whereas other cholinergic neurons have only one type. Not all nicotinic ACh receptors are identical, since ACh nicotinic receptors in the autonomic ganglia and skeletal muscle respond differently to certain stimulating and blocking agents.

Receptors for adrenaline, noradrenaline, and related substances are called adrenergic and, like cholinergic receptors, they can be classified according to their specificity to natural and synthetic compounds. Ahlquist (1948) used a series of agonist drugs and showed that there were two discrete potency sequences for various types of physiological responses. His observation allowed adrenergic receptors to be divided into two distinct classes, called α and β. Subsequent pharmacological studies have further subdivided the β-receptor into two major types, $β_1$ and $β_2$. The present characterization of adrenergic receptors is based on (1) the relative potency of a series of adrenergic agonists for eliciting the specific response and (2) the potency of an antagonist for blocking the response to a given agonist. Table 4.3 summarizes the classification of adrenergic receptors based on ligand specificity and also shows typical physiological actions.

Snake venoms have been particularly useful in the study of excitable membranes. The active components of venoms from the *elapidae* snake, α-bungarotoxin (α-BuTX), which blocks the acetylcholine receptor, has been extensively used. When the toxin is radio-labeled with ^{131}I it is possible to demonstrate specific binding to the end-plate region of the neuromuscular junction. Such studies have allowed the number of toxin–receptor sites to

TABLE 4.3

Some Characteristics of Adrenergic Receptors[a]

	Action	Examples of Specific Agonists/Antagonists
α	Blood vessel constriction, pupil dilation, contraction of smooth muscle in uterus, increased release of acetylcholine in skeletal muscle	Noradrenaline, adrenaline, clonidine, oxymetazoline, naphazoline/phentolamine, phenoxybenzamine, dibenamine, WB-4101, azapetine
$β_1$	Increased myocardial contractivity, increased heart rate, blood vessel dilation	Isoproterenol, adrenaline, noradrenaline, propranolol, dichloroisoproterenol, practolol
$β_2$	Relaxation of respiratory system, changes in twitch tension and increased glycogenolysis in skeletal muscle	Isoproterenol, adrenaline, noradrenaline, salbutamol, terbutaline/propranolol, butoxamine

[a] From Giesecke and Hebert (1979).

be estimated. For the frog neuromuscular end plate, the number of sites estimated by this technique is about 10^5 μm^{-2}. The total number of ACh receptor sites has been found to depend on both the size and the developmental stage of the animal. For example, in the particular case of motor end plates in rats, radio-labeled ^{125}I-α-BuTX studies have demonstrated that the number of receptors per end plate is significantly larger for heavier rats, about 1.4×10^7 for a male rat weighing 100 g compared to 3.98×10^7 for a full-grown male rat weighing 360 g (Fambrough and Hartzell, 1972).

CENTRAL-NERVOUS-SYSTEM NEURONS AND DRUG EFFECTS

The properties of individual central-nervous-system neurons and their pharmacological properties are poorly understood in comparison to end plates. Examples of drugs whose mechanism of action is known and that are also important as research tools are *strychnine* and *picrotoxin*. Both have been shown to selectively block inhibitory pathways in the CNS. Strychnine produces excitation at all portions of the CNS by acting as a competitive antagonist to the inhibitory sites in a manner similar to the way curare blocks ACh activation at the neuromuscular junction. A particular well-studied CNS inhibitory pathway that not only illustrates how strychnine acts but also is an example of feedback inhibition is that formed by the axon collaterals of the *spinal motoneurons* and the *Renshaw* cells. This inhibitory pathway is illustrated in Fig. 4.36, which shows that the axons of moto-

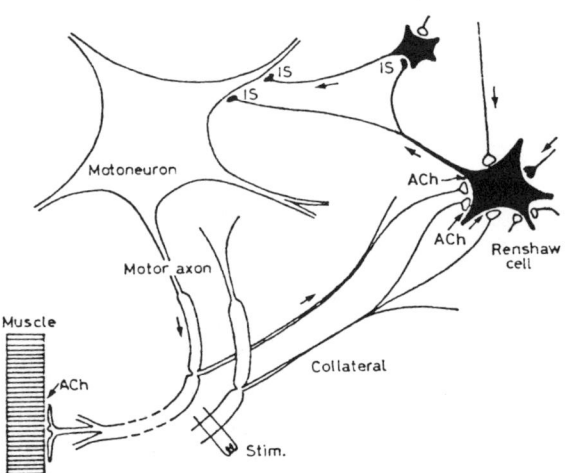

FIG. 4.36 The inhibitory pathways to motoneurons by their collaterals and Renshaw cells. IS: inhibitory synapse; Stim.: stimulation point. [From Eccles (1969).]

neurons have one or more collateral branches as they transverse the spinal cord to emerge in a ventral root (see Fig. 4.12). These collaterals are distributed solely to Renshaw cells where they make *monosynaptic* excitatory synapses with ACh as the transmitter. Each Renshaw cell sends axons to several motoneurons and also to other Renshaw cells. Transmission at synapses formed by Renshaw axons and motoneuron cell bodies are inhibitory, and thus Renshaw cells are sometimes loosely called *inhibitory neurons*. Antidromic stimulation of the proximal end of a severed ventral root as shown (stim.) in Fig. 4.36 activates the Renshaw cell through the release of ACh at the collateral terminals. That acetylcholine is the neurotransmitter at collateral terminals as it is at neuromuscular junction is an example of Dale's principle (p. 95). Those axons from the Renshaw cells terminating on motoneurons produce large and relatively prolonged IPSPs. These IPSPs can be measured by intracellular recordings. Examples of such recordings from a motoneuron generated by antidromic volleys are shown in Fig. 4.37. Strychnine actions can be monitored by observing its suppression of IPSPs. When the Renshaw cell feedback inhibition is blocked, ongoing motorneuronal activity is enhanced and uncoordinated muscular contractions occur. The tetanus toxin also blocks postsynaptic inhibition; however, it acts by preventing the release of the transmitter, presumed to be glycine, from the inhibitory terminals of Renshaw cells, whereas strychnine actions are postsynaptic.

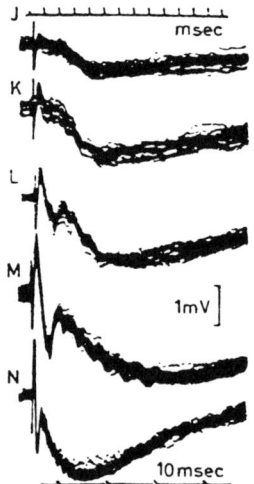

FIG. 4.37 Inhibitory potentials generated in motoneuron from activation of Renshaw cells by antidromic volleys of increasing stimulus magnitude (J to N). Time scale: above J in milliseconds for J–M, and below N in 10 msec for N. [From Eccles *et al.* (1954).]

COMMON DISEASES ASSOCIATED WITH SYNAPTIC FUNCTION

Malfunction of normal receptor activity and specificity and alterations in the amount of neurotransmitters in specific areas of the CNS are known to be implicated in several well-known clinically recognizable disease states. The following discussion is restricted to only two neuronal diseases: (1) myasthenia gravis, which is a neuromuscular disorder characterized by weakness and fatigability of voluntary muscles; and (2) Parkinson's disease, the classic signs of which are rigidity, abnormal slowness of movement, and tremor. However, the number of diseases that are due to improper neuronal functioning is quite large.

The most characteristic feature of myasthenia gravis is progressive muscular weakness during activity. The clinical features have been well known for about a century. The patient is unable to sustain or repeat muscular contractions. For example, if the affected muscle is the flexor muscle of the hand, the first squeeze of a ball will be quite forceful; however, a rapid decrease in forcefulness occurs with increasing numbers of squeezes. With rest the strength will return. Not all muscles in the body are uniformly affected, but the disease is progressive and slowly spreads to other muscles. About 10% of the patients die shortly after becoming afflicted; however, those fortunate enough to survive the first three years have an excellent chance that their condition will stabilize and allow them to live, although with reduced activity. The disease usually starts in the second to fourth decade of life and is more common in females. The basic defect is now known to be a reduction in the number of acetylcholine receptors at the neuromuscular junction, although the cause of this substantial decrease in receptor number is unknown. This reduction in the number of receptors was shown by measuring the amount of bound radio-labeled ^{125}I-α-BuTX on muscle biopsies obtained from myasthenia gravis patients. As discussed earlier, the toxin inhibits ACh-induced depolarization of vertebrate skeletal muscle, and this inhibition is related to the highly specific binding to the ACh receptors. ^{125}I-α-BuTX binds irreversibly and specifically to the ACh receptor, and thus the number of ACh receptors per junction can be easily quantified. Myasthensis muscles have a pronounced reduction in the number of receptors, averaging about 20% of the number occurring in healthy muscle (Fambrough et al., 1973). With this finding, the muscular weakness that characterizes the disease can be easily understood, since the decreased number of receptors reduces the probability of interaction between acetylcholine and its receptor. In some muscle fibers the substantial reduction results in below-threshold end-plate potentials with a corresponding failure to trigger a muscle action potential. Since not all fibers of the muscle are activated at junctions affected by myasthenia gravis, the net power of the muscle is reduced. In normal muscle tissue the safety margin for neurotransmission is large; approximately 10 times as many acetylcholine receptors are activated as are required to initiate an action potential. This observation has been used in diagnostic tests for myasthenia gravis. A small dose of the postsynaptic blocking agent curare is administered intravenously. In a normal patient a small dose, because of the large safety factor, will not result in any changes of muscular strength. However, patients afflicted with myasthenia gravis will show a pronounced increase in weakness since blockage of even a small number of ACh receptors is sufficient for clinical manifestation of fatigue and weakness. This *curare test* entails some risk since artificial ventilation may be necessary, although this problem can be circumvented by limiting the administration of curare to a single limb by use of a tourniquet.

For most patients with myasthenia gravis, the first line of treatment is the application of anticholinesterases. These drugs inhibit the enzymes responsible for the hydrolysis of acetylcholine to choline and acetate and in effect increase the lifetime of acetylcholine in the synaptic cleft. This increased lifetime enhances the number of collisions with the ACh receptors and thereby results in larger end-plate potentials. The most widely used drug is pyridostigmine bromide, which can be administered orally and has an effective duration of about 4 hr.

There are other treatments, the basis of which are not understood. In some patients clinical improvement is evident after the removal of the thymus gland. Younger patients, for reasons unknown, show the greatest improvement after thymectomy. For patients whose muscular weakness cannot be satisfactory controlled by anticholinesterases, steroid administration in the form of prednisone has been found helpful. Steroids are generally used in conjunction with properly controlled amounts of anticholinesterases and, once started, require an indefinite maintenance dose either every day or every other day. The mechanisms of action of corticosteroid agents and thymectomy are not yet understood. However, from general knowledge of their effects, it is suggestive that myasthenia gravis is an autoimmune disease in which endogenous antibodies destroy the receptors [Drachman, 1979].

As noted above, the clinical features of Parkinson's disease include rigidity, tremor, slowness of movement, and speech difficulties. These were first discerned in 1817 by James Parkinson, an English physician. In contrast to myasthenia gravis, the site of the defect is located in the heart of the CNS, namely, the *basal ganglia*. The basal ganglia is a constellation of gray-matter structures lying beneath the *cerebral cortex* and just lateral to the *dorsal thalmus*. Their locations and connections are illustrated schematically in Fig. 4.38. The basal ganglia is divided into several substructures called the *caudate nucleus*, the *putamen*, and the *globus pallidus*. Modern researchers also include within the collective term basal ganglia several nearby neural

COMMON DISEASES ASSOCIATED WITH SYNAPTIC FUNCTION

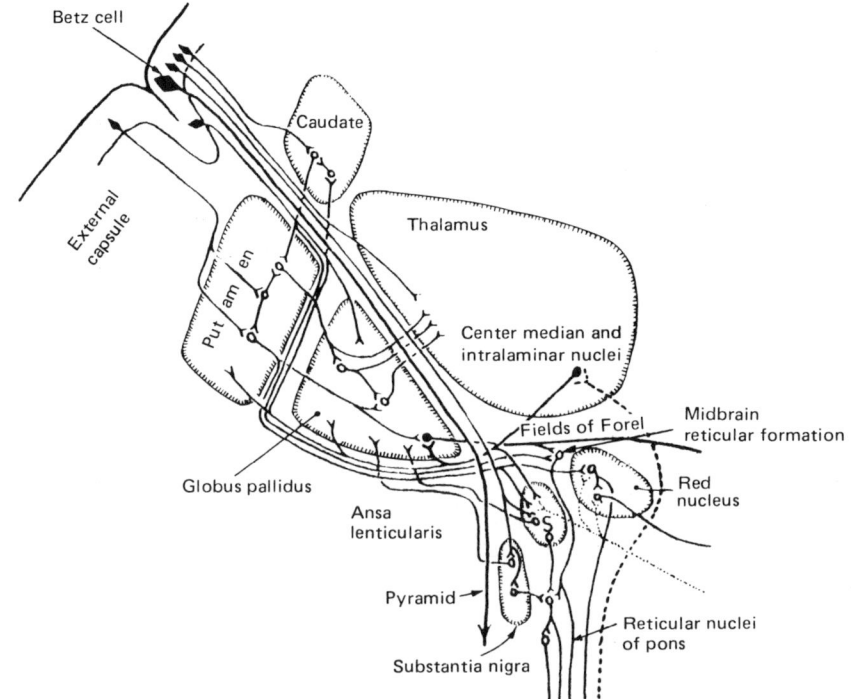

FIG. 4.38 The basal ganglia and their principal connections. [From Denny-Brown (1960).]

structures, namely, the *substantia nigra*, the *subthalamic body*, and the *red nucleus*, since they are intimately associated with the functioning of the caudate nucleus and the putamen. The caudate nucleus and the putamen are collectively called the *corpus striatum*. In lower vertebrates, e.g., birds, the basal ganglia is associated with the highest level of motor integration; however, in the higher vertebrates its functions are incompletely understood. For our purpose, it is sufficient to know that axons of some neurons of the *substantia nigra* terminate on cells in the corpus striatum. Some of these *nigrostriatal* axons form inhibitory synapses with striated neurons and have as a transmitter the *catecholamine dopamine* (see Fig. 4.4).

Table 4.4 shows a comparison of the regional concentration of dopamine and norepinephrine in brains of nondiseased and parkinsonian patients. Examination of the table shows that patients with Parkinson's disease have a pronounced decrease of dopamine in the corpus striatum as compared to normal nondiseased brains. This result clearly indicates some type of dysmetabolism of dopamine, although it does not establish the cause of the loss of brain dopamine. This decrease in dopamine was first observed in

TABLE 4.4

Regional Concentrations of the Catecholamines, Dopamine and Norepinephrine in Brains of Normal and Parkinsonian Patients[a,b]

Brain Region	Dopamine		Norepinephrine	
	Normal	P.D.	Normal	P.D.
Caudate Nucleus	3.50	0.32	0.07	0.03
Putamen	3.57	0.23	0.11	0.03
Globus pallidus	0.30	0.44	0.09	0.11
Substania nigra	0.46	0.07	0.04	0.02

[a] From Sourkes (1972).
[b] Normal ≡ nonneurological case; P.D. ≡ Parkinson's disease; concentration in micrograms per gram wet weight of tissue.

animal-model studies, in particular, in those using monkeys, in which small lesions were placed stereotactically in different regions of the basal ganglia. Such studies demonstrate that when nerve fibers from the substantia nigra to the striatum are interrupted, the animal displays the clinical features of parkinsonism. Histochemical studies show that these clinical signs correlate with reduced dopamine levels in the corpus striatum. In this way the destruction of the substantia nigra and its efferent axons to the corpus striatum has been identified in the CNS malfunction for parkinsonism. The origin of this degeneration of the nigrostriatal dopamine neurons is unknown. Fortunately, the administration of the chemical L-dopa provides some clinical improvement for at least half of the patients with Parkinson's disease. L-Dopa (dihydroxyphenylalanine), whose structure is shown in Fig. 4.39, is the in vivo precursor of the neurotransmitter dopamine and is itself synthesized from the naturally occurring amino acid L-tyrosine by the enzyme tyrosine hydroxylase. Dopamine neurons of the brain are those for which the formation of the dopamine is the terminal step in the biochemical pathway illustrated in Fig. 4.39. Norepinephrine neurons contain the enzyme dopaminebeta hydroxylase. The presence of this enzyme in the synaptic vesicle allows for the conversion of dopamine to norepinephrine and the vesicular storage of the latter. The therapeutic advantage of L-dopa over dopamine resides in its ability to cross the blood–brain barrier (Vol. I, p. 291), whereas dopamine cannot. After administration, L-dopa spreads to all regions of the brain. The ubiquitous enzyme L-dopa decarboxylase converts L-dopa to dopamine, thereby replenishing the supply of dopamine at appropriate deficient sites. This method of treatment for Parkinson's disease constitutes an excellent example of the practical value of elucidating biochemical pathways of neurotransmitters.

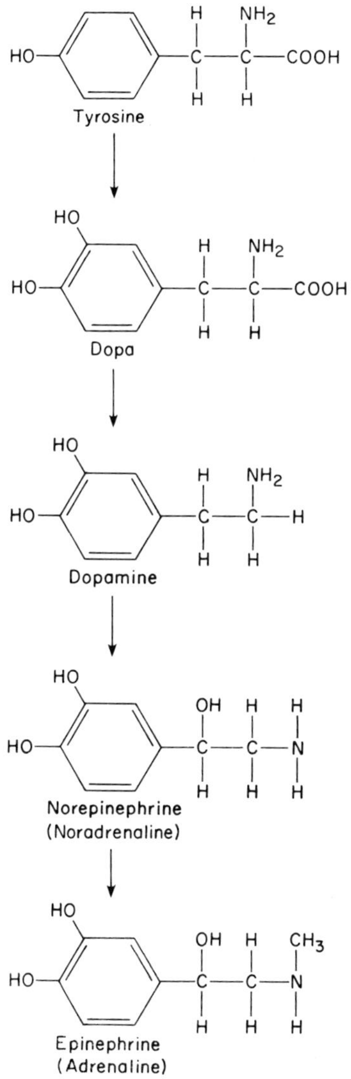

FIG. 4.39 The primary synthesis pathway for tyrosine, dopa, dopamine, norephinephrine, and epinephrine from the dietary amino acid L-tyrosine.

REFERENCES

Ahlquist, R. P. (1948). A study of the adrenotropic receptors. *Am. J. Physiol.* **153**, 586.
Bennett, M. V. L. (1977). Electrical transmission: A functional analysis and comparison to chemical transmission. *In* "Handbook of Physiology." (E. R. Kandel, ed.), Sect. 1, Vol. II, Chapter 11, p. 357. Physiol. Soc., Bethesda, Maryland.
Boyd, I. A., and Martin, A. R. (1956). The end-plate potential in mammalian muscle. *J. Physiol. (London)* **132**, 74.
Brock, L. G., Coombs, J. S., and Eccles, J. C. (1952). The recording of potentials from motoneurons with an intracellular electrode. *J. Physiol. (London)* **117**, 431.
Coleman, P. D., and Riesen, A. H. (1968). Environmental effects on cortical dendritic fields I. Rearing in the dark. *J. Anat.* **102**, 363.
Coombs, J. S., Eccles, J. S., and Fatt, P. (1955a). The specific ionic conductances and the ionic movements across the motoneuronal membrane that produce the inhibitory postsynaptic potential. *J. Physiol. (London)* **130**, 326.
Coombs, J. S., Eccles, J. C., and Fatt, P. (1955b). The inhibitory suppression of reflex discharges from motoneurones. *J. Physiol. (London)* **130**, 396.
Cottrell, G. A. (1976). Does the giant cerebral neurone of Helix release two transmitters: ACh and serotonin? *J. Physiol. (London)* **259**, 44.
Cowley, S., Fuller, N., Rand, R. P., and Parsegian, V. A. (1977). Measurement of repulsion between charge phospholipid bilayers. *Biophys. J.* **17**, 85.
Curtis, B. (1972). Spinal cord. *In* "An Introduction to the Neurosciences" (B. A. Curtis, S. Jacobson, and E. M. Marcus, eds.), p. 120. Saunders, Philadelphia, Pennsylvania.
Dale, H. H. (1935). Pharmacology and nerve endings. *Proc. R. Soc. Med.* **28**, 319.
Del Castillo, J., and Katz, B. (1954). Quantal components of the endplate potential. *J. Physiol. (London)* **124**, 560.
Del Castillo, J., and Katz, B. (1957). La base "quantale" de la transmission neuro-musculaire. *Colloq. Int. C.N.R.S.* **67**, 245.
Denny-Brown, D. (1960). Diseases of the basal ganglia; their relation to disorders of movement. *Lancet* **2**, 1155.
Drachman, D. B. (1979). Immunopathology of myasthenia gravis. *Fed. Proc., Fed. Am. Soc. Exp. Biol.* **38**, 2613.
Eccles, J. S. (1969). "The Inhibitory Pathways of the Central Nervous System." Thomas, Springfield, Illinois.
Eccles, J. S., and Jaeger, J. S. (1958). The relationship between the mode of operation and the dimensions of the junctional region at synapses and motor end-organs. *Proc. R. Soc. London, Ser. B* **148**, 38.
Eccles, J. S., Fatt, P., and Koketsu, K. (1954). Cholinergic and inhibitory synapses in a pathway from motor-axon collaterals to motoneurones. *J. Physiol. (London)* **126**, 154.
Fambrough, D. M., and Hartzell, H. C. (1972). Acetylcholine receptors: Number and distribution at neuromuscular junctions in rat diaphram. *Science* **176**, 189.
Fambrough, D. M., Drachman, D. B., and Sataymurti, S. (1973). Neuromuscular junction in myasthenia gravis: Decreased acetylcholine receptors. *Science* **182**, 293.
Fatt, P., and Katz, B. (1952). Spontaneous subthreshold activity of motor nerve endings. *J. Physiol. (London)* **117**, 109.
Furshpan, E. J., and Potter, D. D. (1959). Transmission at the giant motor synapses of the crayfish. *J. Physiol. (London)* **145**, 289.
Geffen, L. B., Jessell, T. M., Cuello, A. C., and Iverson, L. I. (1976). Release of dopamine from dendrites in rat substantia nigra. *Nature (London)* **260**, 258.
Giesecke, J., and Hebert, H. (1979). The molecular structure of adrenergic and dopaminergic substances. *Rev. Biophys.* **12**, 263.

Gray, E. G., and Whittaker, V. P. (1962). The isolation of nerve endings from the brain: An electron-microscopic study of cell fragments derived by homogenization and centrifugation. *J. Anat.* **96**, 79.

Griffin, D. R. (1962). "Animal Structure and Function." Holt, New York.

Heuser, J. E. (1976). Morphology of synaptic vesicle discharge and reformation at the frog neuromuscular junction. In "Motor Innervation of Muscle" (S. Thesleff, ed.), p. 51. Academic Press, New York.

Heuser, J. E., and Resse, T. S. (1979). Synaptic-vesicle exocytosis captured by quick-freezing. In "The Neurosciences: Fourth Study Program" (F. O. Smith and F. G. Worden, eds.), p. 573. MIT Press, Cambridge, Massachusetts.

Hillman, D. E. (1979). Neuronal shape parameters and substructures as a basis of neuronal form. In "The Neurosciences: Fourth Study Program" (F. O. Schmitt and F. G. Worden, eds.), p. 477. MIT Press, Cambridge, Massachusetts.

Hodgkin, A. L., and Rushton, W. A. H. (1946). The electrical constants of a crustacean nerve fibre. *Proc. R. Soc. London, Ser. B* **133**, 444.

Jones, D. G. (1981). Ultrastructural approaches to the organization of central synapses. *Am. Sci.* **69**, 200.

Korn, H., Triller, A., and Faber, D. S. (1981). Fluctuating responses at a central synapse: n of binomial fit predicts number of stained presynaptic boutons. *Science* **213**, 898.

Loewi, O. (1921). Über humorale überfragbarkeit der herznervenwirkung. I. Mitteilung. *Pfluegers Arch. Gesamte Physiol. Menschen Tiere* **189**, 239.

McClell, A. L. (1963). "Tables of Experimental Dipole Moments." Freeman, San Francisco, California.

MacGregor, R. J., and Lewis, E. R. (1977). "Electrical Signal Processing in the Nervous System." Plenum, New York.

MacIntosh, F. C. (1980). The present status of the vesicle hypothesis. In "Cholinergic Mechanisms and Psychopharmacology" (D. J. Jenden, ed.), p. 297. Plenum, New York.

MacVicar, B. A., and Dudek, F. E. (1981). Electrotonic coupling between pyramidal cells. A direct demonstration in rat hippocampal slices. *Science* **213**, 782.

Marrazzi, A. S., and Lorente de Nó, R. (1944). Interaction of neighboring fibers in myelinated nerve. *J. Neurophysical.* **7**, 83.

Miledi, R. (1966). Strontium as a substitute for calcium in the process of transmitter release at the neuromuscular junction. *Nature (London)* **212**, 1233.

Parsegian, V. A. (1977). Considerations in determining the mode of influence of calcium on vesicle-membrane interaction. *Soc. Neurosc. Symp.* **2**, 167.

Scheibel, M. E. and Schiebel, A. B. (1970). Of pattern and place in dendrites. *Int. Rev. Neurobiol.* **13**, 1.

Sherrington, C. S. (1906). "The Integrative Action of the Nervous System." Yale Univ. Press, New Haven, Connecticut.

Sourkes, T. L. (1972). Parkinson's disease and other disorders of the basal ganglia. In "Basic Neurochemistry" (R. W. Albers, G. J. Siegel, R. Katzman, and B. W. Agranoff, eds.), p. 565. Little, Brown, Boston.

Steinbach, J. H., and Stevens, C. F. (1976). Neuromuscular transmission. In "Frog Neurobiology: A Handbook" (R. Llinas and W. Precht, eds.), p. 33. Springer-Verlag, Berlin.

Steinbach, J. H., Merlie, J., Heinemann, S., and Bloch, R. (1979). Degradation of junctional and extrajunctional acetylcholine receptors by developing rat skeletal muscle. *Proc. Natl. Acad. Sci. U.S.A.* **76**, 3547.

Stevens, C. F. (1974). Kinetics of postsynaptic membrane response at the neuromuscular junction. In "Synaptic Transmission and Neuronal Interaction" (M. V. L. Bennett, ed.), p. 45. Raven Press, New York.

Stevens, C. F. (1979). The neuron. *Sci. Am.* **241** (3), 54.
Takeuchi, A. (1977). Junctional transmission. I. Postsynaptic mechanisms. *In* "Handbook of Physiology" (E. R. Kandel, ed.), Sect. 1, Vol. II, p. 295. Am. Physiol. Soc., Bethesda, Maryland.
Takeuchi, A., and Takeuchi, N. (1959). Active phase of frog's end-plate potential. *J. Neurophysiol.* **22**, 395.
Takeuchi, A., and Takeuchi, N. (1960). On the permeability of endplate membrane during the action of transmitter. *J. Physiol. (London)* **154**, 52.
Takeuchi, A., and Takeuchi, N. (1962). Electrical changes in pre- and postsynaptic axons of the giant synapse of Loligo. *J. Gen. Physiol.* **45**, 1181.
Taxi, J. (1965). Contribution à l'étude des connexions des neurones moteurs du système nerveux autonome. *Ann. Sci. Nat., Biol. Anim. Zool.* [12] **7**, 413.
Volkmar, F. R., and Greenough, W. T. (1972). Rearing complexity affects branching of dendrites in the visual cortex of the rat. *Science* **176**, 1445.
Whittaker, V. P., and Gray, E. G. (1962). The synapse: biology and morphology. *Br. Med. Bull.* **18**, 223.
Zimmermann, H. (1979). Vesicle recycling and transmitter release. *Neuroscience* **4**, 1773.

CHAPTER 5

Neuronal Integration and Rall Theory

INTRODUCTION

In Chapter 4 some of the more quantitative aspects of transmitter release and postsynaptic potentials for transmission at the neuromuscular junction were discussed. The effects of dendrites and the properties of soma potentials in cases where synaptic inputs terminated on dendrites (axodendritic) were not considered. Because of the neglect of dendritic effects, extension of the properties of endplate potentials to synapses formed by two neurons is not straightforward. This should be evident since the electrical responses recorded postsynaptically at the neuromuscular junctions were from membrane locations near the site of the electrical inputs. It might be expected therefore that potentials recorded intracellularly from a neuron for which inputs terminate spatially near or on the soma will display properties analyzable in terms of the model formulated for the neuromuscular junction. This is indeed the case. However, this model, which is sometimes called the Eccles model in honor of his pioneering research, does not provide for a description of the transmembrane potential of the soma in cases for which synaptic inputs are spatially remote from the recording intracellular electrode located at the soma. There are several properties that the potential of the soma must display if only axosomatic synapses are considered. These are the following: (1) The amplitude of the soma's potential (i.e., PSP) should be a linear function of its membrane potential. (2) There should be a unique equilibrium potential at which no transmembrane current flows; that is, the soma can be held at a certain membrane potential for which inputs to the cell do not elicit a potential response. (3) For a fixed input potential, the PSP changes sign when the membrane potential is made to cross the equilibrium potential from either direction. Electrophysiological recordings discussed in Chapter 4 show that end-plate potentials do indeed have these properties. The inhibitory PSPs recorded in cat motoneurons also show similar behavior (see p. 102 and in particular Fig. 4.14). However, most monosynaptic excitatory PSPs recorded from cat motoneurons do not display these properties. For example, the EPSP amplitude does not increase with increasing membrane hyperpolarization in contrast to the expected behavior, property 1, of the Eccles model. In addition, the EPSPs usually do not possess a unique reversal potential (the absence of properties 2 and 3). An example illustrating this lack of a unique reversal potential for alpha motoneurons is shown in Fig. 5.1. This figure shows the EPSP evoked at four different transmembrane potentials. Note that at no potential is the synaptic response null. Instead, with increasing depolarization [from record (d) to record (a)]

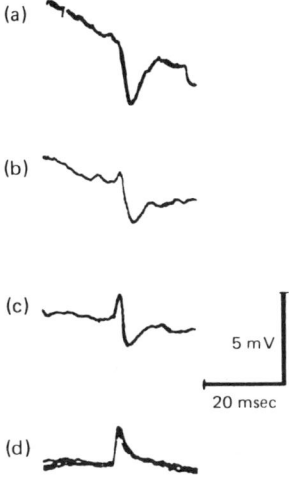

FIG. 5.1 Monosynaptic excitatory postsynaptic potentials (EPSPs) recorded in an alpha motoneuron at different membrane voltages. (d) illustrates the EPSP when the membrane potential was at its resting value. (c), (b), and (a) are records of EPSP at increasing membrane depolarization. [From Smith et al. (1967).]

INTRODUCTION

the EPSP develops a biphasic response. A response of this nature is not explainable within the context of Eccles models in which only the effects of axosomatic synapses are included. Evoked soma potentials that have properties not consistent with the Eccles synaptic model are now understood, largely because of the pioneering theoretical efforts of Rall (1959a, 1960). He and his coworkers showed that the EPSP of cat motoneuron somas contain significant contributions arising from synaptic inputs located on dendrites. The theoretical models and formulas employed in the analysis of dendritic inputs are described in this chapter. An important finding of these investigations is that the soma's potential is strongly affected by the geometry of the dendrites. Since the shape and extent of the dendrites are strongly species specific and are affected by nutritional and environmental factors, as discussed below, this result has important implications regarding learning and memory.

The cat motoneuron is one of the most studied neurons in the central nervous system (CNS), and the results obtained from this neuron should be applicable to CNS neurons in other species. This neuron has served as a model CNS neuron primarily because, as we saw in Chapter 4 (pp. 103, 135), there are several ways in which the cell can be activated. In addition, because of the large size of the soma, about 60–80 μm, it is easy to insert a recording electrode. The dendrites of the motoneuron are passive (i.e., they do not conduct action potentials) and there exists only one region of the neuron, namely, the axon hillock, for the generation of an action potential. This behavior of motoneuron dendrites should be contrasted with that of normal hippocampus pyramidal cells and cerebellar Purkinje cells where dendritic action potentials (spikes) are found.

It is important to emphasize that in both the preceding chapter and this one the treatment of neuronal integration assumes the validity of the following: (1) The morphological unit of the nervous system is the nerve cell with each neuron having identifiable parts, namely, a cell body, an axon, and one or more dendrites. (2) The functional integration (passage of information measured in terms of the membrane polarization) is from dendrites and the cell body to the cell's axon. In a sense, informational flow is unidirectional, i.e., from dendrites to soma to axon. All feedback circuits occur through interneurons. A classic example illustrating neuronal feedback is the Renshaw interneuron shown in Fig. 4.36. (3) Dendrites act as passive elements in the sense that they do not conduct action potentials. (4) Dendrites occur as only postsynaptic elements. These assumptions are quite restrictive and in fact it is known from intracellular recordings of cells in the mammalian olfactory bulb (Rall *et al.*, 1966) that dendrodendritic synapses (dendrite to dendrite) occur. This type of synapse, by allowing dendrites to function as both presynaptic and postsynaptic elements, greatly enhances the type of circuits that can be

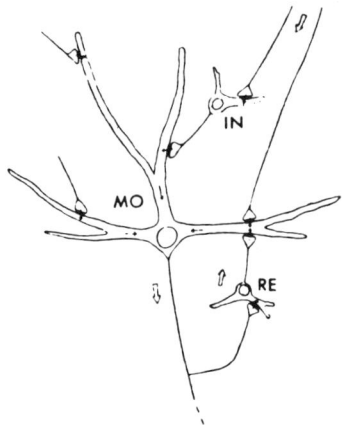

FIG. 5.2 Synaptic organization of spinal motoneuron. MO: motoneuron; IN: interneuron; RE: Renshaw interneuron. Large arrows indicate direction of impulse flow. All synapses are of the classical type, axodendritic (functional polarity of synapse in a single direction and indicated in figure by small solid arrows). There are no dendritic presynaptic elements. [From Shepherd (1972).]

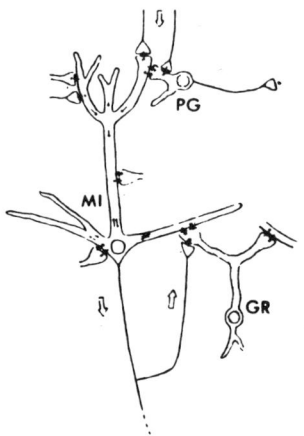

FIG. 5.3 Mitral cell of olfactory bulb synaptic organization. MI, Mitral cell; PG, periglomerular short-axon cell; GR, granule cell. Note that many synapses are bidirectional and that the mitral cell has synaptic outputs originating from its dendritic tree (dendrites are functioning as presynaptic elements at synapses). [From Shepherd (1972).]

EFFECTS OF ENVIRONMENT AND MALNUTRITION ON DENDRITE GROWTH 149

formed among neuronal elements. Figure 5.2 illustrates the synaptic organization of a typical motoneuron whose functional elements are explicable in terms of the assumptions listed above, whereas Fig. 5.3 illustrates some of the functional elements that occur in the mitral cell of the olfactory bulb but that are not explicable under our restricted assumptions. The existence of dendrodendritic synapses increases by a large factor the complexity of the functional properties of neurons (see Shepherd, 1972, 1977 for a discussion of dendrodendritic synapses and their effects on properties of cells in the mammalian olfactory bulb). The model developed in this chapter is restricted to only passive dendrites with no dendrodendritic synapses.

EFFECTS OF ENVIRONMENT AND MALNUTRITION ON DENDRITE GROWTH

The number of dendrite branches, and therefore the associated synaptic connections, increases with higher-order animals (see Fig. 5.12). It is evident that it is these numbers, dendrites and their synaptic connections, that determine the level of complexity of signal processing, which we assign as a level of intelligence (see, for example, Berry et al., 1978). Experiments have shown that the pyramidal cells in the cat cortex associated with vision develop a dendritic structure quite rapidly after birth. Figure 5.4 shows such a cell 1–2, 7–10, and 90 days after birth. That this is not simply a continuation of the

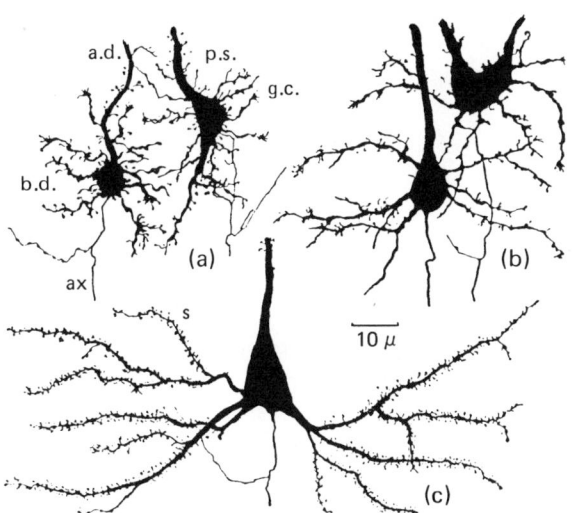

FIG. 5.4 Cell bodies and basilar dendrites of pyramidal cells in the visual cortex of kittens at three postnatal epochs: (a) 1–2 days; (b) 7–10 days; and (c) approximately 90 days. Abbreviations: ax, axon; b.d., basilar dendrites; a.d., apical dendrites; p.s., pseudospines; s, mature spines; g.c., growth cones. [From Scheibel and Scheibel (1971).]

growth process has been demonstrated by the diminution of the visual cortex in growing monkeys that have been visually deprived in one eye (Hubel, 1979).

Ramón y Cajál (1911) first postulated that there could be a modification of existing neuronal connections, or the growth of new connections, as a result of their exercise or use, so that the subsequent passage of neuronal activity along these pathways is facilitated. Conversely, unused connections tend to regress and become nonfunctional. This "use–disuse hypothesis" has been the subject of much investigation ever since (see, for example, Crick, 1982). Although no conclusive evidence of dendritic atrophy has been shown, evidence of growth with use is emerging and a few examples will be briefly discussed.

Rutledge (1976) performed a series of experiments with cats that were trained by electric shock to have a chronic foreleg reflex. Among the subsequent measurements were postmortem microscope examination of the pyramidal cells in the leg motor cortex. A number of changes against controls were reported, illustrated schematically in Fig. 5.5 by the dotted lines, namely, (1) increased length of terminal portions and the appearance of new spines, (2) new spines on new twigs, (3) increases in oblique dendrites, (4) budding of new spines near existing spines, and (5) increase of area of synaptic contact. Rutledge admits that some of these observations may be assumptions because some growth buds are known to exist without a chronic reflex stimulation. However, coupled with observations of other experiments, it is increasingly evident that these assumptions of growth may well be valid.

Having shown evidence of dendritic growth with muscular use, we shall now report evidence of growth from a more subtle source, namely, environ-

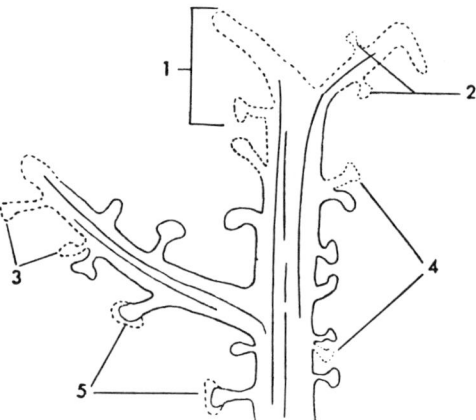

FIG. 5.5 Specific loci and types of morphological changes observed or postulated in apical dendrites of the contralateral cortex of cats with conditioned chronic foreleg reflex. Numbers identified in text. [From Rutledge (1976). Copyright 1976 MIT Press.]

mental stimulation. A wealth of data has been reviewed by Greenough (1976). Briefly, rats were raised in three types of social environments: isolated conditions (IC); social conditions (SC), i.e., housed with other rats; and environmental complexity (EC), often called "environmentally enriched." Usually different toys or mazes are put in the cage each day, and the animals are also put daily in free-exploration boxes. Although these stimuli unboubtedly fall short of the environment of a wild rat, there is significant contrast with rats raised under IC or SC conditions. Following 30 days of the different rearing conditions, Diamond *et al.* (1972) found that the EC rats had both heavier brains and thicker cortices in the somesthetic, temporal, and frontomedial regions. An examination of four occipital-cortex cell types following differential rearing was made by Volkmar and Greenough (1972) and Greenough and Volkmar (1973). They did a statistical study of branching of the apical and basal dendrites in cells from the regions that Diamond *et al.* had shown to undergo the greatest thickness changes. The left-hand side of Fig. 5.6 shows the technique of branch counting of stained cells by concentric rings placed over a drawing of a typical cell. On the right-hand side of Fig. 5.6 is a plot of the mean number of branches per neuron versus the branch order for apical and basal dendrites. It is seen that in both cases the rats raised under environmentally enriched conditions have the more complex dendritic structures. Evidence is also being accumulated showing that increased environmental enrichment can stimulate dendritic growth even in adult rats (Uylings *et al.*, 1978), which has implications against enforced retirement.

Severe, early malnutrition has been shown by many investigators to affect both brain structure and brain function [see reviews by Winick (1980) and McConnell (1980)]. For example, Widdowson and McCance (1960) showed that rats malnourished from birth to weaning had brains that were smaller even into adulthood, no matter how much they were fed after weaning. However, rats malnourished after weaning had smaller brains after the period of malnutrition, but their brains returned to normal after refeeding. A controversy among investigators exists concerning the period of malnutrition from which recovery is possible.

A series of careful experiments by McConnell and Berry (1978) compared the dendritic processes of Purkinje cells in the cerebellum of rats under nutritional restriction with those of controls for periods of 10, 15, 20, and 30 days after birth. An example of Purkinje cells after 30 days is shown in Fig. 5.7. Both a decrease in branching pattern and overall dendritic length is seen for the cell of the undernourished rat. Statistical data of these effects are plotted in Fig. 5.8. On the left is plotted the overall length of the dendrites for experimental and control rats versus days after birth. On the right the number of branches (segment frequencies) are plotted. It is seen that pronounced deviation begins at 20 days.

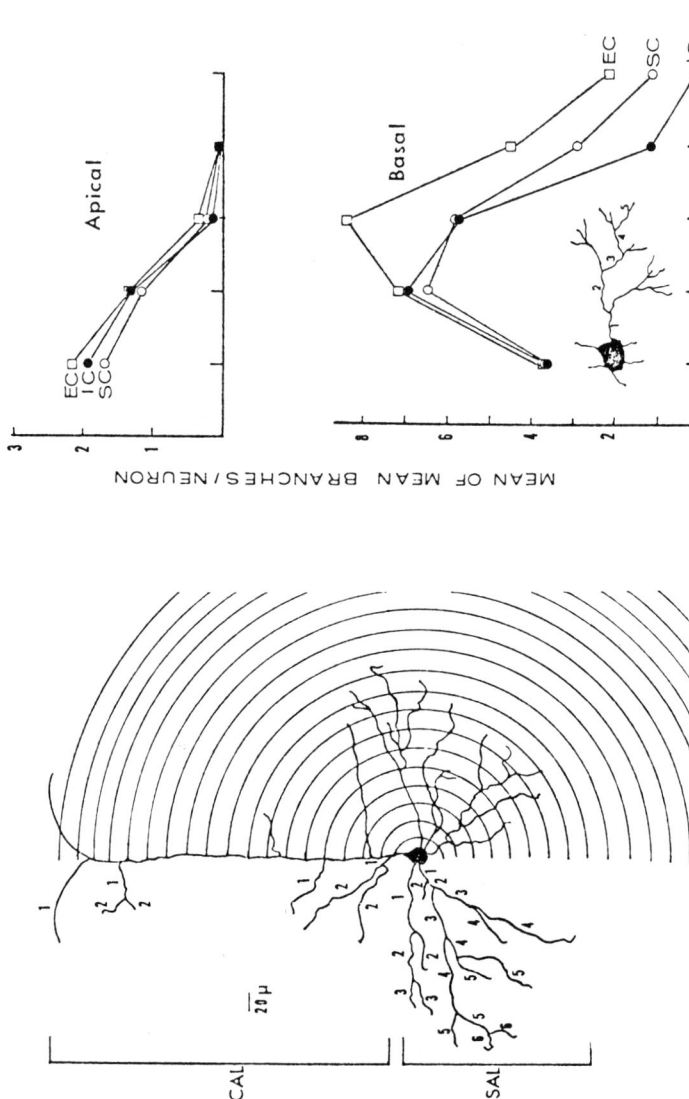

FIG. 5.6 Left: Analysis procedure for counting dendrite branching in stained neural cells of rats. Right: Plots of number of branches versus order of the branch for apical and basal dendrites in layer-4 pyramidal neurons from six triplet sets ($N = 18$) of rats reared for 30 days under isolated conditions (IC), social conditions (SC), and environmentally complex conditions (EC). [From Greenough (1976). Copyright 1976 MIT Press.]

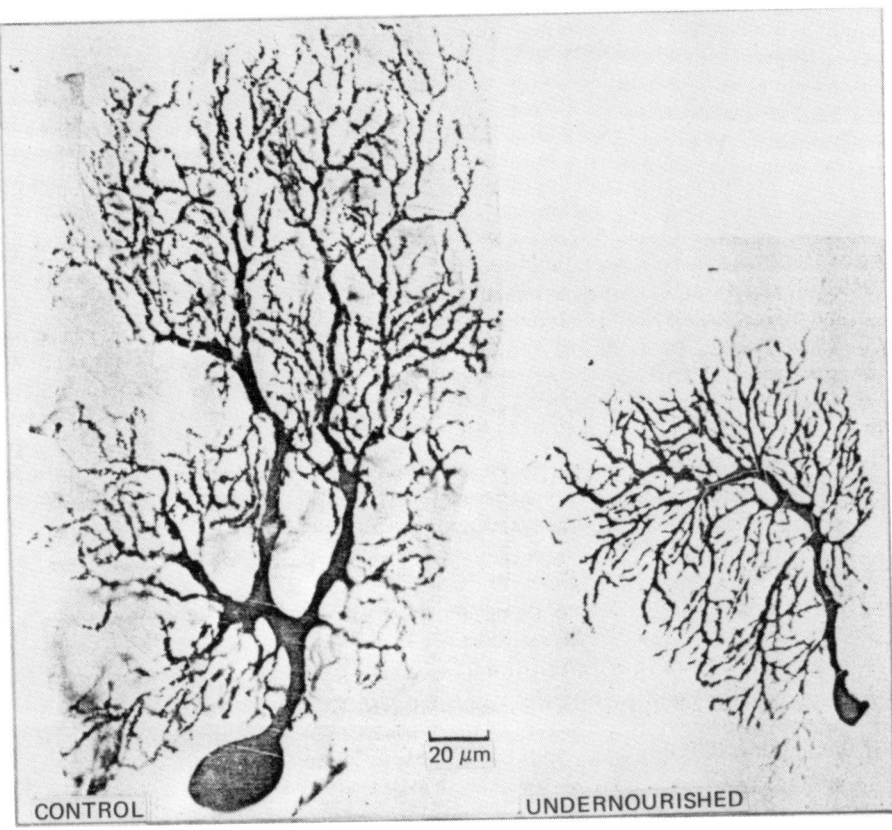

FIG. 5.7 Photomicrographs of stained (Golgi–Cox impregnated) Purkinje cells from control and undernourished rats at 30 days after birth. [From McConnell and Berry (1978).]

It is difficult at present to draw conclusions on the possible recovery by refeeding since other factors have been shown to be important. For example, the stage of human brain growth has been shown to have an effect. There are three defined stages of growth: hyperplasia—during intrauterine life; hyperplasia and hypertrophy—birth to about 18 months; and hypertrophy—18 months to beyond 3 years. In animal experiments by Winick and Noble (1966), hyperplasia, the period of rapid cell division, was shown to be the most vulnerable phase. This is the period of rapid cell division, and malnutrition results in fewer cells, a condition that remains as a permanent deficit. In contrast, malnutrition during hypertrophy, although resulting in a retardation of cell enlargement, appears to be reversible by subsequent feeding. Data from the examination of the brains of children who died of malnutrition during their first year of life also show fewer cells (Winick and

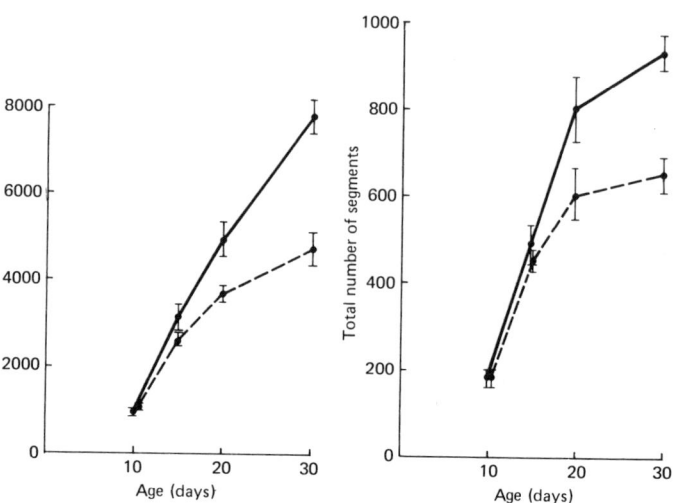

FIG. 5.8 Plots of total dendritic length and segment count against control (—) and undernourished (---) rats at various ages. [From McConnell and Berry (1978).]

Rosso, 1969). From the above animal studies it is inferred that this change is permanent. Since different brain cells undergo division at maximum rates during different periods of development, it is not yet clear what effect selected time of malnutrition has on the developing brain (see McConnell and Berry, 1981).

A number of studies have been carried out on the interactive effects of malnutrition and environmental stimulation. Levitsky and Barnes (1972), among others, showed that animals that were environmentally stimulated during malnutrition did not suffer from the nutritional deficiency. Studies on Korean orphans and Third World siblings also showed that environmental stimulation may counter the effects of malnutrition.

A possible cause of this seemingly inexplicable interaction of environment and malnutrition has recently been proposed by Morgan and Winick (1979). Although numerous biochemical changes in the brain during malnutrition have been reported, these investigators have found that one, n-acetylneurominic acid (NANA), may provide the basis for a further exploration. Their data show that the concentration of NANA within the ganglia of rat cerebrum and cerebellum is reduced as a result of early malnutrition. The concentration of NANA is also reduced in animals subjected to early environmental deprivation. The concentration increases during environmental stimulation and prevents the decrease due to malnutrition. Injection of NANA into brain ganglia of malnourished animals results in its incorporation into the cells and prevents subsequent abnormal behavior. This exciting finding suggests a cause-and-effect relation between the concentration of NANA in brain ganglia and the behavioral abnormalities induced by both malnutritional and environmental deprivation. Although the nutritional and environmental problems of the world may not be solved in our lifetime, this research suggests that there may be a magic pill in the future that may prevent some of the consequences of these problems.

DENDRITE ELECTROTONUS AND THE LINEAR CABLE EQUATION

The gross effects of nutrition and environment on dendritic organization and their functional properties discussed above emphasize the importance of the dendritic field on the integrative properties of neurons. In the remainder of this chapter the cell's response to synaptic inputs terminating on dendrites will be examined. Our treatment is restricted to only passive (electrotonus) dendritic responses, although, (see p. 147), some dendrites are able to produce action potentials. A passive membrane is defined as one whose resistance remains constant with respect to changes in time and transmembrane potential. The mathematical model used to assess the importance of axodendritic synapse is the linear cable theory (see Vol. I, p. 81) within the context of models formulated by Wilfrid Rall at the National Institutes of Health. These models have provided a framework for the correction of the interpretation of earlier experiments and for assuring that data from future experiments will be properly analyzed. The basic cable equation will be considered in this section, and its extension and application to realistic neuronal models will be considered in subsequent sections.

In the calculation of electrical conduction along a nerve fiber, either an axon or a dendrite, it is customary to approximate the fiber by a uniform cylindrical shell. This cylinder is filled with axoplasm, a gel, which has a constant electrical resistivity per unit length. The cylinder itself is the membrane, which has a constant thickness and therefore a constant electrical resistivity and dielectric constant per unit surface area. That the membrane has a finite resistance implies that even in its resting state there is a leakage current across it. Furthermore, in vivo, this cylinder lies within a conducting medium external to the cylinder. If very fine electrical probes (microelectrodes) are used, then three potentials at an instant of time can be measured: (1) at a position inside the cylinder there will be a potential $V_i(x, t)$; (2) a potential V_e in the external medium; and (3) a potential V_r across the membrane when the nerve fiber is at rest. Since potential differences are measured between two probes, excursions to a potential V due to stimulation will be measured against the resting potential difference. These concepts were treated in Vol. I, p. 80, but will be developed here also with some slight changes in notation.

The assumptions involved in the derivation of the cable equation can be summarized as follows:

1. The departure at a time t of membrane potential from its value in the resting state, denoted by the symbol V, is a function of the distance x along the length of the cylinder measured from some reference point $x = 0$ and of the time t, i.e., $V = V(x, t)$.

2. The intracellular fluid provides an ohmic resistance to axial current flow, but offers no resistance to radial current flow. The core, or axial, resistance per unit length will be denoted by r_i (unit: $\Omega\,\text{cm}^{-1}$) and is assumed to be independent of x.

3. The extracellular medium behaves ohmically with a resistance per unit length r_e (unit: $\Omega\,\text{cm}^{-1}$) and, like the core resistance, is taken to be independent of spatial variables.

4. The membrane is electrically equivalent to a linear resistance and a perfect capacitance in parallel. The electrical circuit equivalent of the neuronal membrane is shown in Fig. 5.9. The resistance across a unit length of membrane, r_m (unit: $\Omega\,\text{cm}$), and the membrane capacitance per unit length of cylinder, c_m (unit: $\text{F}\,\text{cm}^{-1}$) are taken to be independent of spatial variables.

In more general treatments of passive membrane responses (see, e.g., Jack, et al., 1975) some of the above assumptions can be removed. None of the assumptions is strictly valid; for example, many cells display a nonlinear

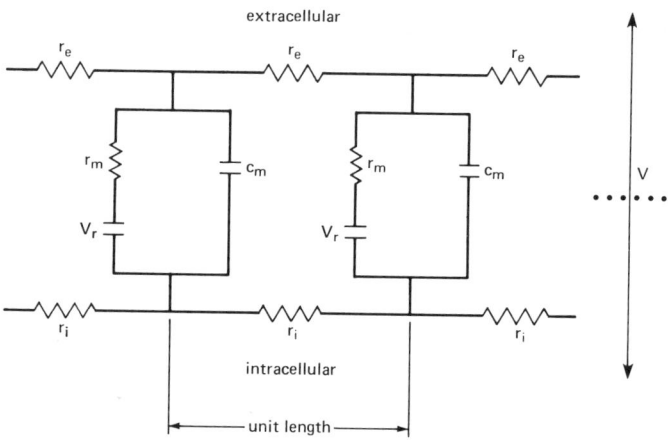

FIG. 5.9 Electrical equivalent circuit for a membrane. V_r denotes the resting membrane potential, i.e., voltage across the membrane; r_m is the membrane resistance for a unit length of membrane; c_m is the membrane capacitance per unit length; r_e is the extracellular resistance per unit length; and r_i is the core resistance per unit length.

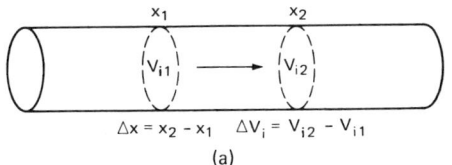

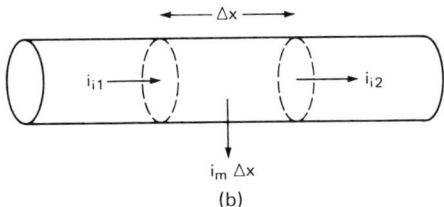

FIG. 5.10 Notation employed in the derivation of the cable equation. [From Rall (1977).]

current–voltage relation for even small depolarizations, thereby negating the validity of assumption (4). These cases will not be considered here.

Assumptions (1)–(4) in conjunction with Ohm's law and the conservation of current are sufficient to derive the basic linear cable equation. The notation given in Fig. 5.9 will be adhered to, and the treatment given by Rall (1977) will be followed.

Consider a small cylinder segment Δx; the resistance it affords to axial current flow is by definition $r_i \Delta x$. Application of Ohm's law between points 1 and 2 as denoted in Fig. 5.10a gives

$$i_i r_i \Delta x = -(V_{i2} - V_{i1}) \equiv -\Delta V_i \tag{5.1}$$

where i_i is the axial core current and V_i is the internal potential, relative to ground or a reference electrode electrically far away, recorded by microelectrodes. The negative sign occurs in Eq. (5.1) because a positive core current flowing in the positive x direction corresponds to decreasing values of the internal potential. In the limit where Δx becomes vanishingly small, Eq. (5.1) assumes the differential form

$$i_i = -\frac{1}{r_i}\frac{\partial V_i}{\partial x} \tag{5.2}$$

The partial derivative signifies that V_i depends also on time. A relation between the membrane current per unit length i_m and the axial current can be obtained by invoking Kirchhoff's current rule, which states that the sum of all electric currents flowing toward a junction is zero. Noting that the net current crossing the membrane of length Δx is $i_m \Delta x$, it follows (see Fig. 5.10b) therefore that

$$i_m \Delta x = i_{i1} - i_{i2} = -\Delta i_i \tag{5.3}$$

which for $\Delta x \to 0$ can be written

$$i_m = -\frac{\partial i_i}{\partial x} \tag{5.3'}$$

With the substitution of Eq. (5.2) this becomes

$$i_m = \frac{1}{r_i} \frac{\partial^2 V_i}{\partial x^2} \tag{5.4}$$

Employing the same arguments as used in arriving at Eq. (5.2) allows the extracellular current i_e to be expressed as

$$i_e = -\frac{1}{r_e} \frac{\partial V_e}{\partial x} \tag{5.5}$$

where r_e is the extracellular resistance per unit length and V_e is the external potential, relative to the same reference electrode as V_i, recorded by microelectrodes.

The current across the membrane i_m, expressed in Eq. (5.4) by the loss of axial current, may also be expressed as a gain in external current. Where there is no applied current, that is, no current-injecting electrodes,

$$i_m = \frac{\partial i_e}{\partial x} = -\frac{1}{r_e} \frac{\partial^2 V_e}{\partial x^2} \tag{5.6}$$

The sign in Eq. (5.6) is opposite that of Eq. (5.4) because one is a loss of current whereas the other is a gain.

The potential difference across the membrane is $V_i - V_e$. When the fiber is at rest electrically, the membrane potential $V_i - V_e$ is constant for all values of x because one of the assumptions has been that r_m and c_m are independent of position along the x axis of the fiber. If we define V_r as the resting potential difference across the membrane and there is subsequently a departure from the resting potential, then the magnitude of this departure, called V, will be the new potential difference minus the resting potential difference, or

$$V = V_i - V_e - V_r$$

If we take the second derivative of V with respect to axial distance, noting that by definition V_r is constant, we obtain

$$\frac{\partial^2 V}{\partial x^2} = \frac{\partial^2}{\partial x^2}(V_i - V_e - V_r) = \frac{\partial^2 V_i}{\partial x^2} - \frac{\partial^2 V_e}{\partial x^2}$$

DENDRITE ELECTROTONUS AND THE LINEAR CABLE EQUATION

Upon substitution of Eqs. (5.4) and (5.6) we may write

$$\frac{\partial^2 V}{\partial x^2} = i_m(r_i + r_e) \tag{5.7}$$

The relation expressed by Eq. (5.7) is independent of any particular molecular membrane model. In the case where the membrane is electrically equivalent to a purely passive resistance in parallel with a perfect capacitor (see Fig. 5.9), the membrane current i_m is a simple sum of a resistance current V/r_m and a capacitance current $c_m \, \partial V/\partial t$. Thus

$$i_m = \frac{V}{r_m} + c_m \frac{\partial V}{\partial t} \tag{5.8}$$

where c_m is the membrane capacitance per unit length (unit: F cm^{-1}). The form of the capacitance current can be rationalized by recalling that the capacitance C of a capacitor is defined as $Q = CV$, where Q is the net charge on one of the plates and V is the potential difference separating the charge Q from its counter charge; therefore, the current is $dQ/dt = C \, \partial V/\partial t$. On combining Eqs. (5.7) and (5.8) one obtains

$$\frac{\partial^2 V}{\partial x^2} = (r_i + r_e)\left[\frac{V}{r_m} + c_m \frac{\partial V}{\partial t}\right]$$

Removing the brackets and cross multiplying results in the cable equation

$$\frac{r_m}{r_i + r_e} \frac{\partial^2 V}{\partial x^2} = V + r_m c_m \frac{\partial V}{\partial t}$$

Customarily the following two constants are defined, namely

$$\lambda = \left(\frac{r_m}{r_i + r_e}\right)^{1/2}, \qquad \tau = r_m c_m \tag{5.9}$$

and the cable equation is written (Hodgkin and Rushton, 1946)

$$\lambda^2 \frac{\partial^2 V}{\partial x^2} - V - \tau \frac{\partial V}{\partial t} = 0 \tag{5.10}$$

In special cases where the extracellular medium is isopotential (i.e., V_e is independent of position), r_e is exactly zero and λ depends on neuronal properties only. In most Ringer solutions, $r_e \ll r_i$, and in this case $\lambda \approx \sqrt{r_m/r_i}$. When comparing results from different cells it is more convenient to use the parameters

$$c_m = 2\pi a C_m, \qquad r_m = R_m/2\pi a, \qquad r_i = R_i/\pi a^2 \tag{5.11}$$

where a is the radius of the cylinder. Thus C_m is the membrane capacitance per unit area (F cm^{-2}), R_m is the passive membrane resistance for a unit area (Ω cm^2), and R_i is the volume resistivity of the intracellular medium (Ω cm). As examples of typical values of these parameters we note that for the squid axon

$$R_i = 30 \quad \Omega \text{ cm}, \quad R_m = 1000 \quad \Omega \text{ cm}^2, \quad C_m \approx 1 \quad \mu\text{F/cm}^2$$
$$\tau = 1 \quad \text{msec}, \quad \text{and} \quad \lambda \approx 0.65 \quad \text{cm}$$

whereas for the cat spinal motoneuron

$$R_i \approx 70 \quad \Omega \text{ cm}, \quad R_m \approx 2500 \quad \Omega \text{ cm}^2, \quad C_m \approx 2 \quad \mu\text{F/cm}^2, \quad \tau = 5 \quad \text{msec}$$

It may be useful before proceeding to consider the physical meaning of the partial derivatives in Eq. (5.10). The term $\partial V/\partial t$ arose from the capacitive membrane current resulting from a change of potential across the membrane, whereas $\partial^2 V/\partial x^2$ arose from Eqs. (5.4) and (5.6) and is proportional to the excess of core current, that is, it is the difference between the current which enters a small length increment of core and that which leaves. This excess escapes the core when no intracellular electrode is present. In the next section we shall explain more fully the physical meaning of the constants in Eq. (5.10).

SPECIAL SOLUTIONS TO THE CABLE EQUATION

A general technique for solving the cable equation (5.10) is discussed later. Here we restrict ourselves to a few special solutions that will help clarify the physical meaning of τ, the membrane time constant, and λ, the membrane length constant.

1. Consider a small patch of a membrane that is initially ($t = 0$) uniformly depolarized and everywhere has a potential $V(0)$, that is, for this small patch there is no x dependence of V. With what rate does this depolarization subsequently decay to its equilibrium value $V = 0$? For the conditions stated, the membrane potential is independent of spatial variables, and therefore Eq. (5.10) reduces to

$$\frac{dV}{dt} = -\frac{V}{\tau} \tag{5.12}$$

the well-known equation governing first-order decay kinetics. The variables are separable and the solution is obtained by simple integration:

$$\int_{V(0)}^{V(t)} \frac{dV}{V} = \ln\left(\frac{V(t)}{V(0)}\right) = -\int_0^t \frac{dt}{\tau} = -\frac{t}{\tau} \tag{5.13}$$

or

$$V(t) = V(0)e^{-t/\tau} \tag{5.14}$$

SPECIAL SOLUTIONS TO THE CABLE EQUATION

It is seen that a plot of ln V versus time of Eq. (5.14) will yield a straight line with a slope of $-1/\tau$. The quantity τ is called the "time constant" or "decay constant" (see Vol. I, p. 365). When $t = \tau$, $V(t) = V(0)/e = 0.37V(0)$ and τ is therefore the time it takes for a departure of the membrane potential from its resting value to decay to e^{-1} of its initial value. Later in this chapter we shall find that in the case where the membrane is not initially uniformly depolarized, Eq. (5.14) is replaced by an infinite sum of exponentials (see p. 178).

The time constant has the dimension of time, as can be seen from the product $r_m c_m$. This product has units of ohm farads. By Ohm's law $V = IR$ and $I = Q/t$; $C = Q/V$ and therefore $RC = V/(Q/t) \times Q/V = t$. Equation (5.11) shows that $r_m c_m = R_m C_m$, and thus the dependence on membrane surface area cancels upon obtaining the product.

2. The physical meaning of λ may be discerned by a different example. Consider a semi-infinite cylinder extending from $x = 0$ to infinity. Suppose a steady-state value of $V = V_0$ is maintained at $x = 0$ by applying a constant current at that point and that no current is applied elsewhere. What is the value of the potential at some other position x after steady state has been achieved? Steady state means that there is no change in V with time; thus

$$\lambda^2 \frac{d^2 V}{dx^2} - V = 0 \tag{5.15}$$

This is known as the Kelvin equation for a leaky cable, and it was considered in Vol. I, p. 81. We shall not show how to solve this second-order differential equation; however, the general solution is well known and can be written

$$V(x) = Ce^{-x/\lambda} + De^{x/\lambda} \tag{5.16}$$

That this is a solution may be readily verified simply by substituting Eq. (5.16) into Eq. (5.15). Equivalently, since the hyperbolic sine and cosine functions are sums of exponentials, $\sinh u = (e^u - e^{-u})/2$ and $\cosh u = (e^u + e^{-u})/2$, they are also solutions of Eq. (5.15). Accordingly, the general solution of the Kelvin equation expressed in hyperbolic functions is

$$V(x) = A \cosh(x/\lambda) + B \sinh(x/\lambda) \tag{5.17}$$

which may also be verified by substitution into Eq. (5.15). The constants A, B, C, and D are determined by the boundary conditions. Consider the solution Eq. (5.16) for the case where $V = V_0$ at $x = 0$ and where $V(x)$ becomes vanishingly small as x approaches infinity. Because $V(x)$ must be zero at $x = \infty$, the constant D must be zero, and because $V(x)$ must be V_0 at $x = 0$, the constant C must be equal to V_0. Therefore, the particular solution is

$$V(x) = V_0 e^{-x/\lambda} \tag{5.18}$$

Thus the constant λ denotes the physical distance from the origin at which the membrane potential has decreased to e^{-1} of the value at the origin. Following the name "time constant" for τ, the comparable name for λ is "length constant."

It is evident from the definition of $\lambda^2 = r_m/(r_i + r_e)$, or in the case of isopotentiality of the external medium where $\lambda^2 = r_m/r_i$, that the length constant does not depend on the capacitance changes of the membrane. Note also that r_m is expressed in Ω cm and r_i and r_e in Ω cm^{-1} and therefore λ^2 has dimensions of cm^2. We may rearrange this definition by dividing by λ and multiplying by r_i and obtain $\lambda r_i = r_m/\lambda$. Thus the length constant λ corresponds to the length of the core conductor for which the core resistance λr_i exactly equals the resistance r_m/λ across the membrane for the same length of membrane cylinder.

When considering the effectiveness of distal synapses (dendrites spatially remote from the soma) on the response of the soma, it is often more convenient to characterize distances in terms of "electrical distance." This is accomplished by defining a new, dimensionless spatial variable $x^* \equiv x/\lambda$ called the *electrotonic* (or electrical) distance. The concept of electrical distance allows investigators to gauge the effectiveness in the propagation of a potential disturbance. The smaller the electrotonic distance between two locations, the more effectively is the electrical perturbance at one end transmitted to the other end; e.g., synaptic input on dendrites having a small electrotonic distance from the soma are more effective in altering the soma's transmembrane potential. For a membrane cylinder characterized by a length constant λ and having a finite physical length L, the term $L^* = L/\lambda$ is often called its *electrotonic length*, and this ratio may be larger or smaller than unity. For the spinal motoneurons of cats the electrotonic length of the dendrites, as inferred from electrophysiological measurements, is about 1.5. This value as indicated above and discussed more fully later indicates that synapses on dendrites are quite effective in controlling the responses of this cell.

3. When the cylinder is of finite length L, the preceding boundary condition $V \to 0$, $x \to \infty$, must be replaced by an appropriate boundary condition at $x = L$. For simplicity, we consider only the case of a cylinder of finite length with a *sealed end* (see footnote, Appendix B). In this approximation the terminal resistance (at $x = L$) is assumed to be so large that the axial core current vanishes at $x = L$ and therefore the potential gradient at $x = L$ is zero. The boundary conditions appropriate for this case are expressed as

$$V = V_0 \quad \text{at} \quad x = 0 \tag{5.19}$$

$$\frac{\partial V}{\partial x} = 0 \quad \text{at} \quad x = L \tag{5.20}$$

SPECIAL SOLUTIONS TO THE CABLE EQUATION 163

The latter boundary condition, Eq. (5.20), follows directly from Eq. (5.2). To obtain a useful solution for $V(x)$ for this case, it is helpful to start with a variation of Eq. (5.17), namely,

$$V(x) = A \cosh\left(\frac{L - x}{\lambda_0}\right) + B \sinh\left(\frac{L - x}{\lambda_0}\right) \quad (5.17')$$

where λ_0 refers to the specific length constant for the cable of length L under consideration. Two mathematical properties of these hyperbolic functions will be used:

$$\sinh 0 = 0, \quad \cosh 0 = 1$$

and

$$\frac{d}{dx}\sinh u = \cosh u \frac{du}{dx}, \quad \frac{d}{dx}\cosh u = \sinh u \frac{du}{dx}.$$

If we first set $V(x) = V(L)$ at $x = L$ in Eq. (5.17'), we obtain $A = V(L)$, and if at $x = 0$ there is a constant voltage V_0, then Eq. (5.17') becomes

$$V_0 = V(L) \cosh\left(\frac{L}{\lambda_0}\right) + B \sinh\left(\frac{L}{\lambda_0}\right) \quad (5.21)$$

Now take the derivative and apply the boundary condition of Eq. (5.20)

$$\left.\frac{\partial V}{\partial x}\right|_{x=L} = 0 = V(L) \sinh 0 + B \cosh 0$$

For this equation to be satisfied, B must be equal to zero. Eliminating $V(L)$ from Eqs. (5.17') and (5.21) results in the relation

$$V(x) = \frac{V_0 \cosh[(L - x)/\lambda_0]}{\cosh(L/\lambda_0)} \quad (5.22)$$

Figure 5.11 illustrates graphically the behavior of $V(x)$ of Eq. (5.22) for different lengths L expressed in terms of λ_0. It is evident from this figure that the electrical potentials of cylinders of finite length decay more slowly with distance than does a cable of semi-infinite length. It is also evident from this figure that it is incorrect to define the length constant by the condition that it is the distance over which the steady-state potential falls by $1/e$ since all four potential curves were calculated using the same length constant λ_0. The profound differences in the shape of the potential are due to the effects of the endpoint boundary conditions.

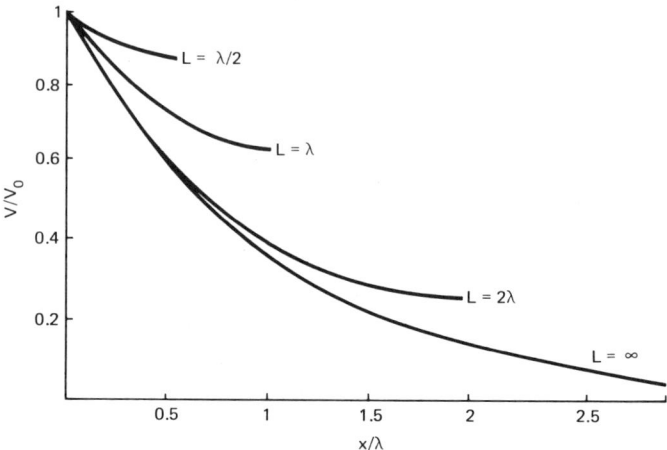

FIG. 5.11 Spatial variation of membrane potential under steady-state conditions for four different cylinder lengths ($L = \lambda/2$, λ, 2λ, and ∞) with $V(x = 0) = V_0$ and sealed-end conditions at $x = L$. λ is the space constant for the cylinders (λ in figure corresponds to λ_0 in text). [Modified from Rall (1959b).]

BOUNDARY CONDITIONS AT DENDRITIC BRANCH POINTS

The dendrites of most neurons branch considerably before terminating, and in many cases this extensive branching imparts treelike structures to the dendrite shafts emerging from the neuron's cell body. Figure 5.12 illustrates this branching in various vertebrate species for cerebellar Purkinje cells. For this particular cell type, the dendritic network develops from only one side of the body, while the axon emerges from the other side. This pattern, however, does not hold for all cell types. Figure 5.12 not only illustrates how well this basic cellular pattern is preserved among different species, but also demonstrates the increase in the number of branches and branch points in the dendritic part of the neuron with increasing complexity of the organism. Presumably the invariance of the basic cellular structure among species is related to the similarity in the function of the cell. The increased branching of the dendrites, by greatly enhancing the neuronal surface area and, correspondingly, the number of dendritic spines and the number of synapses impinging on the cell, provides for greater integrative cellular functioning (see p. 149).

The application of the cable equation for the analysis of current flow in networks such as those illustrated in Fig. 5.12 necessitates considerable mathematical approximations. An extremely useful model that has certain reasonable assumptions, some of which have been shown to be valid by experiment, is the Rall model (1959a,b). His model assumes that each

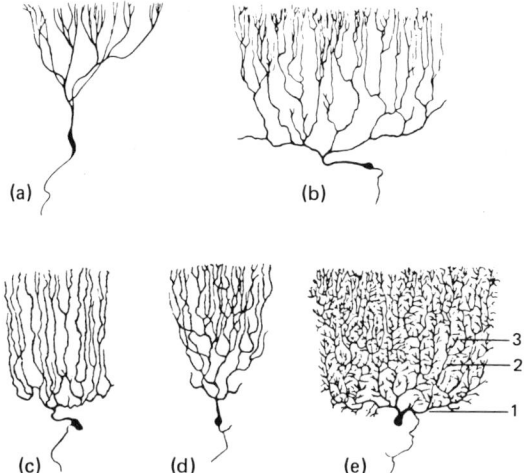

FIG. 5.12 The dendritic patterns in cerebellar Purkinje cells of various vertebrate species: (a) sturgeon; (b) perch; (c) turtle; (d) pigeon; (e) man. In (e), 1, 2, and 3 refer to primary, secondary, and tertiary dendrites, respectively. [From Scheibel and Scheibel (1970).]

dendritic tree consists of an initial cylindrical trunk with cylindrical branches. Each branch is approximated by a cylinder that is assumed not to taper between branch points. In addition, the membrane resistance and capacitance are taken to be independent of position in the dendritic network, and their values are assumed to be the same as those characterizing the cell body. A further simplification is to neglect the effects of the extracellular environment, which is equivalent to saying that the extracellular resistance is very small compared to the resistance of the cell's intracellular medium. Within the context of this model, Rall (1959b) showed that if electric current and potential obey the one-dimensional cable equation [Eq. (5.10)] and if internal current is conserved at branch points, then at each branch point the following relation must hold:

$$B_l d_0^{3/2} = \sum_k B_{lk} d_{lk}^{3/2} \qquad (5.23)$$

where d_0 is the diameter of the parent dendritic branch and d_{lk} is the diameter of its kth daughter branch. The quantities B_l and B_{lk} are constants depending only on the boundary conditions assumed for the terminal ends of the dendritic tree. This equation expresses a relation between the electrical and geometrical properties of a dendrite at a branch point. [See Rall (1962) for a generalization of this equation to a much larger class of dendritic trees.] The derivation of Eq. (5.23) is given in Appendix B. In many cases the ratio of the constants

B_{lk}/B_l can be taken to be unity, and in this case Rall's branching rule assumes a particularly simple form,

$$d_0^{3/2} = \sum_k d_{lk}^{3/2} \tag{5.24}$$

This equation is commonly known as the restricted $\frac{3}{2}$-power rule.

For cat spinal motoneurons, the rule is obeyed quite well. For example, Lux et al. (1970) used autoradiography for anatomic measurements of dendritic branching in motoneurons. They reported $\sum d_{lk}^{3/2}/d_0^{3/2}$ as varying between 0.8 and 1.2, in good agreement with the predicted value of unity if the ratio of the constants B_{lk}/B_l is taken to be unity. The validity of the $\frac{3}{2}$ rule for cat motoneurons has been confirmed by other investigators although Barrett and Grill (1974), using procion dye injections to help them make anatomic measurements, reported a slight taper (a decrease) of $\sum d_{lk}^{3/2}$ with increasing distance from the cell body. These measurements are extremely difficult to make, and the reported taper could be an experimental artifact caused by the incomplete filling of the finer dendritic branches by procion dye. Agreement with Eq. (5.24) is not obtained for pyramidal and Purkinje cells (Hillman, 1979). For these neurons, it is found that

$$d_0^2 = \sum_k d_{lk}^2 \tag{5.25}$$

a result equivalent to the conservation of cross-sectional area at branch points. In contrast, it should be noted that Eq. (5.24) implies that cross-sectional area is not conserved for motoneurons.

The difference between motoneurons and pyramidal and Purkinje cells in the power rule obeyed at dendritic branch points is thought to reflect differences in the internal structure of dendrites. Most proteins needed in neuronal processes have to be synthesized in the nerve cell body and transported into axons and dendrites. This rate of transport from the soma to the periphery has been measured by following the time course of radioactive proteins in the dendrites after the injection of radio-labeled amino acids into the soma. The transport diffusion coefficient inferred from such measurements is quite large, estimated to be 2×10^{-6} cm^2 sec^{-1}, which is about one order of magnitude larger than the diffusion coefficient for proteins in water (Uylings et al., 1975). If nutrients flow along dendrites at a constant current density, i.e., the same rate of flow per unit cross-sectional area, then Eq. (5.25) shows that all branches are served with the same flow rate. However, when the $\frac{3}{2}$ branch rule is obeyed, then the flow rate decreases at the branch points, which is not a tenable result. Electromicrographs show that nutrients are actually transported by small processes called microtubules, which extend along the dendrites. For constant nutrition per unit cross-sectional area to occur in dendrites obeying the $\frac{3}{2}$-power law, the number of microtubules at each branch point must be conserved. If this is the case, then the density of microtubules (number per unit cross-sectional area) must increase with increasing branch number. Electron micrographs have shown this to be the case. It is not unreasonable to suggest the conclusion that if nutrition flow rate is unaffected by the two types of power laws, then these laws must be related to their electrical behavior.

EQUIVALENT CYLINDER APPROXIMATION

An approximation that considerably simplifies the analysis of the effects of dendrites on the electrical response of the soma is that of the equivalent cylinder. It is stated as follows. If (1) boundary conditions at *all* terminal points of the dendritic tree are the same, (2) all terminal points of a dendritic tree are of the same electrotonic distance L^* from the soma, and (3) Rall's branching rule holds at each branch point, namely,

$$d_0^{3/2} = \sum_k d_k^{3/2} \tag{5.26}$$

then the entire dendritic tree reduces to a cylinder with electrotonic length L^*. Furthermore, if *sealed-end conditions* are assumed, then the membrane potential at any point x^*, measured in electrotonic units, can be calculated using the equation

$$V = V_0 \frac{\cosh[(L-x)/\lambda]}{\cosh(L/\lambda)} = V_0 \frac{\cosh(L^* - x^*)}{\cosh(L^*)} \tag{5.27}$$

where V_0 is the membrane potential at the origin. This theorem was initially proved by Rall (1959b), and the proof is shown in Appendix C.

Assumption (3) in the equivalent cylinder theorem was made prior to any experimental guidance or confirmation that dendrites indeed followed this simple power law. It is remarkable that an assumption that lead to an elegant theorem was found to be obeyed by Nature. Without this theorem, however, such behavior probably would not have been looked for by investigators. In fact, its success has led to further research into patterns of dendrites. Although no dendritic tree exactly satisfies the conditions of this theorem, the application has been useful in estimating the effect of distal synapses and has provided a means of inferring dendritic membrane resistance from electrophysiological measurements. The theorem is illustrated schematically in Fig. 5.13a. The terminal ends A, B, and C are at the same electrotonic distance L^* from the cell body, denoted by 0 in the figure. The dashed lines connect points on different branches that are at equal electrotonic distances from the cell body. It is to be emphasized that branches need not have equal physical

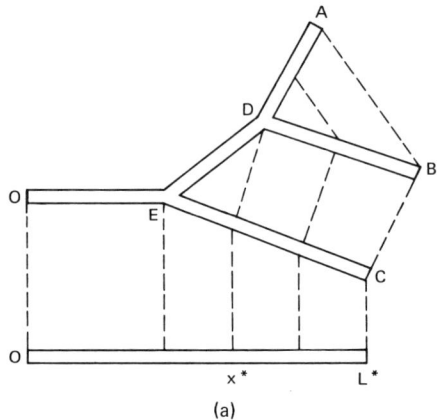

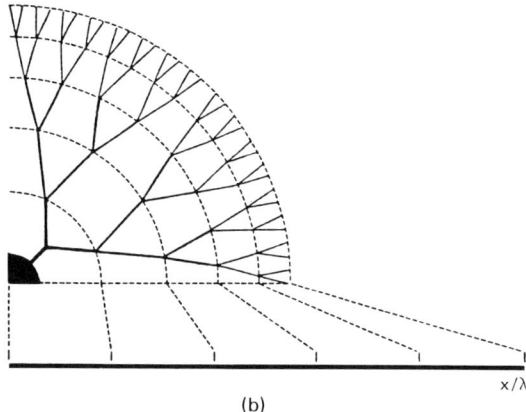

FIG. 5.13 (a) Reduction of the dendritic tree $OABC$ to an equivalent cylinder of electrotonic length L^*. Dashed lines connect points that are at equal electrotonic distances from the soma. (b) Diagram indicating the mathematical transformation of a class of dendritic trees into an equivalent cylinder. [From Rall (1967).]

lengths and diameters. For example, the branch EC in Fig. 5.13a need not have a length equal to the sum of the lengths of ED and DB, nor do the lengths of DA and DB have to be equal, although their electrotonic lengths must be equal. Rall's much-reproduced illustration, Fig. 5.13b, is somewhat deceptive in that all branch points have been placed at common equipotential distances. This placement is not necessary.

This reduction of one dendritic tree to an equivalent cylinder can sometimes be extended to neurons having many dendritic trees. If each dendritic

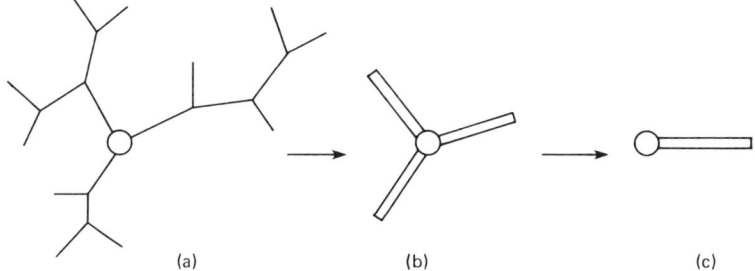

FIG. 5.14 Reduction of (a) a neuron with three dendritic trees to (b) an equivalent neuron with three equivalent cylinders, and finally to (c) a neuron with one equivalent cylinder.

tree can be reduced to an equivalent cylinder and all equivalent cylinders have equal electrotonic lengths, then the entire dendritic network of the cell is equivalent to a single cylinder. Figure 5.14 illustrates this reduction schematically.

PROPERTIES OF SOMATIC POTENTIALS WITH DENDRITIC SYNAPTIC INPUTS

The calculation of the soma's transmembrane potential in response to inputs located at different distances from the cell's body is quite complicated. Even the equivalent cylinder approximation needs to be further simplified in order to perform such a calculation. One method of accomplishing this is to use the compartmental model first introduced by Rall (1964, 1967). The following discussion is limited to the general aspects of this model and a semiquantitative discussion of computed soma potentials. Mathematical details are kept to a minimum.

Compartmental Model

In the compartmental model the cylinder representing the dendritic trees of the neuron is divided into smaller segments or compartments (see Fig. 5.15b). These compartments may or may not be of equal length. Each compartment is taken to be isopotential. This approximation neglects any effects of passive electrotonic spreading within a compartment and, in effect, represents each segment by a membrane circuit as shown in Fig. 5.15c. Although the circuit shown in Fig. 5.15c allows for the neuronal membrane to transport only potassium and sodium ions, the restriction is not essential. In fact, different regions of the neuronal membrane surface, i.e., compartments in the model, could transport different ions. Each compartment is joined to adjacent ones by a series resistance, as Fig. 5.15b illustrates. The

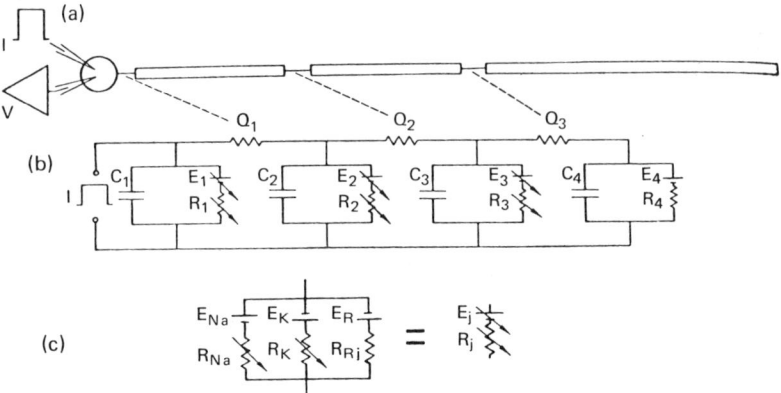

FIG. 5.15 Reduction of equivalent cylinder to compartmental neuronal model. Three dendritic compartments are shown. (a) Equivalent cylinder and cell body. (b) Compartmental model. The Q_i are longitudinal dendroplasmic resistances. (c) Electrical circuit for each compartment. C_j is the membrane capacitance; E_{Na}, E_K, and E_R are, respectively, the sodium, potassium, and resting potentials. R_{Na}, R_K, and R_{R_j} are the sodium, potassium, and resting conductances, respectively. [From Llinas and Nicholson (1976).]

effects of the cell body are included by assuming that it can be approximated by an isopotential sphere, and it is further assumed to be electrically parallel with the dendrites. Synaptic inputs at any location along the cylinder are formally introduced into the model by varying the conductances at that location. Variation of the potential within any one compartment causes current to flow into (or out of) neighboring compartments by way of the resistances Q_i. By this mechanism the transient voltage response spreads to the compartment representing the soma. The time variation of the potential for the soma compartment is the theoretical prediction for the excitatory postsynaptic potential (EPSP), since experimentally the intracellular recording site is generally at or near the cell body. This is because the recording electrodes are comparable in size to the diameters of the dendrites so that when penetration of a cell occurs it is almost always at or very close to the soma.

Shape and Time Behavior of Soma Potentials

Rall (1967, 1977) has performed extensive calculations of EPSPs for a ten-compartmental model of the type shown in Fig. 5.15 for the motoneuron. The following is a summary of his theoretical results.

In the case where synaptic inputs are distributed uniformly over the entire neuronal surface (this includes both the soma and dendrites), there is no electrotonic spreading between different compartments. For these conditions,

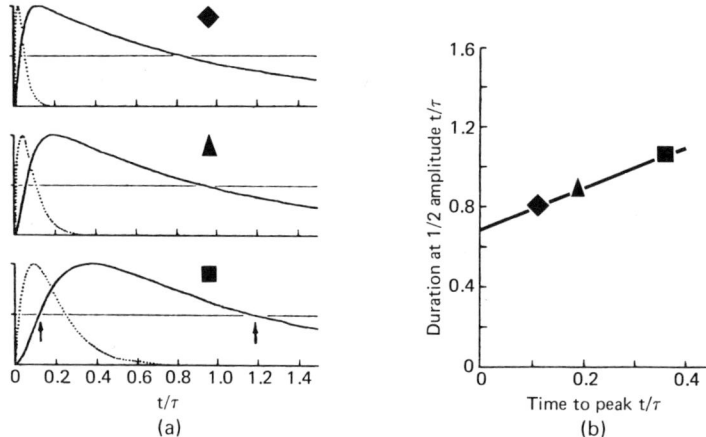

FIG. 5.16 (a) Theoretical EPSPs (solid lines) generated by synaptic currents of different time courses (dotted lines). Synaptic currents were uniformly distributed over the entire neuronal surface. (b) Shape index plot. The three points plotted are for EPSPs illustrated in Fig. 5.16a. Scales are in units of the dimensionless ratio t/τ; τ is the membrane time constant. [From Rall et al. (1967).]

the excitatory postsynaptic potentials (EPSPs) at each location have the same time dependence. Examples of EPSPs generated by uniform synaptic inputs but with varying synaptic current duration are illustrated in Fig. 5.16a. The dashed lines in Fig. 5.16a denote examples of the synaptic current input time course, and the solid lines are the calculated EPSPs. The important point to note here is that the slower synaptic currents (broader dashed lines in Fig. 5.16a) produce EPSPs that attain their peak amplitudes at later times and last longer than EPSPs evoked by faster synaptic input currents. In comparing theoretically calculated EPSPs with experimentally determined EPSPs, Rall (1967) found it convenient to characterize the shape and time course of the potentials by two *shape indices*. These shape indices are (1) the time required for the EPSP to attain its peak value, and (2) the half-width of the EPSP. The half-width is defined as the time duration measured between points on the rising and falling phases of the EPSP that are one-half the peak amplitude. For example, the time interval separating the arrows in the lower curve in Fig. 5.16a is the half-width for that particular EPSP. Having defined the two indices, a given EPSP can be represented as a single point in a shape-index graph. Such a graph is shown in Fig. 5.16b, and the three points plotted in this figure correspond to the three uniformly generated synaptic potentials illustrated in Fig. 5.16a. The introduction of these two shape indices is motivated by the inability to compare a given theoretical EPSP with an experimental result because of the large variability of EPSP experimental data. This variability, which will be shown, arises from the nature of the experiments. An electrical probe is inserted blindly into a region of motoneurons. Penetration of one is indicated by a sudden potential drop. Synaptic inputs are then produced by muscle stimulation of the animal. This crude cause–effect coupling of input to EPSP obviously results in a wide variance of EPSPs. Furthermore, a given Ia fiber (Vol. II, p. 74) can have collateral axons to many different motoneurons (Fig. 5.2). Stimulation of the Ia axon does not in general produce the same EPSP in each motoneuron in cases where the axon collaterals have axodendritic synapses.

When synaptic inputs are theoretically restricted to a single compartment, the resulting theoretical EPSPs differ considerably from those calculated using uniform excitation. Figure 5.17 illustrates the effect of synaptic input location on the shape of the computed synaptic potentials. Curves B, C, and D in Fig. 5.17 are examples of EPSPs computed in the soma, designated as compartment 1 for a chain of ten compartments when synaptic inputs are localized to the first, fourth, and eighth compartment. It is seen that for synapses located spatially close to the soma, the EPSPs rise faster and have a shorter duration than EPSPs generated by more distal synaptic inputs. The slower rise and slower decay of the EPSP evoked by distal synapses,

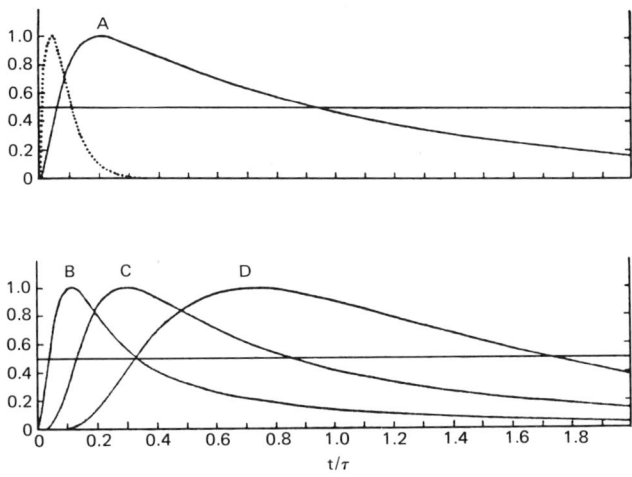

FIG. 5.17 Four EPSP shapes computed for a chain of 10 compartments. Dotted curve shows the assumed synaptic conductance transient. Curve A shows EPSP for equal inputs into all compartments. Curve B shows the EPSP when the synaptic input is restricted to compartment 1 alone. Curve C shows the EPSP when the synaptic input occurs in compartment 4 only. Curve D shows the soma's potential when the synaptic input is in compartment 8 only. Ordinate scale represents amplitudes relative to each peak amplitude. Abscissa in units of the dimensionless ratio t/τ; τ is the membrane time constant. [From Rall (1967).]

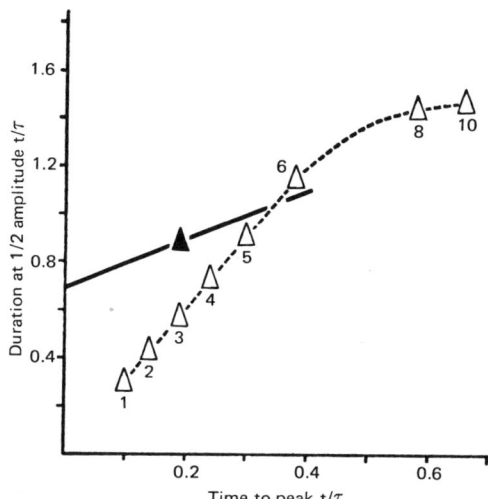

FIG. 5.18 Plot of EPSP shape index values calculated using a 10-compartmental model. Triangles with numbers give indices when synaptic input is restricted to the numbered compartment. 1 labels the soma compartment whereas 10 labels the most distal compartment. Solid line represents indices of EPSP when cell is uniformly depolarized. The solid triangle denotes the EPSP labeled by the same symbol in Fig. 5.16a. Scales are in units of dimensionless ratio t/τ. τ is the membrane time constant. [From Rall et al. (1967).]

illustrated by curve D in Fig. 5.17, can be understood intuitively as the consequence of electrotonic spread; the farther the distance of travel, the slower the rising and falling time. It simply takes a longer time for the distal-evoked potentials to reach the soma, the position where experimental EPSPs are recorded. Thus the shape indices are smaller for more proximal synapses than for distal synapses. This is shown in Fig. 5.18, which gives the values for the shape indices for EPSPs evoked by single synaptic inputs, assuming a ten-compartmental neuronal model. The numbers near the triangles label the compartment that received the synaptic input. The number 1 labels the soma, while compartment 10 represents the most distal part of the dendritic tree. Compartments 2–9 represent divisions of equal electrotonic distance labeled from the dendritic trunk to the periphery. As this figure illustrates, those EPSPs having small shape indices correspond to synaptic input locations proximal to the soma.

The theoretical EPSPs and the shape indices reported in Figs. 5.16–5.18 were calculated using a ten-compartmental model and electrical parameters appropriate for spinal motoneurons (Rall, 1967). Thus the computed EPSP indices can be compared with indices inferred from experimentally observed EPSPs recorded from spinal motoneurons. Such a comparison is shown in

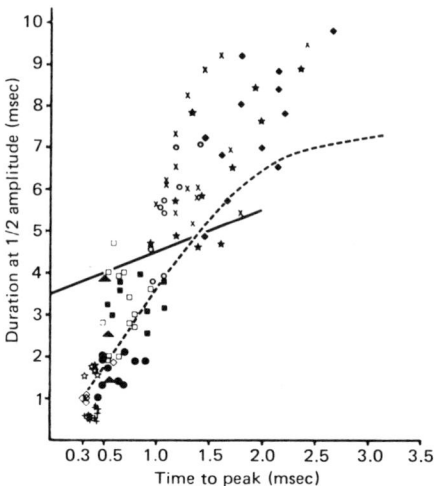

FIG. 5.19 Shape index plot for experimentally observed miniature EPSPs for several different groups of Ia afferent fibers. The dashed line represents the loci of computed EPSP shapes with localized input (see Fig. 5.18). Solid line denotes loci of indices for uniformly depolarized input (see Fig. 5.16). [From Rall et al. (1967).]

Fig. 5.19. The different arrays of symbols are the shape indices for experimentally observed *miniature* EPSPs evoked by several different groups of Ia afferent fibers (see Fig. 5.2). Miniature EPSPs are those soma potentials elicited by the activity of a *single* Ia afferent fiber.

The comparison between theory and experiment shown in Fig. 5.19 necessitated a choice in the value of τ, the membrane time constant, since the theoretical shape indices given in Figs. 5.16 and 5.18 are in dimensionless units t/τ, whereas the experimental EPSP's indices are measured in real time, e.g., in milliseconds. The value of 5 msec was assumed for this conversion because it is a typical value for motoneurons. The scatter in the experimental indices is expected since a single Ia fiber forms many synapses with a motoneuron, and there is no a priori reason to expect all motoneurons to have the same spatial arrangement of synaptic inputs. It is evident from the figure that the experimental results cannot be explained if only somatic inputs (sometimes called axosomatic inputs) are considered, i.e., only compartment 1 has inputs. The experimental EPSP indices are explicable only in terms of large contributions to the soma's potential from synapses on the dendrites. The large half-width indices of Fig. 5.19 can be explained only by having a significant contribution to the soma potential of the motoneuron from distal synapses. This conclusion is opposite to the view held by many electrophysiologists prior to the early 1960s. At that time it was thought that postsynaptic potentials

PROPERTIES OF SOMATIC POTENTIALS WITH DENDRITIC SYNAPTIC INPUTS   175

originating from dendritic inputs were significantly attenuated before arrival at the soma (Eccles, 1964; Coombs *et al.*, 1955).

The concept of a reversal potential was discussed in Chapter 4. It was defined as that membrane potential for which a synaptic input does not evoke a membrane potential change. As shown in Fig. 4.14a, the IPSPs in cat spinal motoneuron have a reversal potential near -82 mV; however, monosynaptic EPSPs, as discussed at the beginning of this chapter and illustrated in Fig. 5.1, do not exhibit a clear reversal potential. It is possible to understand this anomalous behavior of EPSPs when distal synaptic inputs contribute significantly to the soma potential. This can be seen in the following way. When only a single synaptic input is activated the EPSP evoked at the soma will exhibit a unique reversal potential. However, when two or more axodendritic synaptic inputs contribute to the soma response there will not usually be a unique reversal potential. This is because synaptic inputs at different electrotonic distances from the soma have different reversal potentials. This is seen more clearly in Fig. 5.20, which shows the computed synaptic potentials for a four-compartment neuron model. When the synaptic input is restricted to only compartment 1, the EPSP reverses at some value between -10 and -15 mV. This is shown in Fig. 5.20 in the column labeled 1. When

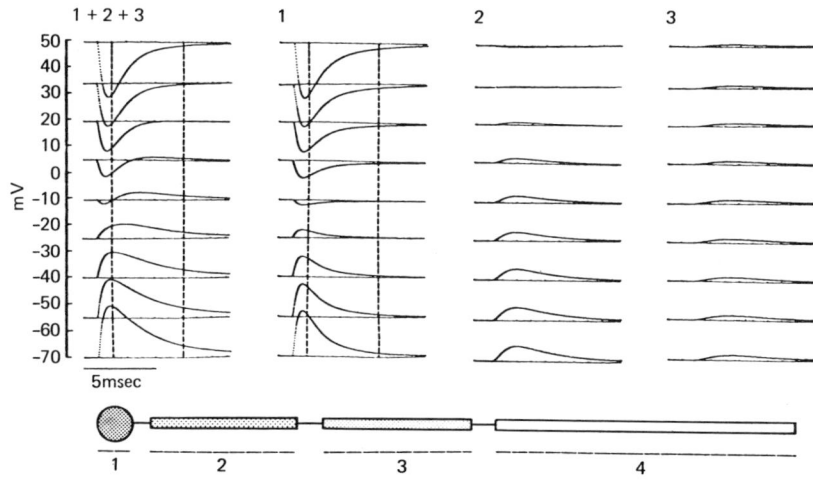

FIG. 5.20   Theoretical somatic potentials following synaptic activation of the three different compartments as a function of membrane potential. Record to left are the EPSPs generated by synaptic activation of compartments 1, 2, and 3. Note that there is no true reversal potential. Record 1 gives synaptic potential when synaptic input is restricted to the soma. In this case the EPSP exhibits a true reversal potential. Record 2 gives somatic potentials when synaptic input is evoked in the first dendritic (labeled 2) compartment. Record 3 gives somatic potentials when input is restricted to compartment 3. [From Llinas and Nicholson (1976).]

inputs are restricted to only compartment 2 the EPSP does not reverse, at least until +40 mV, where it is seen to go to zero, probably prior to reversal. However, when the neuron receives inputs into compartments 1, 2, and 3, the potential recorded at the soma does not have a unique reversal potential. This is illustrated in Fig. 5.20 in the column labeled 1 + 2 + 3. At no membrane potential is the response, the soma potential, zero. In fact the rising part of the soma's EPSP is the first to reverse, whereas the component of the soma's potential due to distal input loci reverses at larger membrane depolarizations. This biphasic reversal of the soma potential with holding potential, as discussed at the beginning of this chapter, is characteristic of EPSPs of motoneurons (see Fig. 5.1). Hence, the absence of a true equilibrium or reversal potential for EPSPs is consistent with large axodendritic synaptic input contributions to the motoneuron soma potentials. The existence of a true reversal potential for inhibitory inputs as illustrated in Fig. 4.14a implies that, in contrast to excitatory inputs, inhibitory inputs synapse on or near the soma.

TRANSIENT PASSIVE MEMBRANE RESPONSES

The cable properties of a neuron, namely, the membrane resistance and the electrotonic length of the cell's dendrites, can, in principle, be determined from an analysis of the voltage transients produced by either hyperpolarizing or depolarizing current pulses. A major advantage of this transient method is that the detailed geometry of the neuron need not be considered. Before discussing how this is accomplished, consider first the charging and decay characteristics of a small membrane segment having a spatially uniform membrane resistance R_m and capacitance C_m. Although the following discussion is restricted to depolarizing current pulses, it should be obvious that the treatment of hyperpolarizing pulses is similar.

Suppose for $t < 0$ the inserted electrode in the membrane injects zero current but that at $t = 0$ a sudden transition to some constant value i_m occurs. The time dependence of the voltage response is found by solving the equation in which the total current is the sum of the ohmic current and the capacitance current across the membrane:

$$i(t) = C_m \frac{\partial V}{\partial t} + \frac{V}{R_m} \tag{5.28}$$

For the conditions stated, the solution to this equation is obtained by the integrating-factor method (Vol. I, p. 365) and has a saturating exponential time dependence:

$$V(t) = i_m R_m [1 - \exp(-t/R_m C_m)] \tag{5.29}$$

where $V(t)$ is measured from its resting value. The quantity $R_m C_m$ has the dimension of time and is called the *passive time constant* of the membrane, $\tau_0 = R_m C_m$. Its value characterizes the rate at which steady state is achieved, and at infinite times $V(\infty) = i_m R_m$. The exponential dependence of the charging rate is illustrated in Fig. 5.21. Note that Eq. (5.28) and its solutions have the same form as those derived for pulsatile arterial blood flow (Vol. I, p. 191).

In a similar manner it can be shown that the decay of the voltage departure to its resting value is exponential in time with the same time constant. For the particular case considered here, the *input resistance* R_N, defined as the ratio of the magnitude of the steady-state voltage deflection, the upper plateau of Fig. 5.21a, to the magnitude of the injected current, is the passive membrane resistance R_m:

$$R_N \equiv \frac{\text{steady-state voltage departure from its resting value}}{\text{current injected}} = \frac{i_m R_m}{i_m} = R_m \tag{5.30}$$

Thus the value of R_m can be obtained from the steady-state voltage deflection, and the value of C_m can be inferred from the time constant. Unfortunately, the values of R_m and C_m so determined are rather poor since the condition assumed in the analysis, namely, uniform polarization, is difficult to achieve experimentally for cells. For realizable cases, where an intracellular microelectrode is employed to apply current between a point inside a nerve cell body and a distant reference electrode in the extracellular conducting medium, the initial polarization is not uniform over the entire neuronal membrane.

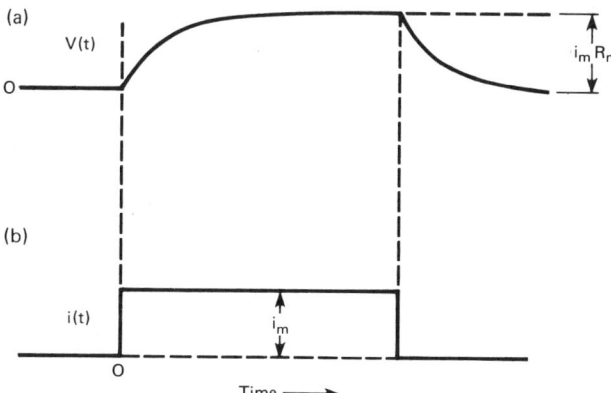

FIG. 5.21 (a) The voltage growth and decay rates characterized by a time constant τ_0: growth rate is proportional to $1 - \exp(-t/\tau_0)$ and decay rate to $\exp(-t/\tau_0)$. The growth and decay are caused by the sudden current pulses (positive and negative) indicated in (b).

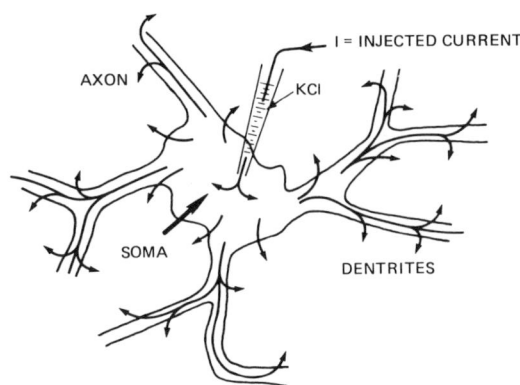

FIG. 5.22 Diagram illustrating how the injected current I flows from a microelectrode whose tip penetrates the soma of a neuron. Full extent of the axon and dendrites not shown. Arrows indicate direction of current flow.

For this reason, the above simple analysis is not generally applicable. Some of the injected current flows across the soma membrane, and some flows into the dendrites and axon as well as across their membrane surfaces. The amount of current that flows along each of these paths is determined by both the geometry of the neuron and the electrical properties of the membranes. Figure 5.22 schematically illustrates the flow of current from the injecting electrode. Because of this redistribution of injected charge, the time dependence of the polarization is not the same at all points on the membrane. Therefore the product $i_m R_m$ in Eq. (5.29), while yielding a voltage, is not the voltage drop across the soma membrane.

The method of calculation of the membrane resistance R_m from the measured input resistance R_N and the form of the passive voltage transient under conditions where the membrane potential is not uniform over the entire neuronal surface were initially given by Rall (1969). He showed that when the entire neuron is approximated by an equivalent cylinder of electrotonic length L^*, the time dependence of the voltage change following the application of a current step to the soma can be written as a linear sum of exponentials,

$$\Delta V(t) = B_0 \exp(-t/\tau_0) + B_1 \exp(-t/\tau_1) + \cdots \quad (5.31)$$

where the time constants are

$$\tau_n = \frac{\tau_0}{1 + n^2\pi^2/L^{*2}} = \frac{R_m C_m}{1 + n^2\pi^2/L^{*2}} \quad (5.32)$$

A proof of this result is given in Appendix D. The exponential terms for $n > 1$ arise from current redistribution within the cylinder from the site of injection and are called *equalization times*. Since τ_n is a decreasing function with increasing n, it follows that at long times only the first term in Eq. (5.31) is significant. This is expected physically since at long times after the onset of the current pulse the charge redistribution has equalized the membrane polarization, thereby imparting to the entire neuronal membrane behavior characteristics of a small, uniformly polarized membrane segment (see p. 176 and Fig. 5.21). Under favorable conditions only the first few terms in Eq. (5.31) contribute significantly to $V(t)$. It is evident that a measurement of τ_0 and τ_1 provides a method for determining the electrotonic length L^* since from Eq. (5.32)

$$L^* = \pi(\tau_0/\tau_1 - 1)^{-1/2} \quad (5.33)$$

The application of this formalism is illustrated in Fig. 5.23 for the hyperpolarizing transients recorded from the cell body of a mouse dorsal root

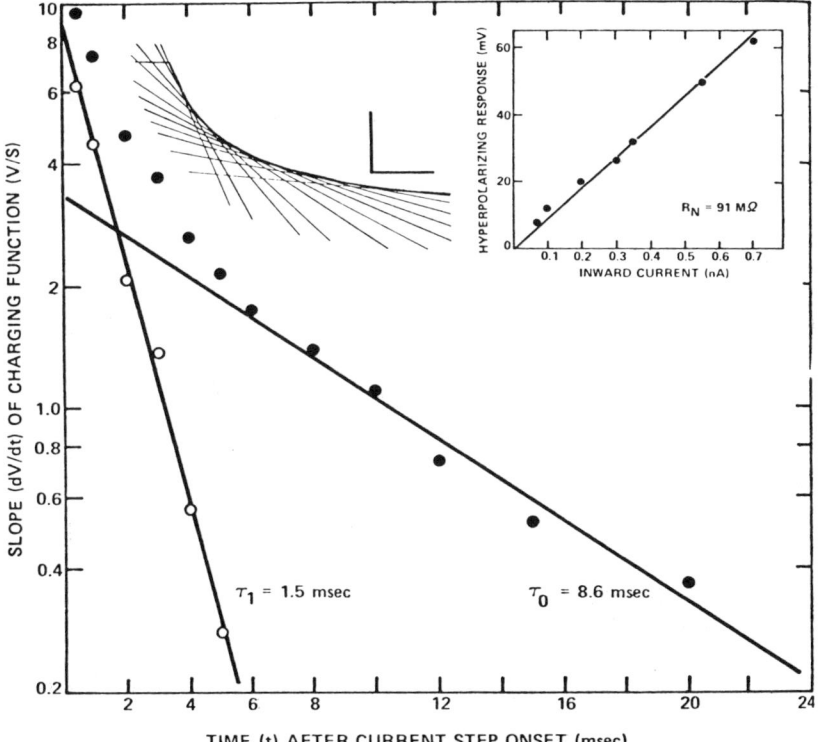

FIG. 5.23 A semilogarithmic (base e) plot of dV/dt versus time. The onset of the hyperpolarizing current pulse denotes the origin of time. The inset at upper right is the steady-state current voltage relation. R_N is the input resistance. The inset at upper left is the voltage trace containing the tangent lines used to obtain the solid-dot data points in the lower curve. [From Brown et al. (1981).]

ganglion neuron. The tangents to a voltage-decay curve for times after the initiation of a hyperpolarizing current pulse were determined graphically from the voltage record as shown in the upper left-hand inset in Fig. 5.23. These values are plotted on a natural-logarithmic scale versus time, as illustrated in Fig. 5.23 (solid circles in lower curve). The nearly linear behavior for $t > 6$ msec corresponds mathematically to the approximation

$$\frac{dV}{dt} \approx \frac{B_0}{\tau_0} \exp\left(-\frac{t}{\tau_0}\right) \tag{5.34}$$

from the first term of Eq. (5.3.1). Thus the slope of $\ln(dV/dt)$ versus t at long times provides an estimate of τ_0 and therefore the product $R_m C_m$. By extrapolating the straight line, which fits the long-time part of the curve, back to earlier times ($t < 6$ msec in the case of Fig. 5.23), a value of B_0/τ_0 can be inferred from the intercept. With the values of B_0/τ_0 and τ_0 determined, the first term in the series

$$\frac{dV(t)}{dt} = \sum_{i=0}^{\infty} \frac{B_i}{\tau_i} \exp\left(-\frac{t}{\tau_i}\right) \tag{5.35}$$

can be subtracted from the experimental data.† Thus

$$\left.\frac{dV}{dt}\right|_{\text{experimental}} = -\frac{B_0}{\tau_0}\exp\left(-\frac{t}{\tau_0}\right) = \frac{B_1}{\tau_1}\exp\left(-\frac{t}{\tau_1}\right) \tag{5.36}$$

From a plot of the logarithm of the left-hand side of Eq. (5.36) versus time, the equalization time constant τ_1 can be estimated. This is illustrated by the solid line through the open circles in the lower curve in Fig. 5.23. Obviously this peeling process relies on favorable values of B_n and τ_n. In many cases the procedure does not work well, and in general it is difficult to deconvolute more than two time constants. Inserting the values of τ_0 (8.6 msec) and τ_1 (1.5 msec) obtained from Fig. 5.23 in Eq. (5.33) gives an electrotonic length $L^* \approx 1.4$. However, estimates based on Eq. (5.32) include the effects of the soma as well as the electrotonic length L_0^* of the cylinder that represents the dendrites.

In a more complete treatment Rall (1969) has shown that

$$L_0^* \cong \left(\frac{\rho}{\rho+1}\right)^{1/2} \pi\left(\frac{\tau_0}{\tau_1} - 1\right)^{-1/2} \tag{5.37}$$

where ρ is defined as

$$\rho \equiv G_D/G_S \tag{5.38}$$

---

† The derivative of Eq. (5.31) differs from Eq. (5.35) by a minus sign. This arises since $\Delta V(t)$ equals the asymptotic final voltage minus the voltage at time t; the latter is $V(t)$.

G_S is the input conductance of the soma, and G_D is the total dendritic input conductance. In the particular example considered above, additional electrical measurements gave $\rho = 1.9$. Thus the electrotonic length L_0^* of the dendrites of mouse dorsal root ganglia neurons is about 1.1. For comparison, the entire dendritic tree of a motoneuron has an electrotonic length of about 1.5 (Rall, 1977). The magnitude of L_0^* shows how far electrically the distal dendritic synapses are from the cell body. The smaller L_0^* is, the more effective synapses on the dendrites are in influencing the output of the neuron. A value of $L_0^* \approx 1.1$ from the above experiment corresponds to an electrically compact cell, i.e., all synapses contribute significantly to the cell's integrative output.

The input resistance R_N, defined in Eq. (5.30) as the steady-state voltage departure from its resting value divided by the current injected, is a major physiologically measured parameter. An example illustrating the experimental measurement of R_N is given in the curve in the upper right-hand inset of Fig. 5.23, i.e., the slope of the curve is the input resistance R_N. When the effects of electrotonic spreading of current into the dendrites are included in the analysis of voltage transients, the equality between input resistance and membrane resistance given in Eq. (5.30) is no longer valid. Instead, it has been shown that (Rall, 1967)

$$R_N = R_m/s(1 + \rho) \tag{5.39}$$

where s is the surface area of the cell body and ρ is defined by Eq. (5.38). Equation (5.39) shows that a measurement of the input resistance does not immediately provide a value for R_m, since the value of R_N depends on s and ρ and therefore on the geometry of the neuron. The cell body surface area s can be directly measured, for example, by use of a phase-contrast microscope, and a value for ρ can be determined from an analysis of voltage transients (Jack et al., 1975). These values with the additional measurement of R_N permit the determination of R_m. With the value of R_m found and with a measurement of τ_0, the capacitance C_m of the neuronal membrane can be determined. Hence, Rall's mathematical model of a neuron allows the investigator to infer the characteristic cable properties of the cell from physiological measurements. The important points to note from the above discussion are that electrophysiological data alone are insufficient and that theoretical considerations are essential.

REFERENCES

Barrett, J. N., and Grill, W. E. (1974). Specific membrane properties of cat motoneurones. *J. Physiol. (London)* **235**, 301.

Berry, M., Bradley, P., and Borges, S. (1978). Environmental and genetic determinants of connectivity in the central nervous system—an approach through dendritic field analysis. *Prog. Brain Res.* **48**, 133.

Brown, T. H., Perkel, D. H., Norris, J. C., and Peacock, J. H. (1981). Electrotonic structure and specific membrane properties of mouse dorsal root ganglion neurons. *J. Neurophysiol.* **45**, 1.

Coombs, J. S., Eccles, J. C., and Fatt, P. (1955). The electrical properties of the motoneurone membrane. *J. Physiol. (London)* **130**, 291.

Crick, F. (1982). Do dendritic spines twitch? *Trends NeuroSci. (Pers. Ed.)* **5**, 44.

Diamond, M. C., Rosenzweig, M. R., Bennett, E. L., Lindner, B., and Lyon, L. (1972). Effects of environmental enrichment and impoverishment on rat cerebral cortex. *J. Neurobiol.* **3**, 47.

Eccles, J. C. (1964). "The Physiology of Synapses." Springer-Verlag, Berlin.

Fox, C. A., and Barnard, J. W. (1957). A quantitative study of the Purkinje cell dendritic branchlets and their relationship to afferent fibres. *J. Anat.* **91**, 299.

Greenough, W. T. (1976). Enduring brain effects of differential experience and training. *In* "Neural Mechanisms of Memory and Learning" (M. R. Rosenzweig and E. L. Bennett, eds.), p. 255. MIT Press, Cambridge, Massachusetts.

Greenough, W. T., and Volkmar, F. R. (1973). Pattern of dendritic branching in occipital cortex of rats reared in complex environments. *Exp. Neurol.* **41**, 371.

Hillman, D. E. (1979). Neuronal shape parameters and substructures as a basis of neuronal form. *In* "The Neurosciences: Fourth Study Program" (F. O. Schmitt and F. G. Worden, eds.), p. 477. MIT Press, Cambridge, Massachusetts.

Hodgkin, A. L., and Rushton, W. A. H. (1946). The electrical constants of a crustacean nerve fibre. *Proc. R. Soc. London, Ser. B* **133**, 444.

Hubel, D. H. (1979). The visual cortex of normal and deprived monkeys. *Am. Sci.* **67**, 532.

Jack, J. J. B., Noble, D., and Tsien, R. W. (1975). "Electric Current Flow in Excitable Cells." Oxford Univ. Press (Clarendon), London/New York.

Levitsky, D. A., and Barnes, R. H. (1972). Nutritional and environmental interactions in the behavioral development of the rat: Long term effects. *Science* **176**, 68.

Llinas, R., and Nicholson, C. (1976). Reversal properties of climbing fiber potential in cat Purkinje cells; an example of a distributed synapse. *J. Neurophysiol.* **39**, 311.

Lux, H. D., Schubert, P., and Kreutzberg, G. W. (1970). Direct matching of morphological and electrophysiological data in cat spinal motoneurons. *In* "Excitatory Synaptic Mechanisms" (P. Andersen and J. S. Jansen, eds.), p. 189. Universitetsforlaget, Oslo.

McConnell, P. (1980). Nutritional effects on non-mitotic aspects of central nervous system development. *Prog. Brain Res.* **53**, 99.

McConnell, P., and Berry, M. (1978). Effects of undernutrition on Purkinje cell dendritic growth in the rat. *J. Comp. Neurol.* **178**, 759.

McConnell, P., and Berry, M. (1981). The effects of refeeding after varying periods of neonatal undernutrition on the morphology of Purkinje cells in the cerebellum of the rat. *J. Comp. Neurol.* **200**, 463.

Morgan, B., and Winick, M. (1979). A possible relationship between brain n-acetylneuraminic acid content and behavior. *Proc. Soc. Exp. Biol. Med.* **161**, 534.

Rall, W. (1959a). Branching dendritic trees and motoneuron membrane resistivity. *Exp. Neurol.* **1**, 491.

Rall, W. (1959b). "Dendritic Current Distribution and Whole Neuron Properties," Res. Rep. NM 01.05 00.01.02 Nav. Med. Res. Inst., Bethesda, Maryland.

Rall, W. (1960). Membrane potential transients and membrane time constant of motoneurons. *Exp. Neurol.* **2**, 503.

Rall, W. (1962). Theory of physiological properties of dendrites. *Ann. N.Y. Acad. Sci.* **96**, 1071.

Rall, W. (1964). Theoretical significance of dendritic trees for neuronal input-output relations. *In* "Neural Theory and Modelling" (R. F. Reiss, ed.), p. 73. Stanford Univ. Press, Stanford, California.

REFERENCES

Rall, W. (1967). Distinguishing theoretical synaptic potentials computed for different soma-dendritic distributions of synaptic input. *J. Neurophysiol.* **30**, 1138.

Rall, W. (1969). Time constants and electrotonic length of membrane cylinders and neurons. *Biophys. J.* **9**, 1483.

Rall, W. (1970). Cable properties of dendrites and effects of synaptic location. In "Excitatory Synaptic Mechanisms" (P. Andersen and J. S. Jansen, eds.), p. 175. Universitetsforlaget, Oslo.

Rall, W. (1977). Core conductor theory and cable properties of neurons. In "Handbook of Physiology" (E. R. Kandel, ed.), Sect. 1, Vol. II, p. 39. Am. Physiol. Soc., Bethesda, Maryland.

Rall, W., Shepherd, G. M., Reese, T. S., and Brightman, M. W. (1966). Dendrodendritic synaptic pathway for inhibition in the olfactory bulb. *Exp. Neurol.* **14**, 44.

Rall, W., Burke, R. E., Smith, T. G., Nelson, P. G., and Frank, F. (1967). Dendritic location of synapses and possible mechanisms for the monosynaptic EPSP in motoneurons. *J. Neurophysiol.* **30**, 1169.

Ramón y Caján, S. (1911). "Histologie du système nerveux de l'homme et des vertébrés," Vol. 2. Maloine, Paris.

Rutledge, L. T. (1976). Synaptogenesis: Effects of synaptic use. In "Neural Mechanisms of Memory and Learning" (M. R. Rosenzweig and E. L. Bennett, eds.), p. 329. MIT Press, Cambridge, Massachusetts.

Scheibel, M. E., and Scheibel, A. B. (1970). Of pattern and place in dendrites. *Int. Rev. Neurobiol.* **13**, 1.

Scheibel, M. E., and Scheibel, A. B. (1971). Selected structural-functional correlations in postnatal brain. In "Brain Development and Behavior" (M. B. Sterman, D. J. McGinty, and A. M. Adinolfi, eds.), p. 1. Academic Press, New York.

Shepherd, G. M. (1972). The neuron doctrine: A revision of functional concepts. *Yale J. Biol. Med.* **45**, 584.

Shepherd, G. M. (1977). The olfactory bulb. A simple system in the mammalian brain. In "Handbook of Physiology" (E. R. Kandel, ed.), Sect. 1, Vol. II, p. 945. Am. Physiol. Soc., Bethesda, Maryland.

Sholl, D. A. (1955). The surface area of cortical neurons. *J. Anat.* **89**, 571.

Smith, T. G., Wuerker, R. B., and Frank, K. (1967). Membrane impedance changes during synaptic transmission in cat spinal motoneurons. *J. Neurophysiol.* **30**, 1072.

Uylings, H. B. M., Smit, G. S., and Veltman, W. A. M. (1975). Ordering methods in quantitative analysis of branching structures of dendritic trees. *Adv. Neurol.* **12**, 247.

Uylings, H. B. M., Kuypers, K., and Veltman, W. A. M. (1978). Environmental influences on the neocortex in later life. *Prog. Brain Res.* **48**, 261.

Volkmar, F. R., and Greenough, W. T. (1972). Rearing complexity affects branching of dendrites in visual cortex of the rat. *Science* **176**, 1445.

Widdowson, E. M., and McCance, R. A. (1960). Some effects of accelerating growth, I. General somatic development. *Proc. R. Soc. London, Ser. B* **152**, 88.

Winick, M. (1980). Nutrition and central nervous system development. *Prog. Brain Res.* **53**, 93.

Winick, M., and Noble, A. (1966). Cellular response in rats during malnutrition at various ages. *J. Nutr.* **89**, 300.

Winick M., and Rosso, P. (1969). The effects of severe early malnutrition on cellular growth of human brain. *Pediatr. Res.* **3**, 181.

Wyckoff, R. W. G., and Young, J. Z. (1956). The motoneuron surface. *Proc. R. Soc. London, Ser. B* **144**, 440.

CHAPTER **6**

Analysis of Membrane Noise at Synaptic Junctions

INTRODUCTION

For all systems, measured variables display some fluctuations. Under certain circumstances it is possible to infer microscopic information about the underlying molecular events involved in a physical process from an analysis of the *random fluctuations* of the variable being measured. For example, membrane noise resulting from the statistical variation in the actual number of open channels around the mean number of open channels provides information about ion channel properties. In this case a mathematical analysis of membrane current noise in terms of the amplitude of the fluctuations and the various frequency components it contains allows for the determination of the ion single-channel conductance and the single-channel lifetime. How this is done will be discussed in this chapter.

The types of noise we concentrate on in this chapter are those due to "spontaneous fluctuations" in the voltage or current or in their analogs in linear systems. The different sources of noise are discussed, and the mathematics needed to characterize fluctuations is developed. Several applications of the technique of noise analysis are given which illustrate its usefulness and also its limitations.

INTRODUCTION

Noise analysis has serious limitations, the most important of which is that the extracted information depends critically on the theoretical model employed in the analysis of the raw data. In general, many different molecular models are able to account for the same observed results. This lack of uniqueness implies that fluctuation analysis is most useful as a supplement to information about the physical system obtained by other methods; models constructed for the quantitative analysis of noise are usually suggested by other experiments. In some instances the analysis of noise allows the investigator to eliminate possible theoretical models. Unfortunately, the analysis rarely, if ever, uniquely defines the molecular mechansim.

The origin of inherent fluctuation in any variable (e.g., pressure, voltage, current) of a system resides in the random nature of the elementary molecular events. Thus noise is a direct manifestation on a macroscopic level of the intrinsic probabilistic nature of individual molecular processes. A simple example will help clarify the meaning of that statement. In elementary physics one learns that the pressure of a gas on the walls of a container results from the momentum imparted to the container's walls by a large number of individual molecular collisions. Each event, which is the collision of a molecule with a wall, is assumed random, the basic assumption of the kinetic theory of gases. Since collisions are random, it follows that the total number of molecules $N(t)$ arriving at the container's wall is a function of time. A typical behavior of $N(t)$ is illustrated in Fig. 6.1a. A function such as $N(t)$ whose value at time t is governed by *probabilistic laws* is called a *stochastic* or a *random function*, and the process it represents is called a *stochastic process*. In a sense a stochastic function is one in which the *dependent* variable, in

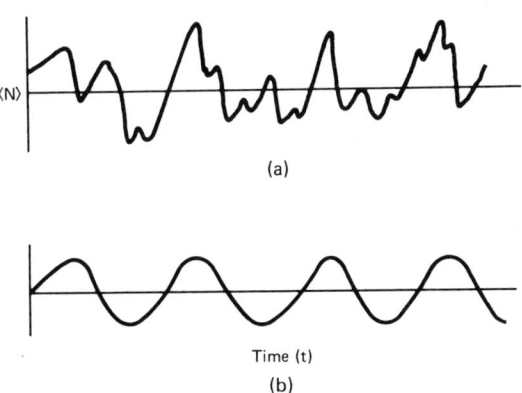

FIG. 6.1 (a) Schematic illustration of the variation of the number of molecules per second colliding with a wall of a container. The variation is measured about the mean number $\langle N \rangle$. (b) The amplitude of a simple pendulum as a function of time.

this case the number of molecules arriving at the container's wall per increment of time, does not depend in a completely defined way on the *independent* variable, here time. Different observations of different incremental times of the same stochastic process generate different functions. In contrast to the function $N(t)$, the amplitude of a simple pendulum, as illustrated in Fig. 6.1b, is a completely determined function of time, having a unique value for each instant. Such a function is not a stochastic function.

In measurements of a macroscopic physical variable, such as the pressure on a container's wall, one generally measures the *time average* or *mean value*, defined formally as

$$\langle N \rangle \equiv \lim_{T \to \infty} \frac{1}{T} \int_0^T N(t)\, dt \tag{6.1}$$

where angle brackets denote the average value. In practice the interval T is taken large enough that the average value is independent of the finite length of the time interval selected. At any instant the actual value of $N(t)$ differs from the mean value by an amount

$$\delta N \equiv N(t) - \langle N \rangle \tag{6.2}$$

Obviously the deviation δN can be positive or negative. A characterization of the amplitude of the fluctuations in a random variable, which eliminates the sign of the deviation, is provided by its mean square deviation or *variance* σ_N^2. The variance is defined as

$$\sigma_N^2 \equiv \langle (N(t) - \langle N \rangle)^2 \rangle \equiv \langle (\delta N)^2 \rangle = \langle N^2 \rangle - \langle N \rangle^2 \tag{6.3}$$

where the angle brackets denote the appropriate time average. The equality on the right-hand side of Eq. (6.3) follows directly by expanding the square of $N(t) - \langle N \rangle$ and noting that $\langle N \langle N \rangle \rangle = \langle N \rangle^2$. The positive square root of the variance is called the *standard deviation* or *root mean square* (rms) value. Returning to the example of the pressure on the wall, we note that because of the intrinsic fluctuations in the number of molecules colliding with the wall the pressure P is expected to display variations similar to those of N. This is evident since P is linearly related to N, and therefore the variance of the fluctuations in pressure, σ_p^2, is proportional to variance of N. If instrumental capabilities are such that the pressure can be measured only to within ΔP, then the intrinsic fluctuations in pressure will go undetected if $\Delta P > \sigma_p$. This example illustrates another major problem associated with noise analysis, namely, the noise in the measuring apparatus must be less than the inherent noise in the physical variable being measured. However, if these difficulties can be surmounted, then the noise in the signals can provide useful insights concerning the underlying fundamental molecular processes. Before directing our attention to biophysical application of noise analysis, it is first necessary to discuss some elementary aspects of mathematics essential for quantifying noise.

ANALYSIS OF RANDOM SIGNALS

A random function $x(t)$ as defined above is a relation for which only the probability of the dependent variable on the independent variable can be specified. If the independent variable t is time, then $x(t)$ is called a *random signal*. For example, the current $I(t)$ through a resistor R subject to a voltage drop V will not have at each instant the time-independent value V/R, but rather $I(t)$ will take on slightly different values at different times due in part to thermal fluctuations. That is, the current will fluctuate about its time average value of V/R. Since the dependent variable does not depend in a completely defined manner on the independent variable, identical experiments do not give the same function. The collection of all possible records of $I(t)$ versus t for the same process is called an *ensemble*. Thus, in our example, the currents for similar circuits will have different current-versus-time behaviors. Three hypothetical members of the ensemble are schematically illustrated in Fig. 6.2. Because of the probabilistic nature of the signals each record for a fixed value of t can have different values, for example, in Fig. 6.2, $^1I(t_1) \neq {}^2I(t_1) \neq {}^3I(t_1)$. This difference in each record of a random process necessitates that one make a distinction between *time* averages and *ensemble* averages. This can be done as follows. Let $^kI(t)$ be the kth member of the ensemble, and let F be some *specific* function of $^kI(t)$. The function F, for example, could be the identity function $F(^kI(t)) = {}^kI(t)$ or a function that

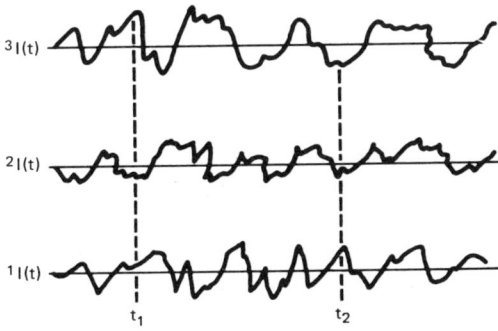

FIG. 6.2 Three hypothetical current records. The collection of all possible random signals for the same physical process is called an ensemble.

squares each value of the signal, i.e., $F(^kI(t)) = (^kI(t))^2$, or any other well-behaved function. The *ensemble average* of F for the random process represented by I at time t_1 is defined as

$$\bar{F}(t_1) \equiv \lim_{N \to \infty} \frac{1}{N} \sum_{k=1}^{N} F(^kI(t_1)) \qquad (6.4)$$

where the sum is over *all* members of the ensemble, although in practice the number of members included in the sum is taken to be large enough that the $\bar{F}(t_1)$ value is unchanged by incorporating new members from the complete ensemble into the average on the right-hand side of Eq. (6.4). Generally, ensemble averages performed at different times do not equal each other, i.e., $\bar{F}(t_1) \neq \bar{F}(t_2)$ if $t_1 \neq t_2$. An important type of random process is the *stationary random process*. These are random processes for which $\bar{F}(t_1) = \bar{F}(t_2)$ for all times t_1 and t_2 and for all functions F. That is, the statistical properties do not change with time.

In contrast to the ensemble average, the *time average* for a given specific function F for the kth representative of the ensemble is defined as

$$\langle F \rangle_k = \lim_{T \to \infty} \frac{1}{T} \int_0^T F(^kI(t)) \, dt \qquad (6.5)$$

where T in practice is made large enough that the value on the right-hand side of Eq. (6.5) is unchanged by further increases of its value. Reference to Fig. 6.2 may help clarify these two types of averaging procedures. Time averages can be thought of as averaging along the horizontal direction, whereas ensemble averaging is averaging in the vertical direction. An important subclass of all stationary random processes is the set of *ergodic processes*. These are random processes for which the ensemble and time averages are equal for all possible functions F. Henceforth, our discussion will be limited to ergodic stationary random processes since this type occurs most often in physical applications. For these random processes, the value of the integral in Eq. (6.5) is independent of the particular ensemble representative, and therefore the subscript k on the bracket expression on the left-hand side can be deleted.

Analysis of ergodic random processes is relatively simple compared to that of nonergodic processes, since for ergodic processes it is possible to construct from a single very long record of the original ensemble a new ensemble that is *statistically equivalent* to the original ensemble. Statistical equivalence means that all properties of the new ensemble are identical to those of the original ensemble. This can be done by cutting the original record into strips T sec long, as illustrated in Fig. 6.3. Each short strip constructed in this manner is considered as a member of the new ensemble. The value of T depends on the physical processes and must be selected large enough that the signal's statistical properties are repeated after T sec. This

ANALYSIS OF RANDOM SIGNALS

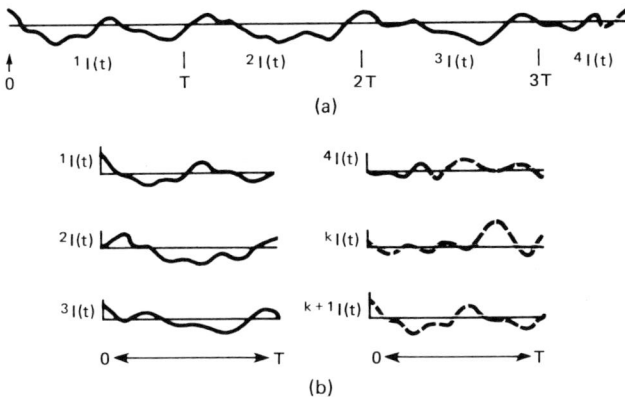

FIG. 6.3 (a) A single random noise record for an ergodic process. (b) The equivalent ensemble constructed from a single noise record.

must occur for some value of T since the original process was assumed ergodic. For nonergodic random processes, this procedure for constructing an equivalent ensemble cannot be done. In many applications experimental conditions may necessitate the use of more than a single record of the original ensemble simply because a sufficiently long single record was not obtainable.

Autocorrelation Function

Intimately related to the selection of the time interval T is the autocorrelation function of a random process. For an ergodic process $I(t)$, where the mean (time-average) value of $I(t)$ is zero, the autocorrelation function $C(\tau)$ is defined as

$$C(\tau) \equiv \langle I(t)I(t + \tau)\rangle \tag{6.6}$$

i.e., the average of the product of the functions at times t and $t + \tau$ (see Appendix E). The average is either the time average (integration over t) of a single record or the average over the ensemble. If $\langle I \rangle \neq 0$, then $I(t)$ and $I(t + \tau)$ in Eq. (6.6) are replaced, respectively, by $I(t) - \langle I \rangle$ and $I(t + \tau) - \langle I \rangle$. If the signal is not continuous but instead is digitized in time at intervals Δt apart, such as occurs in an experimental print-out, and if the record is divided into N segments, then the autocorrelation function $C(\tau)$ at $\tau = l\,\Delta t$, where l is the number of Δts, can be calculated by forming the product of the magnitudes of the signals at time points $m\,\Delta t$ and $(m + l)\,\Delta t$, $I(m\,\Delta t)I((m + l)\,\Delta t)$, where m is some integral multiple of Δt that specifies the initial time, and averaging over all such products separated from the initial time by $l\,\Delta t$. The digitized autocorrelation function is

$$C(\tau) \equiv C(l\,\Delta t) = \frac{1}{N} \sum_{m=1}^{N} I(m\,\Delta t)I((m + l)\,\Delta t) \tag{6.7}$$

6. ANALYSIS OF MEMBRANE NOISE AT SYNAPTIC JUNCTIONS

Equation (6.7) gives a definite prescription for calculating $C(\tau)$ where values of the signal I at times $(m + l)\Delta t > N \Delta t$ are replaced by the signal's value at time $(m + l - N)\Delta t$. The number of intervals N and the interval length Δt actually used are determined by how smoothly behaved the actual autocorrelation function is.

The autocorrelation function at time τ provides a measure of the similarity of the random function to itself after τ seconds. If $I(t)$ is a very irregular function such that after τ sec it is unrelated to its behavior at $\tau = 0$, then $C(\tau)$ will be zero. Hence, the autocorrelation estimates the memory content of the random signal at each time separation. For stationary ergodic processes, the value of $C(\tau)$ for sufficiently large τ must be zero since the stochastic variable takes on values unrelated to the stochastic function initial values. This behavior of $C(\tau)$ therefore gives a criterion for the selection of the time interval T employed in the construction of the *equivalent ensemble*. The time T must be selected such that $C(T)$ is very near zero.

It is evident from the definition of the variance σ^2 of an ergodic process given in Eq. (6.3) that σ^2 is equal to the value of the autocorrelation function at time $\tau = 0$; that is, $C(0) = \sigma^2$. It can be shown furthermore that $\sigma^2 \geq |C(\tau)|$ for all times τ (Bendat, 1958). A central quantity in characterizing stochastic processes is the *correlation time* τ_c. Formally, it is the minimum time for which the autocorrelation function has decreased to $1/e$ of its initial value, i.e.,

$$|C(\tau_c)| = e^{-1} C(0) \qquad (6.8)$$

and where for $\tau \geq \tau_c$, $|C(\tau)| < C(\tau_c)$. For the stochastic processes illustrated in Figs. 6.4a and b, the correlation time τ_c is longer in case (b) than in case (a). This is evident since in case (b) the signal exhibits slower oscillations than in case (a). For nonrandom oscillatory functions, e.g., $I(t) = I_0 \sin \omega_0 t$, the autocorrelation function will itself oscillate. For

$$I(t) = I_0 \sin \omega_0 t, \quad C(\tau) = I_0^2 (\cos \omega_0 \tau)/2$$

and in this case τ_c is infinite.† An example for this case is illustrated in Fig. 6.4d.

---

† This can be derived from Eq. (6.6) by using the trigonometric identity

$$\sin(\omega_0 t + \omega_0 \tau) = \sin \omega_0 t \cos \omega_0 \tau + \sin \omega_0 \tau \cos \omega_0 t$$

and noting that the time average of $\sin^2 \omega_0 t$ and $\cos \omega_0 t \sin \omega_0 t$ over a cycle are $\tfrac{1}{2}$ and 0 respectively. This can be seen as follows:

$$\frac{1}{T}\int_{\text{cycle}} \sin^2 \omega_0 t \, dt = \frac{2}{\omega_0 T}\int_0^\pi \sin^2 x \, dx = \frac{2}{\omega T}\left[\frac{1}{2}\omega t - \frac{1}{4}\sin 2\omega_0 t\right]_0^\pi = \frac{2}{\omega_0 T}\frac{\pi}{2} = \frac{1}{2}$$

since $\omega_0 = 2\pi f = 2\pi/T$. The same result is obtained for the time average of $\cos^2 \omega_0 t$. The time average of $\cos \omega_0 t$ (or $\sin \omega_0 t$) over a cycle is 0,

$$\frac{1}{T}\int_{\text{cycle}} \cos \omega_0 t \, dt = \frac{2}{\omega_0 T}\int_0^\pi \cos x \, dx = \frac{2}{\omega_0 T}[\sin \omega_0 t]_0^\pi = 0$$

and clearly the product of $\sin \omega t$ and $\cos \omega t$ averaged over a cycle is zero.

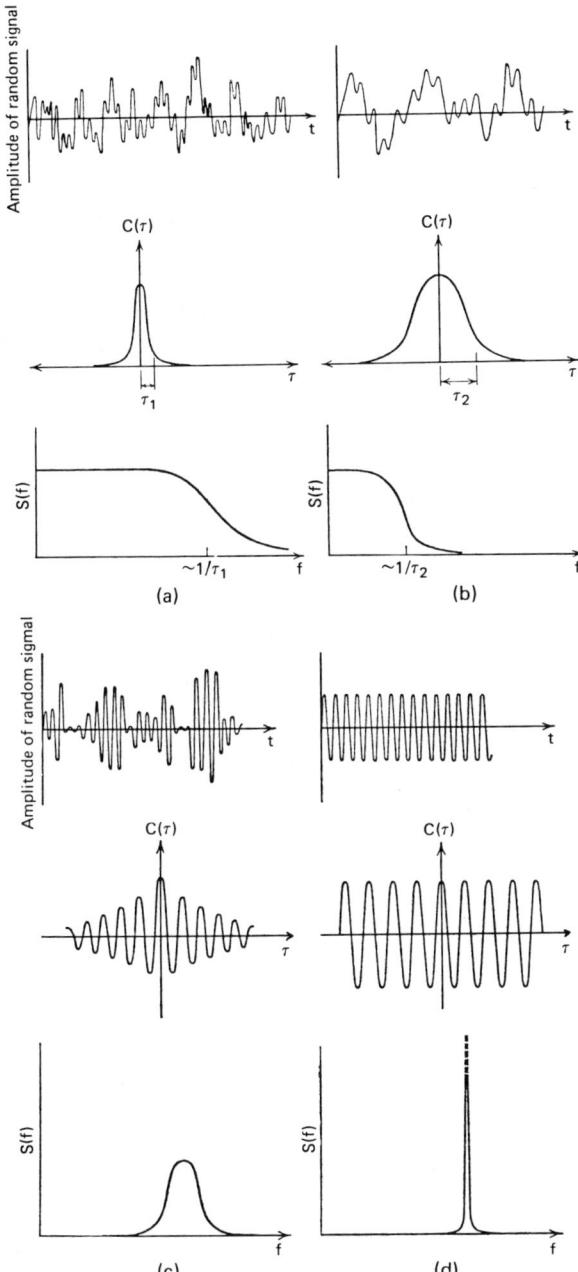

FIG. 6.4 Illustrations of random signals, their autocorrelation function $C(\tau)$ and power spectrum $S(f)$ for various types of statistical fluctuations. (a) Noise that is quite random over time intervals greater than τ_1. (b) Noise that is quite random for times $t > \tau_2$, where $\tau_2 > \tau_1$. (c) Noise generated by a lightly damped resonant system. (d) A sinusoidal wave (nonrandom) of frequency f_0. [From MacDonald (1962).]

Power Spectrum

It is important to emphasize that a stochastic function is not completely characterized by its mean and variance, since these values contain no information about how rapidly the random function varies. For example, the two random functions shown in Fig. 6.5 have the same mean and the same variance although their time course is distinctly different. One varies quite rapidly in time, whereas the other is slowly varying. A distinction between signals such as these is provided by their *power spectrum*, a function which will now be defined. For this, we shall need the concept of *Fourier series*.

Fourier series is a mathematical way of representing a periodic function by an appropriate sum of sine and cosine functions. It will be sufficient here to state Fourier's theorem as follows. If a function $I(t)$ has a periodicity of T [i.e., $I(t) = I(t + T)$], if it is well behaved (i.e., it does not contain singularities), and if the square of the function is integrable, then by proper selection of the coefficients a_n and b_n in the sum expression

$$I_N(t) = \frac{a_0}{2} + \sum_{n=1}^{N} \{a_n \cos 2\pi f_n t + b_n \sin 2\pi f_n t\} \tag{6.9}$$

where $f_n = n/T$ and T is the period, the function $I_N(t)$ will approximate the original function, and this approximation will get better as N gets larger. The frequency f_1 is called the *fundamental* frequency or *first harmonic*. Frequencies f_n with $n > 1$ are called overtones or harmonics, and $f_2, f_3, \ldots$ are referred to as the second, third, ... harmonics. Figure 6.6 illustrates how a sharp-edge square wave can be built up from smoothly varying sine waves. All coefficients (a_n, b_n) have been selected according to Eqs. (6.10) and (6.11).

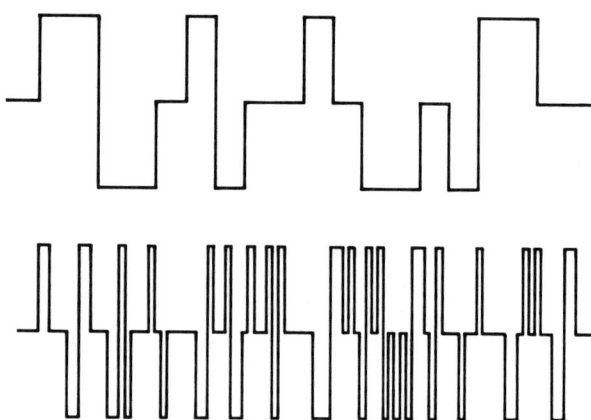

FIG. 6.5 Schematic illustration of two random functions that have the same mean and variance but differ in their time variation.

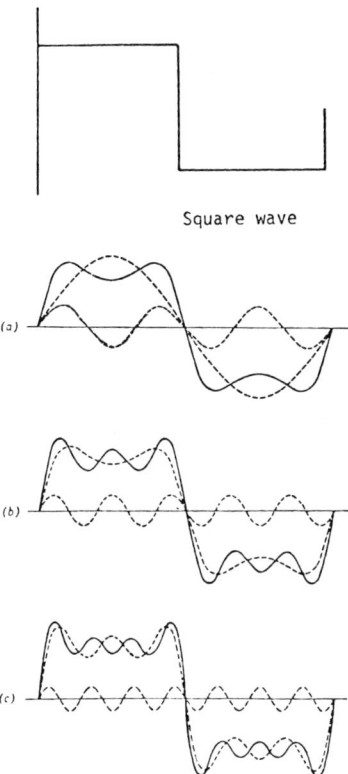

FIG. 6.6 Construction of a sharp-edged square wave with a Fourier series. (a) The addition of a wave having the fundamental frequency and its third harmonic produces a waveform shown by the solid line. (b) The addition of the fifth harmonic wave to results in (a). The solid line denotes the resulting wave due to the first, third, and fifth harmonic waves. (c) Same as (b) but with the addition of the seventh harmonic.

Note that the approximation to the original wave gets better as the number of harmonics included in the sum increases. This example also illustrates that not every harmonic of the fundamental frequency need be included in the wave. That is, some of the a_n and b_n coefficients can be equal to zero. In the case illustrated in Fig. 6.6 all even harmonic coefficients are exactly zero. The best set of coefficients, in the sense that for a fixed N, $I_N(t)$ is closest to $I(t)$, are (see Appendix F)

$$a_n = \frac{2}{T} \int_0^T I(t) \cos 2\pi f_n t \, dt \tag{6.10}$$

$$b_n = \frac{2}{T} \int_0^T I(t) \sin 2\pi f_n t \, dt \tag{6.11}$$

Coefficients for a function $I(t)$ determined in this manner are called Fourier coefficients. If the coefficients are given by Eqs. (6.10) and (6.11), then it can be shown that the mean-square-error difference between $I_N(t)$ and the function $I(t)$ vanishes in the limit of infinitely large N. This is written mathematically

$$\lim_{N \to \infty} \frac{1}{T} \int_0^T |I(t) - I_N(t)|^2 \, dt = 0$$

A convergence of this kind is often called *convergence in the mean*.

Fourier-series representations of periodic functions can also be generalized to treat stationary random processes. In this case the a and b coefficients themselves will be the statistical variables (see Rice, 1944, 1945). We shall disregard these mathematical subtleties and instead treat the random function like an ordinary function in the sense that a random signal $I(t)$ in the equivalent ensemble, which consists of random functions each of length T sec, will be written as a Fourier series, and this series will represent the function in the interval of time $0 \le t \le T$. It is easy to see from Eq. (6.9) and the footnote on p. 190 that $\langle I \rangle$, the time average of $I(t)$, is $a_0/2$; thus

$$I(t) - \langle I \rangle = \sum_{n=1}^{N} [a_n \cos 2\pi f_n t + b_n \sin 2\pi f_n t] \qquad (6.12)$$

The square of each side of this equation, averaged over the time interval T and then averaged over the ensemble gives the variance[†]

$$\sigma^2 \equiv \langle (I(t) - \langle I \rangle)^2 \rangle = \frac{1}{2} \sum_{n=1}^{N} \overline{(a_n)^2} + \overline{(b_n)^2} \qquad (6.13)$$

where the bar signifies the ensemble average.

[†] The result of squaring the sum on the right-hand side can be seen by taking two terms, $n = 1, 2$, as an example:

$$\sum_{n=1}^{2} a_n \cos 2\pi f_1 t + b_n \sin 2\pi f_n t = a_1 \cos 2\pi f_1 t + b_1 \sin 2\pi f_1 t + a_2 \cos 2\pi f_2 t + b_2 \sin 2\pi f_2 t$$

The square of this sum is

$$a_1^2 \cos^2 2\pi f_1 t + a_2^2 \cos^2 2\pi f_2 t + b_1^2 \sin^2 2\pi f_1 t + b_2^2 \sin^2 2\pi f_2 t$$

plus products of sine and cosine functions to the first power. As shown in the footnote on p. 190, the time average over a cycle of any single product of sine and cosine is zero and that of $\sin^2$ or $\cos^2$ is $\frac{1}{2}$. Thus, the result for the example of the sum of two terms is

$$\frac{a_1^2}{2} + \frac{a_2^2}{2} + \frac{b_1^2}{2} + \frac{b_2^2}{2} \quad \text{or} \quad \sum_{n=1}^{2} \frac{a_n^2 + b_n^2}{2}$$

The same type of summation results regardless of the number of terms in a summation.

It was shown in Vol. I, p. 207, that the intensity (or energy) of a wave is proportional to the square of its amplitude. When a series of Fourier components add to represent a wave function, a meaningful relation called the *spectral density* or *power spectrum* can be defined as the contribution to the total intensity (energy, or power if time rate is included) of each frequency component. This function $S(f_n)$ is represented by the square of the amplitude per unit frequency at $f_n = nf_1$, where f_1 is the fundamental frequency given by $f_1 = 1/T$, where T is the period. Note that for an ergodic process this definition gives the intensity of each frequency per unit bandwidth, which, by definition, is the fundamental f_1. Therefore $S(f_n)$ may be written

$$S(f_n) = (\overline{|a_n|^2} + \overline{|b_n|^2})/2f_1 \tag{6.14}$$

where the factor of 2 occurs from the time average as shown in the footnote on p. 194 and the bar signifies an appropriate ensemble average. If $S(f_n)$ is inserted into Eq. (6.13), then

$$\sigma^2 = \sum_{n=1}^{N} S(nf_1)f_1 \tag{6.15}$$

which in the limit, as f_1 goes to zero and $nf_1 \to f$, i.e., the time interval T of each record becomes infinitely large is, by the definition of an integral [see Eq. (G.8) in Appendix G]

$$\sigma^2 = \int_0^\infty S(f)\,df \tag{6.16}$$

Thus, the variance of an ergodic stochastic process is equal to the integral over all frequencies of the spectral density function. The spectral density function for each of the stochastic processes illustrated in Fig. 6.4 is also shown there. Note that in the case of a periodic function (Fig. 6.4d) the spectral density function is sharply peaked about the frequency of the signal and is identically zero at other frequencies. For random processes characterized by an autocorrelation function proportional to $e^{-t/\tau}$, the spectral density function is almost independent of f for frequencies less than τ^{-1} and decreases rapidly to zero for $f > \tau^{-1}$. Figures 6.4a and b illustrate random signals with this type of behavior. In the more complex case where the autocorrelation function has damped oscillator behavior $C(\tau) \propto e^{-t/\tau} \cos \omega_0 t$, as shown in Fig. 6.4c, the spectral density function is peaked about $f_0 = \omega_0/2\pi$ with a width proportional to $(2\pi\tau)^{-1}$.

Formally, the summations in Eqs. (6.12), (6.13), and (6.15) should be over an infinite number of terms, although in practice the summation is restricted to a moderately small number. This number is determined both by the length of the time interval T in each sample record and by the *sampling rate*, i.e., the time interval Δt between the points sampled in a single record. It is evident from the above discussion that the lowest frequency f_1 is determined by the length of the records, i.e., $f_1 = 1/T$. However, the highest frequency is generally determined by experimental design; by this is meant that there may be physical reasons for probing frequencies up to some prescribed maximum frequency. This maximum frequency determines the sampling rate as discussed below.

DETERMINATION OF POWER SPECTRUM
(SPECTRAL DENSITY FUNCTION)

The relation between the maximum frequency sampled and the sampling rate can be illustrated as follows. Suppose a function is sampled at equal time intervals

$$\Delta t = 1/2f_s \qquad (6.17)$$

where f_s is the sampling frequency. If there are higher frequencies $f' = f + nf_s$, where n is an integer greater than zero, and $f \leq f_s$, then this higher frequency will appear as a lower "alias" frequency in the spectral density function. This can be seen by noting that for $t = m \, \Delta t$,

$$\sin[2\pi f' t] = \sin[2\pi(f + nf_s)m/2f_s] = \sin[2\pi f t + \pi m n] = \pm \sin 2\pi f t \qquad (6.18)$$

$$\cos[2\pi f' t] = \cos[2\pi(f + nf_s)m/2f_s] = \cos[2\pi f t + \pi m n] = \pm \cos 2\pi f t. \qquad (6.19)$$

The sign on the right-hand side is determined by both the time interval t and the magnitude of the higher frequency relative to the sampling frequency. In the Fourier sum, Eq. (6.12), the terms

$$a_f \cos 2\pi f t + b_f \sin 2\pi f t + a_{f'} \cos 2\pi f' t + b_{f'} \sin 2\pi f' t$$

where f and f' have been used for notational reasons instead of subscript n, can now be rewritten with the help of Eqs. (6.18) and (6.19)

$$(a_f \pm a_{f'}) \cos 2\pi f t + (b_f \pm b_{f'}) \sin 2\pi f t$$

This demonstrates explicitly that a frequency $f' = f + nf_s > f_s$ will contribute to the Fourier components of frequencies f and thereby distort the power spectrum. A visual illustration showing schematically how *aliasing* errors can arise is shown in Fig. 6.7. To avoid this type of distortion of the spectral density function by higher-frequency components, it is essential to select the sampling frequency greater than or equal to the maximum frequency, i.e., $f_s \geq f_{\max}$. Thus the time interval should be selected such that

DETERMINATION OF POWER SPECTRUM

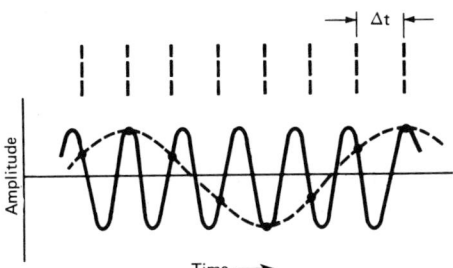

FIG. 6.7 Illustration of how slow sampling of a fast sine wave (solid line) can give the appearance of a slow wave (dotted line). Δt denotes the sampling interval. [From Bendat (1958).]

$\Delta t \leq 1/2f_{max}$. The *sampling theorem*, which is derived in Appendix H, states that the stochastic function is completely determined for all values of time by samples at the rate f_{max}, i.e., taking the fundamental interval $\Delta t = 1/2f_{max}$. A proof of this theorem is given in Appendix H. An interval selected according to this prescription is called a *Nyquist interval*. In most experimental studies one generally selects a maximum frequency to which one wishes to determine the spectral density function and then filters out all frequencies greater than this value prior to analysis of the data. This technique ensures that aliasing errors do not occur.

The spectral density function can also be determined directly from the autocorrelation function by use of the Wiener–Khintchine theorem (Kittel, 1958; Wiener, 1930). This theorem, whose proof is beyond the scope of this book, states that the power spectrum and autocorrelation of a stochastic process are Fourier cosine transforms of each other. (Fourier transforms are discussed in Appendix G.) Thus

$$S(f) = 4 \int_0^\infty C(\tau) \cos(2\pi f \tau) \, d\tau \qquad (6.20)$$

and

$$C(\tau) = \int_0^\infty S(f) \cos(2\pi f \tau) \, df \qquad (6.21)$$

As an example, note that if $C(\tau)$ is exponential, i.e., $C(\tau) = \exp(-\tau/\tau_c)$, where τ_c is the correlation time, then the Wiener–Khintchine theorem tells us that

$$S(f) = 4 \int_0^\infty \exp\left(-\frac{\tau}{\tau_c}\right) \cos(2\pi f \tau) \, d\tau = \frac{4\tau_c}{1 + (2\pi f \tau_c)^2} \qquad (6.22)$$

This equation is often written in a slightly different form. If $S(0)$ is the value of the spectral density function at $f = 0$, then Eq. (6.20) gives for this value

$$S(0) = 4 \int_0^\infty \exp\left(-\frac{\tau}{\tau_c}\right) d\tau = 4\tau_c, \qquad (6.23)$$

and thus Eq. (6.22) may be alternatively written

$$S(f) = S(0)/[1 + (2\pi f \tau_c)^2] \qquad (6.24)$$

A spectrum of this form is called *Lorentzian*. This functional form for the power spectrum is quite common in physical application since most physical systems approach their equilibrium state exponentially after having suffered a slight perturbation from their steady state. Figure 6.8 illustrates schematically the power spectrum and the autocorrelation for this case. Note that for $f < f_c$, where $f_c \sim 1/2\pi\tau_c$, $S(f)$ depends weakly on f since the exponential dominates in the integral of Eq. (6.20). A spectral density function that is independent of frequency over a fixed frequency domain is called *white* over this region. Hence a Lorentzian spectrum is white for $f < f_c$, e.g., the flat portion in the upper curve of Fig. 6.8.

Before presenting applications of noise analysis for biological systems it is helpful to examine a few classic examples that fostered the development of the technique.

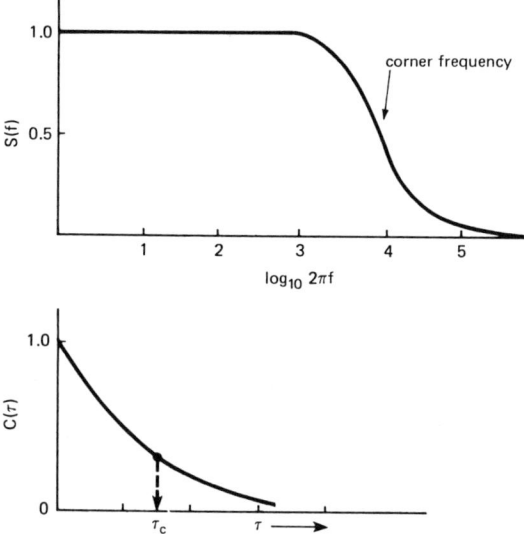

FIG. 6.8 The autocorrelation function $C(\tau)$ and its spectral density function $S(f)$ in the case where $C(\tau)$ decreases exponentially in time with a correlation time τ_c. See p. 210 for definition of corner frequency.

THERMAL NOISE

By far the most common type of noise is thermal noise, which arises from the thermal agitation of the elementary particles involved in the random process. It is always present, although it may under some circumstances make only a small contribution to the overall noise of the system. Its existence is clearly a macroscopic manifestation of the molecular behavior of matter. Thermal noise is known to be associated with all passive components and regions of electronic devices through which a conduction current flows. The Nyquist theorem (Nyquist, 1928), which gives a quantitative expression for the thermal noise generated by a system in thermal equilibrium, is of great practical utility in experimental research. In particular, it allows the experimenter to estimate the limiting signal-to-noise ratio of his experimental apparatus. The Nyquist theorem, in its simplest form, states that the variance of the *open circuit* voltage across a passive resistor of resistance R in thermal equilibrium at temperature T is given by

$$\langle (\delta V)^2 \rangle = 4RkT\,\Delta f \tag{6.25}$$

where k is Boltzmann's constant and Δf is the frequency bandwidth within which the voltage fluctuations are measured. The angle brackets denote the time average. This theorem is a quantitative statement and generalization of what many amplifier technicians had observed experimentally, namely, that "noise" increases as the input resistance is made larger. The validity of Eq. (6.25) was confirmed by Johnson (1928). Figure 6.9 illustrates the dependence of the apparent voltage variance with resistance for several different conductors. Since $\langle (\delta V)^2 \rangle / \Delta f$ is independent of frequency, the spectrum is white. The terms Johnson noise, Nyquist noise, and thermal noise are synonyms for the process described by Eq. (6.25).

The classic example of thermal noise is Brownian motion. Brownian motion is random motion of small macroscopic particles and arises from repeated bombardment of the observed particle by molecules of the surrounding medium. This type of motion is named for Robert Brown, who, in 1828, first reported on the random motion of grains of pollen dispersed in water. The correct explanation of this phenomenon was provided more than 70 years later by Albert Einstein, who showed that a macroscopic displacement could arise from the random molecular bombardment of the grains by faster molecules. Einstein showed that the average square of the displacement obeys the relation (Appendix I)

$$\overline{x^2} = 2\mu kTt \tag{6.26}$$

Here t is time, T is temperature, k is Boltzmann's constant, and μ is the mobility of the particle, i.e., μ is the limiting velocity per unit applied force.

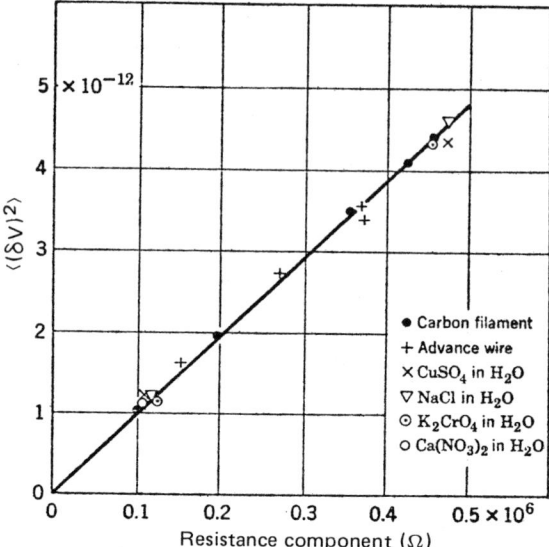

FIG. 6.9 A plot of the average square of the voltage fluctuation versus resistance for several different types of conductors. [From Johnson (1928).]

The bar signifies the ensemble average. Any particular observation of the particle position need not conform to this relation; however, the average of a sufficiently large number of independent observations does conform to this relation. An important point to note in Eq. (6.26) is that for unpredictable, random motion the root mean square displacement (standard deviation) varies as the square root of the time, whereas for motion describable by Newton's laws the displacement varies linearly with time.

All electrical conducting systems have a temperature dependence in conductivity. This dependence arises from a variety of factors such as the scattering of the charge carriers by thermal fluctuations in the medium through which they pass. In the case of electrons in a resistor, the scattering is caused by thermal vibrations of the lattice atoms in the solid through which they pass. In the case of ions in a medium, thermal effects arise both by scattering (Brownian motion) and by the fact that their diffusion velocity is also temperature dependent. Shortly after Einstein derived the Brownian-motion equation he generalized it to all types of behavior in which there is thermal motion. For example, in a solid, local groups of atoms may, for a short time, be statistically hotter than the adjacent atoms and vibrate with greater amplitudes and thereby cause more local scattering of the conducting electrons. The same is true for ion motion in a medium that may also statistically develop short-time local hot spots. Such deviations give rise to a thermal noise in any electrical measurement.

SHOT NOISE

In cases where the current is nonzero the voltage fluctuations are larger than those predicted by Eq. (6.25). A component of this enhanced noise is called *shot noise*. This type of noise, in the case of electrical circuits, arises from the finite size of the electronic charge. Consider a diode in the special case where all electrons emitted by the cathode are collected at the anode, i.e., a temperature-limited diode. In this case the random emission of electrons from the cathode will result in a corresponding random arrival of electrons at the anode. This statistical variation in the number of electrons arriving at the anode results in fluctuations in the diode current. By arguments similar to those developed in Appendix A (see also Appendix L) the emission of electrons from the cathode obeys Poisson statistics, i.e., the probability P_K of K electrons emitted in a time t is given by the relation

$$P_K = [(vt)^K/K!]e^{-vt} \tag{A.17}$$

where v denotes the statistical average number of electrons emitted per second [see Eq. (A.3)]. Thus the variance $\langle(\delta n)^2\rangle$ in the number of carriers during a time t is vt, and the charge fluctuation in the diode circuit in a time t is

$$\langle(\delta q)^2\rangle_{\text{shot}} = e^2\langle(\delta n)^2\rangle = e^2 vt = eIt \tag{6.27a}$$

where $I = ev$ is the average (direct) current. The linear dependence of the variance on time expressed by Eq. (6.27a) is quite similar to that given for thermal diffusion [see Eq. (6.26)]. For comparison, we note that the variance in electric charge due to thermal agitation (not derived) in a current-carrying element having a resistance R is given by a relation quite similar to that in Eq. (6.27a), namely,

$$\langle(\delta q)^2\rangle_{\text{thermal}} = (2kT/R)t \tag{6.27b}$$

Although both $\langle(\delta q)^2\rangle_{\text{shot}}$ and $\langle(\delta q^2\rangle_{\text{thermal}}$ have the same functional dependence on time, the origins of the fluctuations are distinctly different. Thermal noise is dependent on temperature rather than on current. The linear dependence of $\langle(\delta q)^2\rangle_{\text{shot}}$ on the mean (direct) current generally implies that shot noise is larger than thermal noise above a certain threshold current. In semiconductors shot noise arises since most of the carriers appear (detrapping) and disappear (trapping) in the sample itself instead of at the electrodes, as is the case with shot noise in vacuum tubes. In general, it might be said that thermal noise occurs from the granulated nature of the environment through which the electrons pass, while shot noise arises from the statistical fluctuations of the number of the electrons.

FLICKER NOISE

All electronic devices carrying a current, in addition to exhibiting shot and thermal noise, display noise at low frequencies in excess of that given by the above two mechanisms. A characteristic of this excess noise is that the intensity of the mean square noise current has frequency components in the power spectrum that vary approximately as the inverse of the frequency. Noise having a power spectrum varying as $f^{-\alpha}$, where $2 > \alpha \approx 1$, is called $1/f$ *noise*. The term *flicker noise* is also used to denote $1/f$ noise because of its resemblance to flicker noise occurring in vacuum tubes. Flicker noise in vacuum tubes apparently arises from sporadic variation in the average rate of electron emission from the cathode. Although fluctuations of the $1/f$ type were first observed in electric current passing through vacuum tubes, it now appears that noise having a $1/f$ spectra is ubiquitous. Spectra of this form appear not only in the frequency fluctuations of the alpha brain wave and in the fluctuation of the human heartbeat period, but also in the analysis of highway traffic flow. The diversity of phenomena displaying a $1/f$ spectra illustrates an important characteristic of noise analysis, namely, that different sources of noise can have very different origins and properties even when the spectra are apparently similar. $1/f$ noise was the first type of noise observed in biological membranes (Verveen and Derksen, 1965). In the case of voltage fluctuations in node of Ranvier membrane of myelinated nerves it appears that flicker noise is associated with the passive transport of K^+ through the membrane.

CONDUCTANCE NOISE

Conductance noise is noise in biological membranes; its origin is in the random opening and closing of individual ion channels. Measurement of conductance fluctuations have allowed for estimates of the conductance of a single open channel. This is discussed later in the chapter.

In general the measured noise in a given circuit or device is due to the combined effects of all sources of noise. Thus, in analyzing the origin of noise for a given system, it is essential to know or to be able to estimate which sources are contributing. For sources of noise that are uncorrelated, the corresponding current and voltage variances combine linearly. This can be seen as follows. Consider the case of current noise in membranes with only two sources. The fluctuations in the total current I from its mean value $\langle I \rangle$ can be written

$$\delta I(t) \equiv I - \langle I \rangle = \delta I_1(t) + \delta I_2(t) \tag{6.28}$$

where $\delta I_1(t)$ and $\delta I_2(t)$ denote fluctuations due to sources 1 and 2, respectively. The variance of the total current is

$$\langle(\delta I(t))^2\rangle = \langle(\delta I_1(t))^2\rangle + \langle(\delta I_2(t))^2\rangle + 2\langle\delta I_1(t)\delta I_2(t)\rangle \quad (6.29)$$

where the angle brackets denote the appropriate time averages as defined in Eq. (6.1). For uncorrelated noise sources, the time average of their product $\langle\delta I_1(t)\delta I_2(t)\rangle$ equals zero, and thus the additions of the variance of each noise source gives the total variance in the signal.

CURRENT NOISE AT THE END PLATE

An example illustrating the usefulness of fluctuation analysis is the well-studied current noise at the neuromuscular end plate induced by the iontophoretic application of the neurotransmitter, measured under voltage-clamped conditions (see p. 105). Even in the absence of externally applied acetylcholine, there will be some small spontaneous changes in the recorded end-plate current when a constant membrane voltage is maintained. This noise is in addition to the spontaneous miniature end-plate currents discussed in Chapter 4. This extraneous noise is primarily instrumental, arising from the electrodes, the voltage clamp, and the associated electronics, with only a small contribution from the spontaneous opening and closing of single channels. This noise varies with the membrane potential both in magnitude and frequency composition. Therefore in any proper analysis of the induced current noise, other sources of noise must be reduced to a small value compared to that to be analyzed. With modern electronics and good experimental technique these extraneous noises have been reduced to a small value compared to the current and associated noise arising from the iontophoretic flow of ACh. Thus these noise spectral components can be measured and analyzed for each spectral component. The following discussion applies to the frog neuromuscular end plate, although it could equally well be applied at other synapses.

The application of a constant amount of acetylcholine increases the number of open channels because of the enhanced collision rate of the transmitter molecules and its receptors. Concurrent with this increase in the number of open channels there is an increase in the amount of noise. Experimetal data illustrating this noise increase with the iontophoretic current are shown in Fig. 6.10. This figure shows a comparison of the membrane current in the absence (indicated by the label "Rest") and presence of acetylcholine. In the case shown here a constant iontophoretic current of 10 nA applied near the postsynaptic membrane produced a mean current of 120 nA (see lower portion of Fig. 6.10) and increased the root mean square noise from 0.07 nA, measured in the absence of ACh, to 0.25 nA. This increase in the

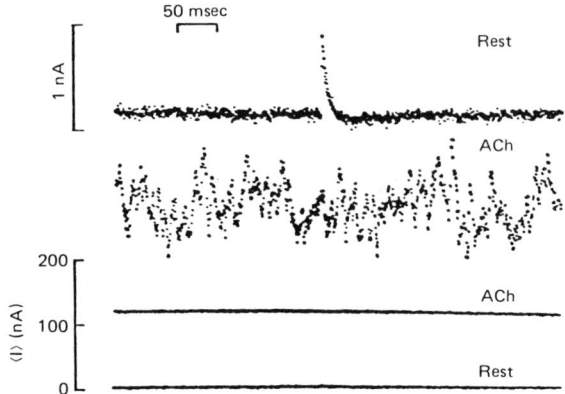

FIG. 6.10 Membrane current in ethylene glycol treated muscle at 8°C measured under voltage clamped conditions. Upper traces: high gain with 1 nA scale as indicated. Lower trace: low gain. The control traces upper and lower denoted by Rest is the current record in the absence of iontophoretically applied ACh. The upper high-gain trace (at rest) shows a spontaneous miniature end-plate current. The two traces labeled ACh are the current record when a small subthreshold (below the amount needed to initiate muscle contraction) amount of acetylcholine is applied. The high-gain record demonstrates the increase in the rapid, irregular fluctuation of membrane current in the presence of ACh. [From Anderson and Stevens (1973).]

amplitude of the variations in the membrane current arises from the random variation in the number of channels open at a given time. To analyze this observation in greater detail, we note that the fixed holding potential (i.e., membrane potential V) and the fixed iontophoretic current I_p do not completely specify the microscopic state of the membrane. Some microstates (i.e., different configurations) will have the same total number of open channels, but will differ in which channels are actually open. Cases (a) and (b) in Fig. 6.11 are examples of this particular situation. Other microstates will have slightly different numbers of open channels, as illustrated in Fig. 6.11c. This stochastic variation in channel number as a function of time is schematically illustrated in Fig. 6.12a. For simplicity, this diagram has been drawn as though the open and closing times are infinitely fast. For a given number K, a function P_K can be defined giving the probability that a membrane segment having a total of N channels will have exactly K channels open. Figure 6.12b is a schematic plot of how the number of open channels is distributed about the average number $\langle K \rangle$ of open channels. P_K is often called the *probability density function*, i.e., P_K is the probability distribution of the deviations (δN) about the mean number of open channels. In Appendix A it is proved that

$$P_K = \frac{N!}{K!(N-K)!} p^K (1-p)^{N-K} \qquad (A.1)$$

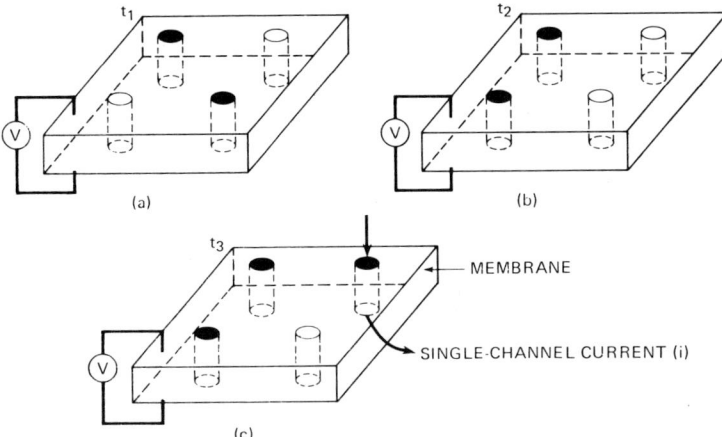

FIG. 6.11 Three different microstates for a membrane having a holding potential V and a constant iontophoretic current I_p (not indicated). Channels indicated by small cylinders in the membrane. Those channels having solid top are open, the remaining channels are closed. (a) and (b) have the same number of open channels but are different microstates because of their different spatial arrangement. Configuration labeled (c), occurring at time t_3, differs because it has three open channels as compared to two in cases (a) and (b).

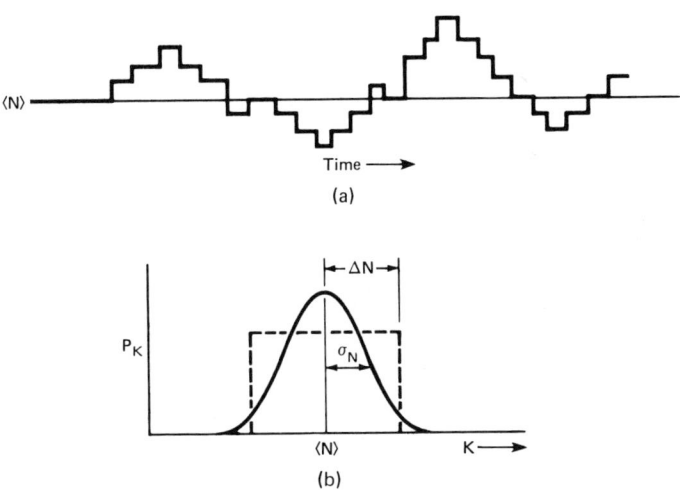

FIG. 6.12 (a) Schematic illustration of the variation in the number of open channels as a function of time for a given applied iontophoretic current. $\langle N \rangle$ denotes the mean number of channels open. (b) Distribution function P_K of the deviations about the mean number of open channels. The standard deviation of the distribution is denoted by σ_N. N is the total number of channels, and $\langle N \rangle$ describes the mean number of open channels. ΔN is a hypothetical measured error (see p. 208).

where p denotes the probability that a single channel is open. If, however, N is very large and p is finite and constant, then the binomial distribution is well approximated by a *Gaussian* distribution, i.e.,

$$P_K = \frac{1}{\sqrt{2\pi}\,\sigma_K} \exp\left\{-\frac{(K - \langle K \rangle)^2}{2\sigma_K^2}\right\} \tag{6.30}$$

where σ_K^2 is the variance and $\langle K \rangle$ is the mean of the distribution. This approximation becomes an identity in the limit as N goes to infinity as follows from the *central limit theorem* (p. 61). This theorem states that the probability distribution of the sum of statistically independent random variables appropriately normalized, tends to become Gaussian as the number of independent variables is increased without limit (Feller, 1966, p. 253). Hence the solid line in Fig. 6.12b is well approximated by a Gaussian. Random processes for which the deviations are distributive in a Gaussian manner about the mean are called *Gaussian random processes*.

Consider a membrane held at a fixed potential V and where a fixed current I_p is applied to the iontophoretic electrode. Let N be the total number of channels in the membrane. If each channel has only two states, open and closed, and the open state has a conductance γ, then the mean transmembrane current is equal to the average number of open channels times the channel conductance times the driving potential, $(V - E_s)$, where E_s is the reversal potential (see p. 104). Thus the mean current $\langle I \rangle$ can be written

$$\langle I \rangle = \langle K \rangle \gamma (\text{driving potential})$$
$$\langle I \rangle = \langle K \rangle \gamma (V - E_s) \tag{6.31}$$
$$\langle I \rangle = N p \gamma (V - E_s)$$

where V is the potential at which the membrane is clamped and p, the probability that a given channel is open, is by definition equal to the mean number of open channels $\langle K \rangle$ divided by the total number of channels N, i.e., $p = \langle K \rangle / N$. If channels are considered as noninteracting, then from the preceding discussion the distribution function obeys binomial statistics (see Vol. I, p. 360). Hence the variance in the distribution of open channels is

$$\sigma_N^2 = \langle (\delta N)^2 \rangle = \langle K \rangle (1 - p) = Np(1 - p) \tag{A.12}$$

as shown in Appendix A [Eq. (A.12)]. V and E_s are constant in any given experiment, and the instantaneous current–voltage relation is linear, i.e., $I = G(V - E_s)$, where G is membrane conductance. It follows therefore that the relation between the variance in current and conductance is linear, i.e.,

$$\sigma_I^2 = \sigma_G^2 (V - E_s)^2 \tag{6.32}$$

The assumption of independent channels implies that the conductance at any instant of time t is $G(t) = N(t)\gamma$, where $N(t)$ represents the number of open channels at time t. Thus it follows that the variance in the conductance

$$\sigma_G{}^2 \equiv \langle(\delta G)^2\rangle = \langle(\delta N)^2\rangle\gamma^2 \tag{6.33}$$

$$\sigma_G{}^2 = Np(1-p)\gamma^2 \tag{6.34}$$

The mean conductance according to Eq. (6.31) is $Np\gamma$, which in conjunction with Eq. (6.34) gives

$$\sigma_G{}^2 = \gamma(1-p)\langle G\rangle \tag{6.35}$$

For $p \ll 1$, a limit that is valid if the current applied to the iontophoretic electrode is small and therefore realizable since it is under the control of the experimenter, we may write

$$\sigma_G{}^2 = \gamma\langle G\rangle. \tag{6.36}$$

The validity of this linear relation for the frog end plate for a wide range of voltage-clamped membrane potentials, sometimes called *holding potentials* since a feedback electronic circuit keeps the membrane potential constant, is shown in Fig. 6.13. The slope of the curve provides a value of the single-channel conductance. Not only does this experiment give a value for the single-open-channel conductance γ equal to $20 \times 10^{-12} \, \Omega^{-1}$, but also shows that γ is voltage independent at least over the ranges of voltages studied. This

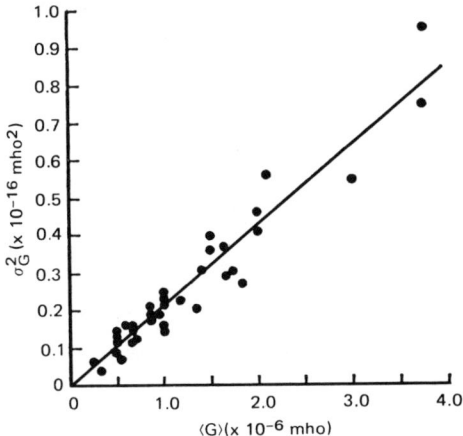

FIG. 6.13 Variance of conductance fluctuations ($\sigma_G{}^2$) as a function of mean end-plate conductance ($\langle G\rangle$) produced by iontophoretic application of acetylcholine. Holding potentials were between -140 and 60 mV. Solid dots indicate experimental points. The slope of the curve gives a single channel conductance of 20×10^{-12} ohm^{-1}. [From Anderson and Stevens (1973).]

can be taken to indicate that after a channel has opened it does not open any more by altering the membrane potential.

Equation (6.36) is remarkably simple, and for this reason it might be thought that its applicability to physical systems is straightforward. Unfortunately this is not the case, since as indicated on p. 186 it is essential that the instrumental noise be less than the noise generated by the random opening and closing of channels. This can be made more apparent as follows. From Eqs. (6.31) and (6.34) it follows that

$$\frac{\sigma_G}{\langle G \rangle} = \frac{\sqrt{Np(1-p)}}{Np\gamma}\gamma \approx \frac{1}{\sqrt{Np}} = \frac{\sigma_N}{\langle N \rangle} \quad (p \ll 1) \quad (6.37)$$

Thus if the instrumental noise produces an error of ΔG in the measurement of conductance, then, for observation of noise generated by the statistics of channel openings, it is essential that $\Delta G < \sigma_G$. This relation implies that

$$\Delta G/\langle G \rangle < \sigma_G/\langle G \rangle \approx 1/\sqrt{N} \quad (6.38)$$

where N is the total number of channels, must be valid. Hence a large N means that instrumental noise, noise other than conductance noise, must be exceedingly small. For example, if conductance can be measured to only 0.1%, then observation of conductance noise will be possible only if

$$\Delta G/\langle G \rangle = 0.1\% < 1/\sqrt{N} \quad (6.39)$$

Thus $\sqrt{N}$ must be less than 10^3, which implies that the total number of channels in the system must be less than one million. Fortunately, membranes being two-dimensional physical systems enhances the likelihood of a small total number of channels. The dotted curve in Fig. 6.12b illustrates a situation where uncertainty (the measured variance) ΔN in channel number is so large that the inequality in Eq. (6.38) does not hold. In this case one would not be able to infer the channel conductance γ from noise measurements.

FREQUENCY COMPOSITION

The variance of a random function provides information only about the amplitude of the fluctuations; it does not contain information about the rate at which departures from the mean value occur. For this reason, measurements of the mean conductance and its variance provide information only about time-independent properties of channels, e.g., its conductance; it does not allow information to be inferred about rates of reactions. Properties of time-dependent rates can, however, be deduced from temporal properties of conductance noise. As discussed earlier in this chapter, the temporal characterization of the fluctuations is provided by its frequency composition, namely, its power spectrum.

FREQUENCY COMPOSITION

In a typical experiment a current record such as shown in Fig. 6.10 might be recorded for a long time. Such a record technically represents only one member from an ensemble of all possible records. However, if the random process is *stationary* and *ergodic*, it is possible to form a statistically equivalent ensemble from a single record by cutting the original record into a sequence of strips of length T sec. The strip length must be selected large enough that the record's behavior during the first portion of the strip, times near zero, is uncorrelated with events for times greater than T (p. 188). This can be stated more formally by demanding that the autocorrelation function be zero or nearly zero for times larger than T. For acetylcholine-activated channels at end plates, this correlation time is the order of 1 msec, as shown below. In fact, in the conformation molecular model (see p. 125) for the generation of end-plate currents, the current correlation time is identical to the mean channel lifetime. Thus a current record 10 sec long is more than sufficient. Having constructed an ensemble of such records, each 10 sec long, a computer samples each record at a finite number of points. We shall, for convenience, take the number of sample points to be 10,000. This corresponds to a sampling rate of 1 kHz. Algorithms used in the evaluation of Fourier coefficients work most efficiently when the number of points is 2^n, so N (the total number of data points) is invariably selected to be 4096, 8192, etc. This is because the most frequently used algorithm for computing Fourier coefficients is that of Cooley and Tukey (1965), which is basically a clever mathematical technique for obtaining Fourier transforms in a minimal number of computer steps. It works best where the number of points is equal to a power of 2. This selection is generally referred to as 4K, 8K, etc. Our choice of 10,000 is for purposes of illustration only. For a sampling rate of 1 kHz the highest-frequency component of the spectral function sampled is 500 Hz, i.e., one-half the sampling frequency, and the lowest frequency component sampled is 0.1 Hz or the reciprocal of the sample record length. Frequencies greater than 500 Hz must be filtered from the original current records before sampling so as to eliminate *aliasing errors* (p. 196). For each current record the 10^4 points determine the coefficients in the Fourier expansion of the current deviations [see Eqs. (6.10)–(6.12)]. This is written formally

$$\delta I(t) = I(t) - \langle I \rangle = \sum_{n=1}^{5000} \{a_n \cos 2\pi f_n t + b_n \sin 2\pi f_n t\} \qquad (6.40)$$

where $\langle I \rangle$ is the average current, $f_n = nf_1$, $n = 1, 2, \ldots$ and f_1 is the lowest frequency sampled. Computer programs using both the intrinsic properties of digital computers and general mathematical relation of Fourier series (Cooley and Tukey, 1965) known as *fast Fourier transforms* (FFT) allow experimenters to calculate quickly the coefficients a_n and b_n for each record of the ensemble. The power spectrum, or spectral density $S(f_n)$, at frequency f_n is then calculated by performing the appropriate ensemble averages. Defining the frequency bandwidth Δf_n as equal to the separation between two adjacent frequencies

$$\Delta f_n = f_{n+1} - f_n = \frac{n+1}{T} - \frac{n}{T} = \frac{1}{T} \quad (\text{sec}^{-1}) \tag{6.41}$$

allows the power spectrum to be written

$$S(f_n) = \frac{\overline{|a_n|^2} + \overline{|b_n|^2}}{2 \Delta f_n} \quad (A^2 \text{ sec}) \tag{6.42}$$

where the bar signifies the average over all members of the ensemble, i.e.,

$$\overline{|a_n|^2} = \frac{1}{M} \sum_{i=1}^{M} |a_n^i|^2 \tag{6.43}$$

where a_n^i is the a_n coefficient for the ith record and M is the number of records in the ensemble. A similar equation holds for $\overline{|b_n|^2}$. M must be taken large enough in order to obtain reliable estimates of $\overline{|a_n|^2}$ and $\overline{|b_n|^2}$. Fortunately in many biological applications M is actually a small number; in some experiments $M = 20$ is sufficient. Representative examples of acetylcholine-induced end-plate current power spectra are shown in Fig. 6.14. Solid dots denote the experimental points, and the smooth curves superimposed on the experimental data were calculated using the formula given by Eq. (6.24):

$$S(f) = S(0)/[1 + (2\pi f/\bar{\alpha})^2] \tag{6.44}$$

where $\bar{\alpha} = 2\pi f_c$, f_c is called the cutoff frequency or corner frequency and is defined as that frequency at which $S(f)$ has decreased to one-half of its zero frequency value,

$$S(f_c) = \tfrac{1}{2} S(0) \tag{6.45}$$

Interpretation of the zero-frequency spectral component $S(0)$ and the parameter $\bar{\alpha}$ depend on the theoretical model employed to derive Eq. (6.44). A function having the general form given in Eq. (6.44) is called Lorentzian, and such a functional dependence unfortunately does not uniquely specify the molecular events involved in a physical process. In fact a large number of quite different molecular models can have a Lorentzian power spectrum. Therefore, inferences about the underlying mechanisms from the power spectrum depend critically on the assumptions of the system's physical properties.

A particularly useful molecular model in interpreting the power spectrum is the conformational transmitter–receptor model of Anderson and Stevens (1973). As discussed in Chapter 4, this model assumes that the binding of

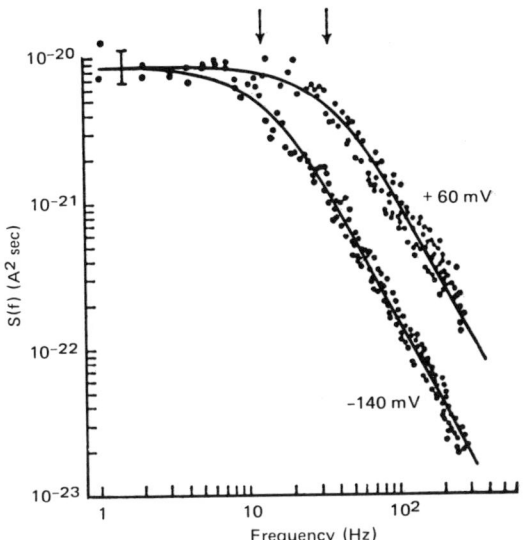

FIG. 6.14 Current spectral function measured at two different membrane potentials for ethylene glycol treated muscle at 8°C. Vertical axis label applies to a holding membrane potential of −140 mV. The 60-mV curve was shifted to facilitate comparison. Arrows denote cutoff frequencies $f_c = \alpha/2\pi$. [From Anderson and Stevens (1973).]

ACh to the receptor is rapid and voltage independent. The rate-limiting step in Eq. (4.2) is the closing of the open channels. This particular model gives theoretical values for the parameters $S(0)$ and $\bar{\alpha}$ consistent with other known experimental data on end-plate potentials. This consistency offers some support for the correctness of the model. The mathematical details involved in deriving the spectral density function for the conformation model are given in Appendix J, where it is shown that

$$S(f) = \frac{2\langle I \rangle \gamma (V - E_s)/\alpha}{1 + (2\pi f/\alpha)^2} \tag{6.46}$$

Here α is the channel closing rate, γ the open single-channel conductance, $\langle I \rangle$ the mean membrane current, and E_s the reversal potential. Comparison of Eq. (6.46) with Eq. (6.44) shows that for the conformational model of Stevens, the following two relations are valid: (a) average closing channel rate

$$\alpha = 2\pi f_c = \bar{\alpha} \tag{6.47}$$

and (b)

$$S(0) = 2\langle I \rangle \gamma (V - E_s)/\alpha \tag{6.48}$$

From the relations given in Eqs. (6.47) and (6.48), a value of γ can be inferred that is in good agreement with that determined from a plot of conductance variance versus the mean conductance (Fig. 6.13). In addition, since the closing channel rate α is proportional to e^{AV}, where A is some constant and V is the membrane potential [see Eq. (4.38)], it follows from the relation given in Eq. (6.47) that a similar dependence of the cutoff frequency on membrane potential exists. This prediction is in accord with experimental results.

CHANNELS AT THE NEUROMUSCULAR JUNCTION

In Chapter 4 the end-plate currents generated at the frog neuromuscular junction were discussed. Arguments were presented demonstrating that acetylcholine acting on the postsynaptic membrane opened transient membrane channels through which primarily sodium and potassium ions flowed. The nature of the chemically gated endplate channels was not examined. Whether the channels for K^+ and Na^+ are kinetically independent or whether these ions are transported by the same channel was not addressed. The distinction between these two possible cases, however, can be made quite simply by employing the techniques of fluctuation analysis to iontophoretically induced endplate currents (Dionne and Ruff, 1977). The essential idea is to apply a constant amount of acetylcholine iontophoretically to the postsynaptic membrane and then to use voltage-clamp techniques to measure the mean current $\langle I \rangle$ and its variance σ_I^2 for a range of different membrane potentials V. Fortunately the functional behavior of $\langle I \rangle$ and σ_I^2 on the holding potential is different in the two cases. Let us consider each case separately.

Case 1. If both K^+ and Na^+ pass through kinetically indistinguishable channels, then the mean current is, according to Eq. (6.31),

$$\langle I \rangle = Np\gamma(V - E_s)$$

where E_s is the reversal potential, γ the conductance of the open channel, N the number of channels, and p the probability that a single channel is open. The parameter p is a complex function of the acetylcholine concentration whose exact dependence need not concern us here. For the single channel, the variance in the current, by combining Eqs. (6.32) and (6.34), is simply

$$\sigma_I^2 = \gamma^2 p(1 - p)N(V - E_s)^2 \tag{6.49}$$

The important point to note from the above two equations is that for the single-channel model both the mean current and its variance vanish at the membrane reversal potential E_s. This is an important conclusion since if the potassium and sodium ions are transported by separate and distinct

CHANNELS AT THE NEUROMUSCULAR JUNCTION 213

channels whose gating properties are different, this relation no longer holds, as seen below.

Case 2. In the two-channel model, where each distinct type of ion has its own channel, the mean total current is simply equal to the sum of the currents carried by each channel. Thus

$$\langle I \rangle = \langle I_K \rangle + \langle I_{Na} \rangle \qquad (6.50)$$

$$\langle I \rangle = \gamma_{Na} p_{Na} N_{Na}(V - E_{Na}) + \gamma_K p_K N_K(V - E_K) \qquad (6.51)$$

where γ_{Na} and γ_K are the open-channel conductances for each ion, p_{Na} and p_K the probabilities that the sodium and potassium channels are open, and N_{Na} and N_K the total number of sodium and potassium channels. E_K and E_{Na} are the equilibrium (Nernst) potentials for K^+ and Na^+ in frog muscle and are known from other experiments to have values of -100 and $+40$ mV, respectively. For noninteracting channels, the variance of the total current is simply the sum of the variances for each current component. This follows from Eq. (6.29) since the random current variables $I_K(t)$ and $I_{Na}(t)$ are uncorrelated for noninteracting channels (see p. 203). Thus the total current variance is

$$\sigma_I^2 = \sigma_K^2 + \sigma_{Na}^2 \qquad (6.52)$$

where

$$\sigma_K^2 = \gamma_K^2 N_K p_K (1 - p_K)(V - E_K)^2$$

and

$$\sigma_{Na}^2 = \gamma_{Na}^2 N_{Na} p_{Na} (1 - p_{Na})(V - E_{Na})^2$$

From these equations it is evident that in the independent two-distinguishable-channel model the variance in the total current is nonzero at the reversal potential since $E_s \neq E_K$ and $E_s \neq E_{Na}$ (see p. 109) and both terms in Eq. (6.52) are positive.

The experimental mean end-plate current (EPC) and its variance σ^2 as a function of the membrane potential V measured in muscle cells of *Rana pipiens* (a particular type of small frog), is shown in Fig. 6.15. Note that for $V = E_s$, which has a value near zero for this particular end plate, both the mean current and its variance become equal to zero. This result is inconsistent with a model of the postsynaptic membrane in which Na^+ and K^+ ions pass through separately gated distinguishable end-plate channels. The result does not, however, rule out the possibility of K^+ and Na^+ ions being transported by different pores that are gated the same. Of the models illustrated in Fig. 6.16 the analysis of the data in Fig. 6.15 eliminates only case

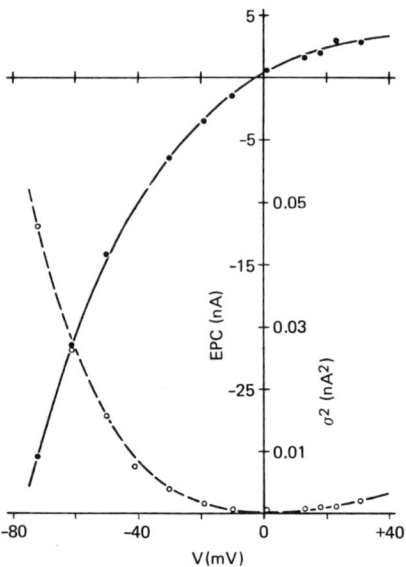

FIG. 6.15 The mean end-plate current EPC (solid dots) and the variance, σ^2 (open circles) as a function of the membrane potential V. [Reprinted by permission from Dionne and Ruff, *Nature* **266**, 363 (1977). Copyright © 1977 Macmillan Journals Limited.]

(c). This property of ion channels at the end plate is quite different from the gating property of K^+ and Na^+ channels in the generation of an action potential. For the axon, each different type of ion channel has a different time and voltage dependence (see Vol. I, Chapter 3). In addition, each channel responds differently to a variety of drugs. For example, tetrodotoxin (TTX), a poisonous substance in the ovary and liver of the pufferfish (Vol. I, p. 85), blocks the Na channel in axons but is ineffective in modifying the trans-

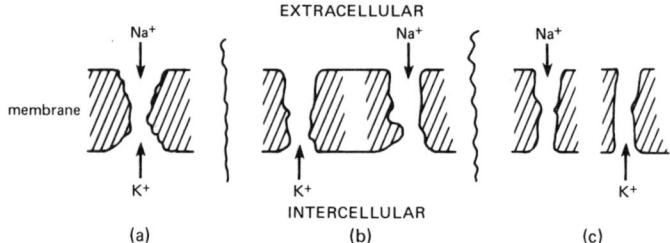

FIG. 6.16 (a) Ions pass through the same pore. (b) Different ions through different pores but gated the same. (c) Different ions through different pores but gated differently.

membrane transport characteristics of potassium ions. In fact, it is this difference in the behavior of K^+ and Na^+ channels in axons with respect to TTX that allowed for a pharmacological separation of the axon membrane currents into various ion components. For K^+ and Na^+ end-plate currents, such a separation appears not to be possible.

The experimental data reported in Fig. 6.15 can also be used to infer a value for the open-channel conductance. A relation to calculate γ can be derived by taking the ratio of Eqs. (6.49) and (6.31). Doing this and assuming $p \ll 1$, a condition that can be satisfied if the concentration of acetylcholine is low, gives

$$\sigma_I^2/\langle I \rangle = \gamma(V - E_s) \tag{6.53}$$

For $V = -70$ mV, $E_s \approx 0$ mV and the experimental values of $\sigma_I^2 \approx 0.047 \times 10^{-18}$ A^2 and $\langle I \rangle \approx -30 \times 10^{-9}$ A at this holding potential Eq. (6.53) gives $\gamma = 22 \times 10^{-12}$ Ω^{-1}, a value in accord with that determined from the slope of σ_G^2 versus $\langle G \rangle$ shown in Fig. 6.13.

The reasonableness of this value for γ is supported by calculating the average number of ions transported per second by a single channel and comparing its value with that expected for the arrival of ions at the mouth of a pore using Fick's law of diffusion (see Vol. I, p. 61). The average number of ions transported per sec $\langle M \rangle$ is simply

$$\langle M \rangle = \frac{\langle I \rangle}{e} = \frac{\gamma}{e}(V - E_s)$$

$$\approx \frac{2.2 \times 10^{-11} \Omega^{-1} \times 90 \times 10^{-3} \text{ V}}{1.6 \times 10^{-19} \text{ C}} \approx 1.2 \times 10^7 \text{ sec}^{-1} \tag{6.54}$$

for a membrane potential of -90 mV. An estimate for the ion arrival rate at the mouth of a channel owing to purely diffusional motion involves solving the diffusion equation for the ions and calculating the flux of ions through a hemispherical sink (S) at one end of the pore. If the ionic channel is taken to be a cylindrical pore 3 Å in radius (R), then the flux of ions at the channel entrance is

$$\varphi = 2\pi RDn \tag{6.55}$$

where D is the ion diffusion coefficient, typically on the order of 10^{-5} cm^2 sec^{-1}, and n is the ionic concentration at S, which is at least 10^{-4} mol cm^{-3}. Thus $\varphi \approx 1.3 \times 10^8$ ions sec^{-1}, a rate ten times that estimated using Eq. (6.54) and the experimental value of the open-channel conductance. Hence the random motion of the ions in the medium alone is sufficient to supply the entrance of a conducting channel. No special mechanisms are needed.

EFFECT OF PORE STRUCTURE

Although we have represented the channel as an aqueous pore, the actual situation is considerably more complex. The microenvironment surrounding the path that the ion transverses is quite heterogeneous, having regions of charged and uncharged molecular moieties. Therefore, the interior of each type of channel presents a different local environment to each type of ion. As a consequence a "free" ion at each spatial region along the membrane channel passageway has a different energy, some regions being local free-energy minima and other regions exhibiting local free-energy maxima. Permeating ions therefore do not just glide freely through a channel subjected to a constant force, as originally assumed in the constant-field equation of Goldman (see Vol. I, p. 67), but instead, the ion hops from one energy minimum to another, a mode of transport quite analogous to ion and electron movement in a disordered solid. Figure 6.17 illustrates schematically an energy-versus-distance profile for a two-barrier model. Obviously models with three, four, or more barriers could be envisioned, although experimental data are currently lacking to support these more complicated models. The actual heights of the barrier, as suggested by experimental studies of Lewis and Stevens (1979), are strongly dependent on the ion being transported. For example, the solid line in Fig. 6.17 could represent the energy profile that a calcium ion experiences in the channel, whereas the dotted line could represent the energy profile for a sodium ion. The rate of transfer between free-energy minima depends exponentially on the barrier height. The arguments for this exponential dependence on barrier height are similar to those developed on p. 128 for the dependence of the channel-closing rate α on

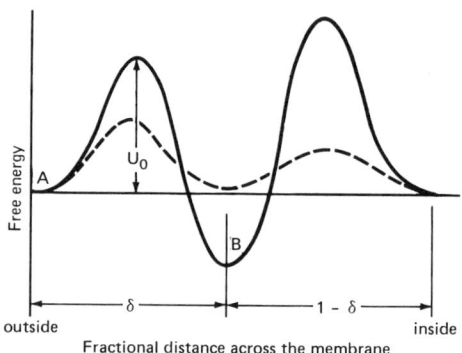

FIG. 6.17 The free-energy profile for a hypothetical channel with two barriers and an energy well located a fraction δ units in distance across the membrane. Solid line denotes energy profile for one type of ion whereas the dashed line is the energy profile for a different ion. [Modified from Lewis and Stevens (1979).]

membrane potential. For calcium ions, therefore, the rate of transfer from site A to site B is proportional to $\exp(-U_0/kT)$, where k is Boltzmann's constant, T the temperature, and U_0 the barrier height, as shown in Fig. 6.17. Thus lower-energy barriers imply faster transport. For the case illustrated in Fig. 6.17, sodium ions pass more readily through the channel than do calcium ions. The chemical microenvironment of the channel therefore determines the single-ion channel conductance, whereas ion selectivity of a channel is thought to be determined not only by the chemical environment at the entrance of a channel but also by the geometrical size of the smallest opening within the channel. Large ions and molecules are excluded by size alone from being transported, as is the case with local anesthetics.

The selectivity of a channel is usually expressed as a ratio of the substituted ion's permeability P_X to that of sodium, P_{Na} (see Vol. I, p. 62, for the definition of ion permeability). One method of determining this ratio is by measurement of reversal potential changes after substituting Na^+ with another monovalent cation X^+ and using the relation

$$E_r(X) - E_r(Na) = \frac{RT}{F} \ln\left[\frac{P_X[X^+]_0}{P_{Na}[Na^+]_0}\right]$$

an equation derivable from the Goldman equation (see Vol. I, p. 71). Here $E_r(X) - E_r(Na)$ is the difference in the reversal potential after cation substitution and $[X^+]_0$ and $[Na^+]_0$ are the extracellular ion concentrations (activities). The ion selectivity for the Na channel for three different preparations is shown in Table 6.1. The table illustrates the remarkable similarity of the sodium channel filtering properties among the three preparations

TABLE 6.1

Selectivity of Sodium Channel Expressed as the Permeability Ratio P_X/P_{Na}[a]

X(ion)	Frog nerve	Frog muscle	Squid axon
Sodium	1.0	1.0	1.0
Hydroxylammonium	0.94	0.94	—
Lithium	0.93	0.94	1.1
Ammonium	0.16	0.11	0.27
Guanidinium	0.13	0.093	0.25
Potassium	0.086	0.048	0.083
Cesium	<0.013	—	0.016
Rubidium	<0.012	—	0.025
Choline	<0.007	—	0.014

[a] From Ulbricht (1977).

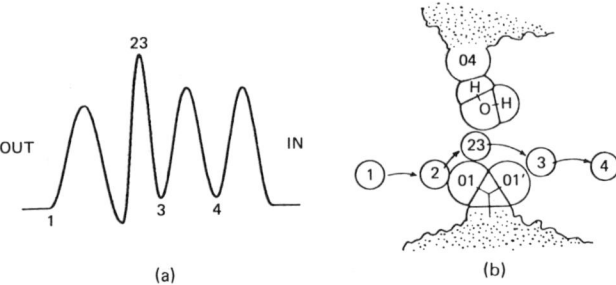

FIG. 6.18 A molecular model for the selectivity of the sodium ion channel at the nodes of Ranvier (Vol. I, p. 79). (a) Energy profile of the sodium channel. (b) Molecular description of the channel. Numbers in (b) correspond to the energy configurations in (a). Labels O1, O1', and O4 denote oxygen atoms. [Reproduced from Hille, *J. Gen. Physiol.* **66**, 535 (1975) by copyright permission of the Rockefeller University Press.]

and the large differences among different cations. The most extensively studied channel is the sodium channel of frog single myelinated nerve fibers. Experiments for this preparation have shown that as the pH of the extracellular medium is lowered, the channel is reversibly blocked. From such studies and the channel ion-selective properties, Hille (1975) has proposed a rather detailed molecular model of the channel's interior, illustrated in Fig. 6.18. The channel is thought to be lined with eight oxygen atoms and has a cross-sectional area at its narrowest opening of 3.1 × 5.1 Å. Figure 6.18b shows an ion of crystal size of Na^+ as it transverses the channel, whereas Fig. 6.18a gives the corresponding free energies of these molecular positions. In the external solution (position 1) the ion is fully *hydrated*. By hydration one means the shell of water molecules that are complexed to the metal ion and that move with the metal ion through the aqueous medium. For the sodium ion, roughly 4.5 water molecules (on the average) form its hydration shell (see Table 6.2). Hydration of an ion increases its size. The greater the number of bound H_2O

TABLE 6.2

Mobility and Size of Some Monovalent Ions[a]

Ion	Ionic crystal radius (Å)	Absolute mobility in H_2O (μm/sec)/(V/cm)	Hydration number
Li^+	0.6	4.01	6
Na^+	0.95	5.2	4.5
K^+	1.33	7.64	2.9

[a] From Conway (1952).

molecules, the larger the ion size. On passage through the channel this hydration shell is selectively removed. For example, at the carboxylic acid oxygen O1 (site 2 in Fig. 6.18b) the ion loses a few water molecules and presumably sheds the remaining H_2O molecules at site 23. The fully hydrated state is returned when the "free" intracellular state (site 4) is reached. Some justification can be made of this behavior by the examination of the data in Tables 6.1 and 6.2. Let us compare what is shown for sodium and potassium. It is seen in Table 6.1 that the permeability of potassium through the nerve wall, i.e., the channels, is smaller than that of sodium by a factor of about 12. This would be expected from the data of Table 6.2, which show that the unhydrated ion of potassium has a diameter about 40% larger than that of sodium. If one examines the hydrated behavior of these ions in Table 6.2, it is seen that the mobility of potassium is $7.64/5.2 \simeq 1.5$ greater than that of sodium. This is explicable by noting that the hydration number of potassium is considerably smaller, which means less water is bound to it than to sodium; more water molecules create a larger collision cross section with other water molecules, thereby reducing the Na^+ mobility through water. The strength of the hydration is electrostatic in that the polar water molecules are attracted to the positive nucleus of the ion. Since Coulomb's law is obeyed, the bonding strength changes inversely with the square of the separation. Because the potassium ion is much larger than the sodium ion, the bonding strength of the water molecules is considerably smaller, which not only results in the lower hydration number but also enables potassium to shed its hydrated molecules more readily. It is this type of argument that is invoked by Hille in this model of the channel interior discussed above. Although these models explain the experimental observations, there is no confirming evidence for the state of hydration within the channels.

SINGLE-CHANNEL EVENTS

The preceding development of current fluctuation explicitly assumed that the underlying conductance changes arise from the random opening and closing of a large number of uncorrelated individual ion channels. This interpretation of macroscopic fluctuations observed in either the current or voltage has, however, received strong support from the direct observation of single channels and from measurements of their electrical characteristics using the *patch-clamp* method of Neher and co-workers (see, for example, Neher et al., 1978). This novel technique involves the electrical isolation of a very small piece of membrane of a neuron (see Fig. 6.19). With sensitive electronic measuring devices the currents flowing through individual ion channels can be measured with a resolution of the order of 10^{-13} A. The

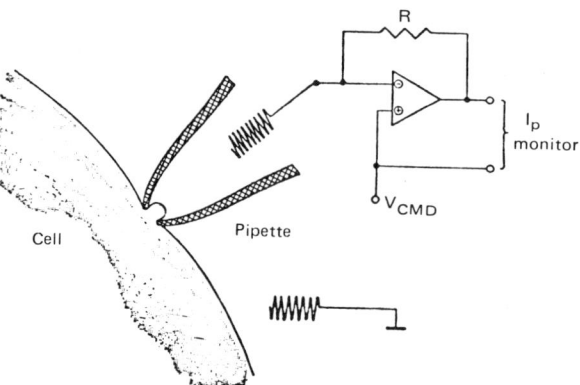

FIG. 6.19 Experimental setup for extracellular patch clamp circuit used in recording membrane currents through a small membrane patch. The membrane spans the electrode (pipette) tip. [Reprinted by permission from Sigworth and Neher, *Nature* **286**, 448 (1980). Copyright © 1980 Macmillan Journals Limited.]

patch-clamp method involves forming a high-resistance seal between the cell's (neuron) membrane and a small, very clean, and properly shaped glass pipette. This high-resistance seal between the outer rim of the electrode and the membrane (resistances on the order of $10^9\ \Omega$) ensures that all currents traversing the small membrane patch are registered by the Ringer solution-filled glass microelectrode, i.e., only small leakage currents exit through the rim of the seal. If high-resistance seals are achieved, then different transmembrane voltages can be applied to the small patch and the corresponding currents recorded, these currents being either spontaneous current produced by the random opening and closing of the individual ion channels within the patch or their modification by agonists and antagonists placed in the Ringer solution of the electrode. Unfortunately the method does not have universal applicability, since for most cells the high-resistance seals cannot at present be formed. In some preparations adherence between the glass electrode's tip and the cell's membrane is sufficiently strong that the seal is maintained even when the electrode is withdrawn from the cell. This causes a small part of the membrane to be torn from the cell. In such cases the membrane patch spanning the electrode's tip behaves as a single bilayer, and in some cases the bilayer's physiological electrical properties can be maintained for tens of minutes. When this fortunate situation exists the ion composition on either side of the membrane can be altered easily. This provides an extremely sensitive method of measuring the selectivity properties of channels. When the number of channels in the patch is sufficiently small that simultaneous open channels are infrequent, the unit currents arising from the opening and closing of single channels are rectangular pulses of constant amplitude for a

FIG. 6.20 Idealized single channel current traces recorded from a membrane under voltage clamp conditions. Absicca: time; ordinate: current. Inward positive current directed downward.

fixed transmembrane potential. The pulses are typically 1–5 pA in amplitude with variable time durations from 1 to 100 msec. Three single-channel openings, each having different open times (t_0), are illustrated in Fig. 6.20. From the duration of the open channel pulse (the lifetime of the open state) t_0 and its amplitude i_0, the net number of ions N_0 transported by the channel can be estimated. This follows since

$$N_0 = Q/Ze = i_0 t_0/Ze \tag{6.56}$$

where Z is the ion's valence, e the electronic charge, and Q the total charge transported. For typical values of $i_0 \simeq 2 \times 10^{-12}$ A, $t_0 = 10$ msec, and for monovalent ions ($Z = 1$), this relation implies that approximately 10^5 ions are transported across the membrane per channel during one opening of a channel.

Both the open-channel lifetimes (denoted by t_0 in Fig. 6.20) and the time intervals between open states are variable. Experimentally it is found that a histogram plot of the frequency of occurrences versus open (closed) lifetimes is exponentially distributed. This is illustrated schematically in Fig. 6.21. The solid line in the figure denotes the exponential function $A \exp(-t/\tau_0)$, where A is a normalization constant determined by the absolute number of open channels sampled and τ_0 is a time constant selected to

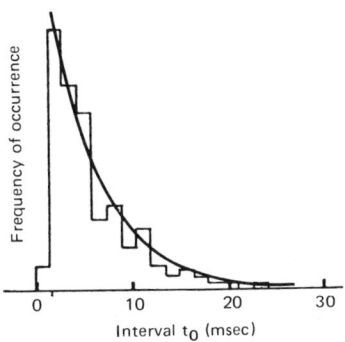

FIG. 6.21 Histogram of the frequency of occurrences of open channel lifetimes (t_0). The variable t_0 is indicated in Fig. 6.20. Solid line denotes the exponential function $A \exp(-t/\tau_0)$, where τ_0 is the average open channel lifetime.

provide a best fit (i.e., a least-squares fit). From the definition of the mean of a random variable (see p. 186) it follows that the average channel lifetime τ is given by

$$\tau \equiv \int_0^\infty tA \exp(-t/\tau_0)\, dt \bigg/ \int_0^\infty A \exp(-t/\tau_0)\, dt = \tau_0 \quad (6.57)$$

This relation states that the time constant of the exponential function providing the best fit to the frequency histogram of open-channel lifetimes is equal to the average lifetime of the open-channel state. A proof that open (or closed) time intervals are exponentially distributed is given in Appendix L. The assumptions involved in the proof are that the probability of a channel making either a transition from the open to the closed state or the reverse transition during the time interval t to $t + dt$ is proportional to dt and it is independent of events outside the interval t to $t + dt$. A similar assumption was previously used in Chapter 4 (see p. 119) to prove that the distribution of time intervals between mepps is exponential. Since the random opening and closing of single channels constitutes the underlying microscopic event responsible for the macroscopic current fluctuations described in the previous sections of this chapter, the characterization of their fluctuations should be derivable from the statistical properties of single channel events. This is indeed the case, as will now be shown.

The autocorrelation function for a sequence of random single-channel openings as illustrated in Fig. 6.20 is shown in Appendix L to be given by

$$C(t) = i_0^2 p(1 - p) \exp(-t/\tau_0) + (i_0 p)^2 \quad (6.58)$$

where i_0 is the amplitude of the single-channel current pulses, τ_0 the average channel lifetime, and p the probability that a channel is open. Setting $t = 0$ gives the variance (σ_i^2) of the single channel record. Thus $\sigma_i^2 = C(0) = i_0^2 p$, a result consistent with the definition of variance as equal to the mean square deviation, i.e.,

variance $\equiv \sigma_i^2 =$ (probability of being open) $\times$ (amplitude of signal)^2
$+$ (probability of being closed) $\times$ (zero)$^2 = p i_0^2$

The equivalence between microscopic and macroscopic measurement is most easily made by transforming Eq. (6.58) to the frequency domain. Using the Wiener–Khintchine theorem [Eq. (6.20)] gives the power spectrum for single channel records, i.e.,

$$S_1(f) = 4 \int_0^\infty C(T) \cos(2\pi f T)\, dT \quad (6.20)$$

$$S_1(f) = \frac{4 i_0^2 p(1 - p)\tau}{1 + (2\pi f \tau)^2} + i_0^2 p^2 \delta(f) \quad (6.59)$$

where $\delta(f)$ is the Dirac delta function.† The second term in Eq. (6.59) arises from the dc component in the autocorrelation function, i.e., the term $i_0^2 p^2$. Although this term will not concern us here, its contribution to the spectral function is essential in order that $S_1(f)$ integrated over all frequencies, including $f = 0$, equal the variance.

The above equation for $S_1(f)$ can be simplified if p is quite small since in this limit terms the order of p^2 can be neglected. With this approximation Eq. (6.59) can be written

$$S_1(f) = \frac{4i_0^2 p\tau}{1 + (2\pi f\tau)^2} \qquad (6.60)$$

This equation reduces to the spectral function for a system of N independent channels by using the superposition principle, which in this case states that the power function $S(f)$ for a system of N channels can be obtained from the power function $S_1(f)$ for a single channel simply by multiplying $S_1(f)$ by N. Thus

$$S(f) = NS_1(f) = \frac{4Ni_0^2 p\tau}{1 + (2\pi f\tau)^2} \qquad (6.61)$$

$$S(f) = NS_1(f) = \frac{4\langle I \rangle \gamma_0 \tau (V - E_s)}{1 + (2\pi f\tau)^2} \qquad (6.62)$$

which is identical to Eq. (6.46) after setting $\tau = \alpha^{-1}$, where α is the channel closing rate and noting that Eq. (6.62) is a one-sided power spectral function (i.e., $f \geq 0$) whereas that given in Eq. (6.46) is two-sided (i.e., defined for all frequencies, both positive and negative). In arriving at Eq. (6.62) the current–voltage relation for a single channel has been assumed to be linear,

$$i_0 = \gamma(V - E_s) \qquad (6.63)$$

and Eq. (6.31) for the mean total current $\langle I \rangle$ was used.

The relation between macroscopic measured variables and the underlying microscopic events is more apparent in the case where p, the probability

---

† The Dirac delta function $\delta(f)$ can be defined by the following properties:

$$\delta(f) = 0 \quad \text{if } f \neq 0$$

and for any function $G(f)$,

$$\int_{-\infty}^{\infty} G(f)\delta(f)\, df = G(0)$$

Effectively, the introduction of a Dirac delta function gives value to the integral only at a given point in space.

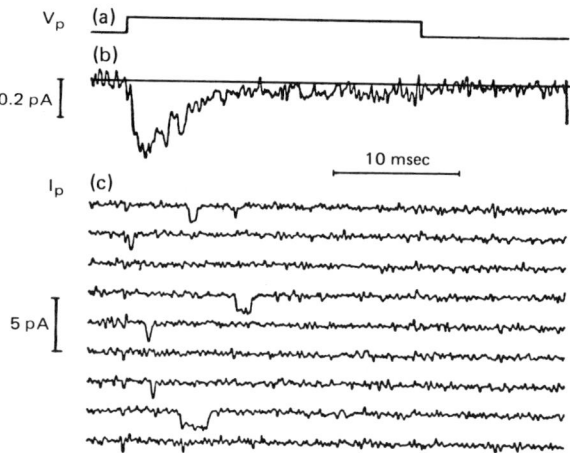

FIG. 6.22 Single channel currents during a 10-mV depolarization recorded in cultured rat muscle cells. (a) A 10-mV depolarization pulse. (b) Average of a set of 300 current records elicited by depolarization pulses shown in (a). (c) Example of nine successive individual single-channel records. [Reprinted by permission from Sigworth and Neher, *Nature* **286**, 448 (1980). Copyright © (1980) Macmillan Journals Limited.]

that a channel is open, is itself a function of time. Consider the case of current transported across the membrane by sodium channels. As shown in Vol. I, p. 77, the time evolution of the sodium current is quite complex in that for a fixed membrane depolarization that is above threshold, the macroscopic current has an activation (rising) phase followed by an inactivation or declining phase. Figure 6.22b illustrates the typical behavior of the sodium current. Deflections directed downward signify positive current entering the cell. The essential concept on how macrocurrent properties, such as the sodium current, are derivable from properties of individual single-channel properties and their statistics can be appreciated by use of the experimental data recorded from cultured cat muscle cells given in Fig. 6.22. Record a of Fig. 6.22 shows the applied depolarization pulse. This pulse activates the opening of the sodium channels. Their opening is evident by the short rectangular current pulses illustrated in Fig. 6.22c. These records were obtained using the patch-clamp technique. Because of the statistical nature of channel opening, each current record in response to a given depolarization step will be different. This is evident in Fig. 6.22c, which shows nine different records. For times much longer than several milliseconds after initiation of the abrupt membrane voltage depolarization pulse, the sodium channel undergoes inactivation as was discussed in Vol. I, p. 60, in the treatment of the Hodgkin–Huxley theory of the action potential. In terms of single-channel events sodium inactivation is equivalent to a decrease in the frequency of occurrences of single-current pulses for long times. This suppression of single-channel openings (i.e., a decrease in p, the probability a channel is open) is evident in the records illustrated in Fig. 6.22c. Currents measured when a large number of channels are involved, what we have loosely called the macrocurrent $I(t)$, can be related to single-channel currents, i.e., the microcurrent events, by noting that $I(t)$ should be proportional to the ensemble average (symbolized by an overbar) of the individual-channel records, $i_k(t)$. Mathematically this is written

$$I(t) \propto \frac{1}{N} \sum_{k=1}^{N} i_k(t) \equiv \overline{i_k(t)} \tag{6.64}$$

Figure 6.22b shows the ensemble average of a large number of current records ($N = 300$) similar to those shown in Fig. 6.22c. The resulting average, $\overline{i_k(t)}$, displays the expected macroscopic kinetic behavior; it has both a rapid activation phase followed by a much slower inactivation phase. This experimental reconstruction of the current $I(t)$ from individual-channel events unfortunately cannot be done theoretically, because of our present lack of understanding of how the probability of being open, p, changes both with time and with transmembrane potential. This example illustrates both the usefulness and the limitations of single-channel recording.

CONDUCTANCE FLUCTUATIONS IN THE PRESENCE OF LOCAL ANESTHETICS

Probing membrane properties by measuring current fluctuations has not only increased our understanding of synaptic and axonal transmission, but has also been helpful in the development of microscopic models of how exogenous chemicals, such as anesthetics and poisons, affect the nervous system. As discussed in Vol. I, p. 84, some nervous-system poisons exert their effect by prohibiting the passage of selective ions through a particular type of membrane channel. For example, tetrodotoxin (TTX) and the related compound saxitoxin (STX) block the sodium channel at the nodes of Ranvier by inserting their positively charged guanidinium group into the aqueous channel pore. Hydrogen bonds are formed between the molecules and the oxygen atoms lining the interior of the channel to help secure the fastening of the poison to the channel. This plugging of the Na channel results in the complete stoppage of transport of sodium ions across the membrane. The effects of anesthetics on neuronal activity and their mechanisms of action are, however, considerably more complex. In many cases no single microscopic mechanism explains their actions.

Anesthetics are defined as chemicals that reduce sensations, in particular the sensation of pain. Anesthetics are usually divided into two groups, *general* and *local*. General anesthetics are those that affect the entire body, causing not only a loss of sensation but also a loss of consciousness. Organic gases like cyclopropane, the inorganic gas nitrous oxide, xenon gas, ether, and chloroform are typical examples of general anesthetics. These examples amply demonstrate the absence of molecular structural similarity among general anesthetics. It is this lack of common molecular features that makes it unlikely that general anesthetics exert their effects by acting at specific sites, e.g., receptors; rather, their interactions with membranes must be highly nonspecific. There exists no universally accepted theory on how general anesthetics exert their influence, although extensive studies have shown an excellent correlation between anesthetic potency and the anesthetic's solubility in lipids, the latter measured by its lipid–water partition coefficient. The more potent the anesthetic, the easier it dissolves in a lipid environment. This empirical observation is generally referred to as the *Meyer–Overton rule* of anesthesia and in more quantitative terms states that the potency of an anesthetic is directly proportional to its lipid–water partition coefficient, with general anesthesia occurring when a general anesthetic attains a critical concentration in the cell's plasma membrane, the order of 30–60 mmol/liter of oil. With few exceptions the effect of general anesthetics on nerve and muscle cells is a depression of both axonal conduction and synaptic transmission, the latter occurring at high concentrations. A noted exception to this general observation is the anesthetic ethanol, which is a *central* depressant. For example, the postsynaptic voltage response to acetylcholine at the vertebrate neuromuscular junctions is potentiated by ethanol, whereas in molluscan neurons ethanol depresses the excitatory postsynaptic potentials. The microscopic mechanisms underlying these observations are still unclear, although it is thought that ethanol enters the membrane and perturbs the nature and state of the lipids surrounding the channel (Gage and Hamill, 1981). A similar process in all likelihood occurs for other general anesthetics.

Local anesthetics, in contrast to general anesthetics, act locally by blocking nerve conduction when applied in appropriate amounts. Local anesthetics do not cause general anesthesia when administered systemically, and with time their action is decreased by bodily processes. It is this property of elimination by the body and their localness of action that accounts for their wide clinical use. Typical local anesthetics are *procaine*, illustrated in Fig. 6.23; cocaine, which was the first local anesthetic discovered and was introduced into clinical use by Sigmund Freud; and lidocaine. Because their interaction is highly specific, most local anesthetics have molecular features quite similar to procaine, having a terminal hydrophilic group and a terminal lipophilic aromatic residue with a connecting intermediate group.

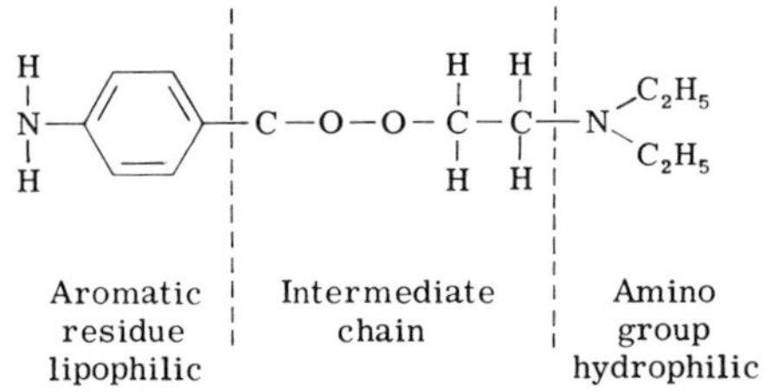

CONDUCTANCE FLUCTUATIONS IN THE PRESENCE OF LOCAL ANESTHETICS   227

$$H_2N-\langle\text{ring}\rangle-C-O-O-CH_2-CH_2-N(C_2H_5)_2$$

| Aromatic residue lipophilic | Intermediate chain | Amino group hydrophilic |

FIG. 6.23 Structural formula of the local anesthetic procaine.

Electrophysiological experiments have provided a wealth of information and support the concept of a specific interaction between a local-anesthetic molecule and the sodium channel. The most extensively studied system has been the synapse at the vertebrate neuromuscular junction. At this synapse the effects of local anesthetic on endplate currents and miniature end-plate currents are quite complex and still poorly understood. Even their effects on squid axon currents, described in Vol. I, p. 84, are incompletely understood. It appears that some local anesthetics exert their effects by *entering* the open-channel state. This results in a channel state having a considerably lower conductance compared with the open-channel state. An open channel plugged by a local anesthetic is called a *blocked channel*. This model is illustrated in Fig. 6.24. A considerable body of experimental data can be explained by assuming that transitions from the closed channel to the blocked channel cannot occur. The kinetic equations for this sequential blocking model, where an open channel can either close or reversibly bind the local anesthetic, are

$$nT + R \rightleftarrows (T_nR) \rightleftarrows (T_nR)^* \tag{6.65}$$

$$(T_nR)^* + Q \rightleftarrows (T_nR)^*Q \tag{6.66}$$

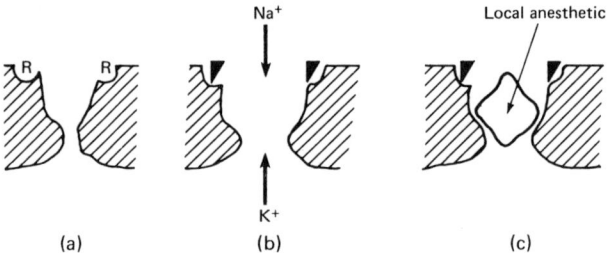

FIG. 6.24 Schematic model of (a) closed, (b) open, and (c) blocked channel. When the transmitter, denoted by ▼, occupies receptor binding sites R, the channel opens. Blockage of channel by a local anesthetic occurs by it binding reversibly to sites within the ion channel.

where T represents the neurotransmitter, R signifies the receptor, Q is the local anesthetic, n is the number of transmitters necessary to open a channel, (T_nR) is the receptor–agonist complex associated with a closed channel, $(T_nR)^*$ is the open conformation of the complex, and $(T_nR)^*Q$ is the transmitter–receptor–local-anesthetic complex corresponding to the blocked state. Generally, the blocked state has a negligible conductance compared to the open-channel configuration. In the absence of local anesthetics the end-plate currents decay exponentially, whereas the kinetic scheme given by Eqs. (6.65) and (6.66) predicts a double exponential decay of the end-plate current, one faster than normal and one slower. Functions that depend exponentially on time correspond to Lorentzian functions in the frequency domain (see p. 188). Hence the power spectrum of the microscopic current fluctuations when the concentration of the local anesthetic is nonzero is given as a sum of two Lorentzian functions. An example of this modified functional behavior of the power spectrum in the presence of a local anesthetic is shown in Fig. 6.25. In the case illustrated the current fluctuations were produced by the ionotophoretic application of the ACh agonist suberyl-

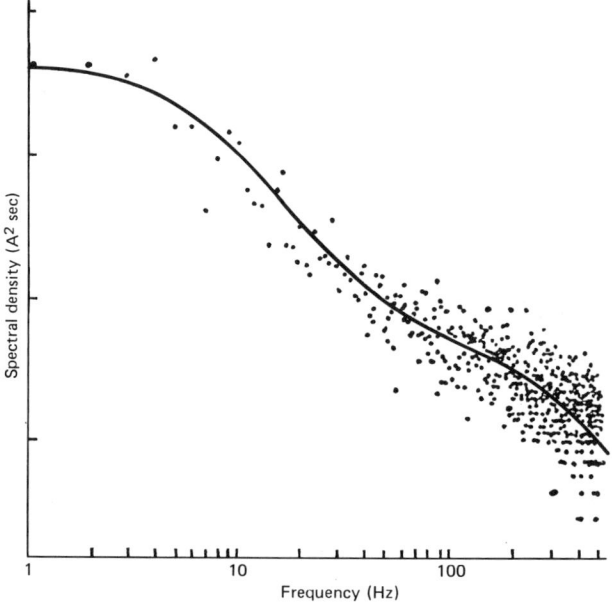

FIG. 6.25 Power spectrum of iontophoretically induced membrane current fluctuations. Double logarithmic plot. Solid dots are experimental points. Solid line is a two-Lorentzian fit. [From Neher and Steinbach (1978).]

CONDUCTANCE FLUCTUATIONS IN THE PRESENCE OF LOCAL ANESTHETICS   229

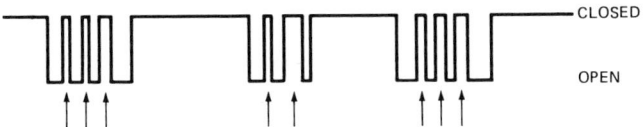

FIG. 6.26 Idealized current trace with three channel bursts. Arrows denote locations within bursts corresponding to the blocked state.

dicholine to the extrajunctional regions of denervated frog muscle in the presence of the local anesthetic lidocaine derivative QX-222. The solid line is a two-Lorentzian fit to the data.

A more direct experimental probe of how local anesthetics exert their effects is to examine changes in the single-ion channel current in the presence of a local anesthetic. In the absence of a local anesthetic single-channel openings give rise to rectangular current pulses with varying opening times, as illustrated in Fig. 6.20. In the presence of a local anesthetic, however, the rectangular current pulses divide into clusters or bursts of rapid activity. Three such bursts are shown schematically in Fig. 6.26. In terms of the sequential blocking model these bursts are interpreted as follows. The burst begins when a channel opens, that is, when the transition $(T_n R) \rightarrow (T_n R)^*$ occurs. The rapid random switching within the burst arises from transitions between the open and blocked states. There are no transitions between the blocked state and the closed state in the sequential model. The transition $(T_n R)^* \rightarrow (T_n R)$ in Eq. (6.65) terminates the burst (Gage and Hamill, 1981). In a sense the drug "enters" and "leaves" a channel many times before the channel closes. The important point to note is that single-channel records show no evidence of a two-channel population as one might infer from the double-Lorentzian fit to the current–power spectrum. Hence, the two decay constants characterizing the end-plate current in the presence of a local anesthetic are not average lifetimes of channels but are only parameters of the kinetic formalism. Other experimental observations such as the temperature, membrane voltage, and local-anesthetic concentration that effect single-channel events are consistent with the simple model of local-anesthetic compounds binding to sites within ion channels. The depression of electrical activity corresponding to the reversible blockage of ion channels occurs for both charged and uncharged anesthetics. Unfortunately, not all local anesthetics conform to such a simple model, and other mechanisms such as alteration of surface charge near the channel or lipid disordering have been proposed. Anesthetics also alter electrical transmission between electrotonic coupled neurons; however, electrical synapses are less sensitive to anesthetics than are chemical synapses and axonal membrane (Johnston et al., 1980).

REFERENCES

Anderson, C. R., and Stevens, C. F. (1973). Voltage clamp analysis of acetylcholine produced endplate current fluctuations at frog neuromuscular junction. *J. Physiol. (London)* **235**, 655.
Bendat, J. S. (1958). "Principles and Application of Random Noise Theory." Wiley, New York.
Conway, B. E. (1952), "Electrochemical Data." Elsevier, Amsterdam.
Cooley, J. W., and Tukey, J. W. (1965). An algorithm for the machine calculation of Fourier series. *Math. Comput.* **19**, 297.
Dionne, V. E. (1979). Modulation of conductance at the neuromuscular junction. *Membr. Transp. Processes* **3**, p. 123.
Dionne, V. E., and Ruff, R. L. (1977). Endplate current fluctuations reveal only one channel type at frog neuromuscular junction. *Nature (London)* **266**, 363.
Feller, W. (1966). "Probability Theory and its Applications," Vol. II. Wiley, New York.
Gage, B. W., and Hamill, O. P. (1981). Effects of anesthetics on ion channels in synapses. *Int. Rev. Physiol.* **25**, 1.
Hille, B. (1968). Pharmacological modifications of the sodium channels of frog nerve. *J. Gen. Physiol.* **51**, 199.
Hille, B. (1975). Ionic selectivity, saturation, and block in sodium channels: A four-barrier model. *J. Gen. Physiol.* **66**, 535.
Johnson, J. B. (1928). Thermal agitation of electricity in conductors. *Phys. Rev.* **32**, 97.
Johnston, M. F., Simon, S. A., and Ramón, F. (1980). Interaction of anaesthetics with electrical synapses. *Nature (London)* **286**, 498.
Kittel, C. (1958). "Elementary Statistical Physics." Wiley, New York.
Lewis, C. A., and Stevens, C. F. (1979). Mechanism of ion permeation through channels in a postsynaptic membrane. *Membr. Transp. Processes* **3**, 133.
MacDonald, D. K. C. (1962). "Noise and Fluctuations: An Introduction." Wiley, New York.
Magleby, K. L., and Stevens, C. F. (1972). A quantitative description of end-plate currents. *J. Physiol. (London)* **223**, 173.
Neher, E., and Steinbach, J. H. (1978). Local anesthetics transiently block currents through single acetylcholine receptor channels. *J. Physiol. (London)* **277**, 153.
Neher, E., and Stevens, C. F. (1977). Conductance fluctuations and ionic pores in membrane. *Annu. Rev. Biophys. Bioeng.* **6**, 345.
Neher, E., Sakmann, B., and Steinbach, J. H. (1978). The extracellular patch clamp: A method for resolving currents through individual open channels in biological membranes. *Pfluegers. Arch.* **375**, 219.
Nyquist, H. (1928). Thermal agitation of electric charge in conductors. *Phys. Rev.* **32**, 110.
Rice, S. O. (1944). Mathematical analysis of random noise. *Bell Syst. Tech. J.* **23**, 282.
Rice, S. O. (1945). *Bell Syst. Tech. J.* **24**, 46; also reprinted *in* "Selected Papers on Noise and Stochastic Processes" (N. Wax, ed.), p. 133. Dover, New York, 1954.
Ruff, R. L. (1977). A quantitative analysis of local anesthetic alteration of miniature endplate currents and endplate fluctuations. *J. Physiol. (London)* **264**, 89.
Seeman, P. (1972). The membrane actions of anesthetics and tranquilizers. *Pharmacol. Rev.* **24**, 583.
Sigworth, F. J., and Neher, E. (1980). Single Na^+ channel currents observed in cultured rat muscle cells. *Nature (London)* **287**, 447.
Stevens, C. F. (1972). Inferences about membrane properties from electrical noise measurements. *Biophys. J.* **12**, 1028.
Stevens, C. F. (1975). Principles and applications of fluctuation analysis: A nonmathematical introduction. *Fed. Proc., Fed. Am. Soc. Exp. Biol.* **34**, 1364.

REFERENCES

Stevens, C. F. (1977). Study of membrane permeability changes by fluctuation analysis. *Nature (London)* **270**, 391.
Ulbricht, W. (1977). Ionic channels and gating currents in excitable membranes. *Annu. Rev. Biophys. Bioeng.* **6**, 7.
Verveen, A. A., and Derksen, H. E. (1965). Fluctuations in membrane potentials of axons and the problem of coding. *Kybernetic* **2**, 152.
Wiener, N. (1930). Generalized harmonic analysis. *Acta Math.* **55**, 117.

CHAPTER 7

New Techniques of Brain Studies: Autoradiography, Positron Annihilation, and Nuclear Magnetic Resonance

INTRODUCTION

In the past 50 years neurophysiologists have done a remarkable job in mapping the central nervous system (CNS) and many of the pathways in the brain. Knowledge of these pathways is only part of the information required for understanding. Tissues that do physical or chemical work, such as skeletal muscle, heart, and kidney, exhibit a close relation between functional activity and energy metabolism. The brain is much more complex. Not only is there a resting metabolic rate, but in such activities as sensory perception or muscular control there is an increase of metabolism associated with that part of the brain's activity. Kety and Schmidt (1945) determined the blood flow rate to the entire brain in man as (50 ml/100 g)/min by measuring the saturation and desaturation of NO_2 by following both the arterial and venous concentration curves. Since then, the development of radioisotopes has led to a revolution in the study of metabolic rates through the detection of emissions from the decay of radionuclides introduced into the body. When such a nuclide is combined with a chemical that has a preferential location in a part of tissue or organ, a photographic film, if placed close to the area under consideration, will produce an "autoradiograph." This technique is frequently used in animal studies in which the animal is sacrificed after the radionuclide uptake and cross-sectional slices of tissue or of an organ are autoradiographed. In other techniques of diagnostics a bolus containing a radionuclide is injected, and its course through the body is followed by means of external counters. An example of this use in the measurement of cardiac performance was described in Vol. I, pp. 237–242. When a γ-ray-emitting isotope is injected into the body in a chemical compound that will concentrate in a brain tumor, the tumor becomes a source of γ rays radiating equally in all directions. Kuhl and Edwards (1970) developed a method of transverse section scanning of these emissions and the employment of a computer program to reconstruct the location of the center or centers of radiation from the angular intensity of emission. This transverse section scanning is called tomography and is described in Vol. I, Chapter 10.

Concurrent with the development of radionuclide tomography was that of x-ray computerized tomography, now known as CT (or sometimes CAT), in which the x-ray source is external. The essential differences between radionuclide tomography and x-ray tomography are twofold. (1) Radionuclide tomography requires an injection into the body so that the brain becomes the emitting source, whereas x-ray tomography uses an external source. (2) The radionuclide technique uses a calculation that triangulates the data to pinpoint the source. The early x-ray CT method compares the differences in absorption coefficients μ of the x rays of different tissues of the brain with that of water. That is, a differential absorption $e^{-\mu x}$ is obtained by allowing part of the x rays to pass through a known and comparable quantity of water, where the μx for water is known. Thus, small deviations in absorption can be obtained and reconstructed into a picture by a suitable computer program. X-ray computer-assisted tomography has revolutionized imaging of the brain, as well as that of other organs of the body. The ability to image with the readily available reconstruction algorithms has effectively opened floodgates for other scientific ideas. These ideas center on the proposition that since imaging exists, what other signals may be imaged and what are their relative advantages and disadvantages? This chapter discusses two of the more advanced techniques, positron annihilation (PETT) and nuclear magnetic resonance (NMR). Both are so promising that they will probably be commercially available soon, although positron annihilation can be employed only where a cyclotron or similar isotope source is available. For this reason, it is highly likely that the NMR device will be the more widely used for clinical examinations.

AUTORADIOGRAPHIC DETERMINATION OF REGIONAL BRAIN METABOLISM

Before discussing the techniques of PETT and NMR imaging, we shall first consider pioneering work in brain metabolism studies by autoradiography and by focused scintillation counters.

The most direct measurement of energy metabolism using a radioactive tracer would be oxygen consumption. However, oxygen and its products are washed out of brain cells quite rapidly, and isotopes of oxygen have short half-lives. Both factors preclude autoradiographic measurements. A second consideration is the rate of glucose consumption. In most cases glucose is almost the sole substrate for cerebral oxidative metabolism, and its consumption is stoichiometrically related to oxygen. Glucose is transported across the blood–brain barrier and is phosphorylated by hexokinase to glucose 6-phosphate. This is further metabolized to CO_2 and water. However, the brain cells do not store for long the glucose 6-phosphate, and the time for metabolic breakdown and elimination is also too short for autoradiographic measurements.

What is clearly required is an analog of glucose whose passage into the cell has essentially the same rate as that of glucose but whose exodus from the cell is slow compared to the measuring time. Sokoloff et al. (1977) used the radioactive isotope carbon-14-labeled analog of glucose 2-deoxy-D-[^{14}C]glucose, for which we shall use the shorthand notation [^{14}C]DG. This substance is taken into the cell by the same carrier as glucose and metabolized within the cell at a definable rate relative to that of glucose. 2-Deoxy-D-glucose differs from glucose only in the substitution of the hydroxyl group on the second atom by a hydrogen. The remainder of the molecule is identical to glucose. It therefore is phosphorylated in the same way as glucose, but its product deoxyglucose 6-phosphate cannot be isomerized into fructose 6-phosphate because of the lack of a hydroxyl group on its second carbon atom. It therefore cannot complete the metabolic pathway and remains trapped within the tissue (see Appendix K). This trapping time (45–60 min) is sufficiently long to perform the experiments.

If a bolus of [^{14}C]DG is injected into the bloodstream of an animal, at any time thereafter the total content of ^{14}C per unit mass of tissue i may be called C_i. This is equal to the sum of the concentrations of free [^{14}C]DG in the precursor pool in the tissue C_E and its product deoxyglucose 6-phosphate C_M in that tissue, or

$$C_i = C_E + C_M \tag{7.1}$$

The rate of change dC_E/dt of free [^{14}C]DG is the difference between its rate of transport into the tissue from the plasma, k_1, and two loss terms, the rate of transport out of the tissue back into the plasma k_2 and the rate of phosphorylation of [^{14}C]DG in the tissue k_3. If C_p is the concentration of [^{14}C]DG in the plasma, the kinetic equation is

$$\frac{dC_E}{dt} = k_1 C_p - k_2 C_E - k_3 C_E \tag{7.2}$$

There is an obvious assumption in this equation that first-order kinetics are being obeyed by all terms. Sokoloff et al. combined this equation with Fick's zeroth law (Vol. I, p. 181) for indicator flow

$$F(C_A - C_V) = \frac{dC_i}{dt} \tag{7.3}$$

where F is the rate of blood flow per unit mass in the ith tissue and C_A and C_V are the arterial and venous concentrations of ^{14}C draining that tissue, respectively, and obtained a general solution for dC_i/dt. The rate constants were determined by separate experiments. By obtaining relative autoradiographic counts in separate portions in the brain, Sokoloff et al. were able to determine the relative metabolic rate of that tissue. A possible criticism is that the rate constants are not determined in the same animal at the same time as are the experiments to infer dC_i/dt. These constants also vary among different brain tissues, with different species, and, although not determined, they may vary with regional cerebral activity and disease. Nevertheless, within these limitations, striking data were obtained for relative regional blood flow in different regions of the brain and under different stimuli.

Experiments were performed on both rats and monkeys. An injection of [^{14}C]DG was given, and after a few minutes the animal was sacrificed and beheaded. The head was preserved in liquid nitrogen, and then 20-μm slices of the brain were made and a photographic film was placed over the slices for an autoradiograph. The resulting films were examined by a densitometer with spots of 25–100-μm size being selected, although the claimed resolution is 100–200 μm. These experiments were repeated with sensory stimuli being either applied or deprived prior to sacrificing the animal.

Example results of these experiments are shown in Fig. 7.1, in which autoradiographs of coronal slices at the same position through the visual cortex (Vol. II, p. 228) for three animals under three conditions are compared. Figure 7.1a shows an animal with normal binocular vision. A laminar distribution of the density is apparent; the dark band corresponds to visual layer IV. In Fig. 7.1b the animal has been deprived of vision in both eyes, with a resulting decrease in density and virtual disappearance of layer IV. In Fig. 7.1c the right eye has been occluded. The left half of the figure is the left

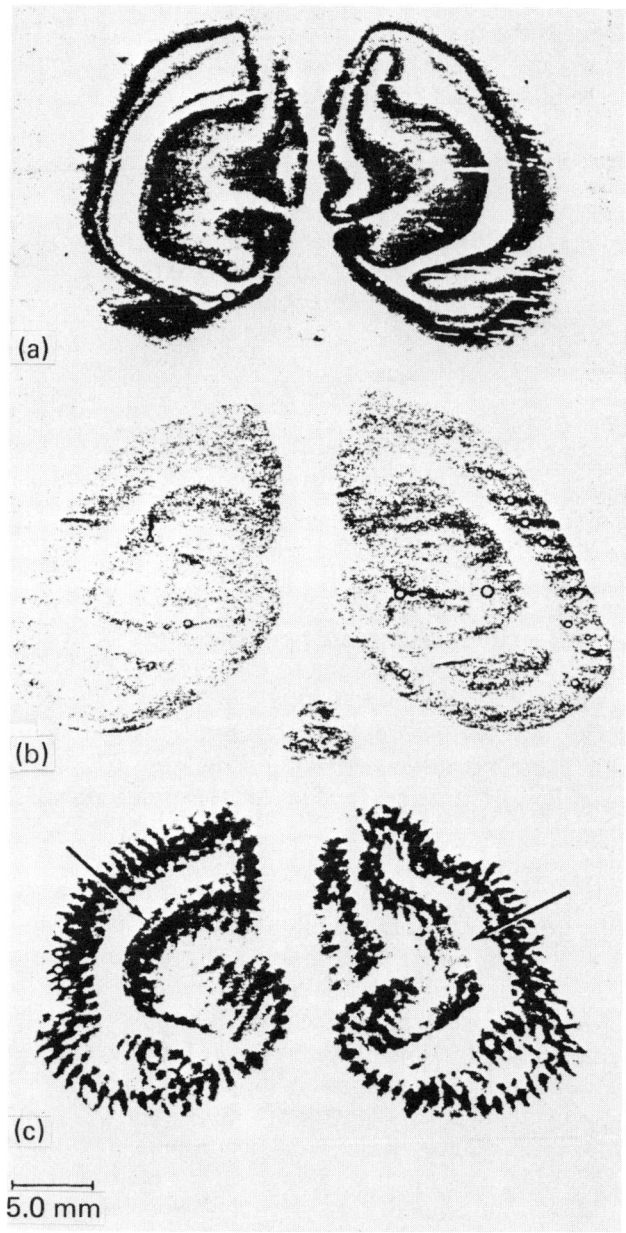

FIG. 7.1 Autoradiography of coronal brain sections of monkeys at the level of the visual cortex; right side of figure corresponds to right side of brain. (a) Animal with normal binocular vision. (b) Animal with bilateral visual deprivation. (c) Animal with right eye occluded. The arrows point to regions of bilateral asymmetry. [From Kennedy *et al.* (1976).]

hemisphere. Alternate light and dark striations are evident, which represent the ocular dominance columns. These columns are most apparent in the dark band corresponding to layer IV. The arrows point to regions of asymmetry where the ocular dominance columns are absent. These are regions in which there is normally input from a single eye on the same side of the cortex as the eye. On the right-hand in this region there is no glucose metabolism, because there is complete sensory deprivation. The arrows indicate regions that are characterized by an absence of ocular dominance columns and are believed to be the cortical representations of the blind spots of the eyes (Vol. II, p. 155).

An alternative compound 2-[^{18}F]fluorodeoxyglucose, labeled with the radioactive isotope fluorine 18, has been used in similar glucose-uptake studies (Reivich et al., 1979). This chemical is also taken into the cell by the same carrier as glucose and it is also phosphorylated. Its subsequent product is also trapped in the cell, not being able to be metabolized to CO_2 and water. Its time of trapping is believed to be longer than that of deoxyglucose, and it offers an alternative emission energy.

DYNAMIC RADIOGRAPHIC STUDIES OF BRAIN METABOLISM

The Kety–Schmidt method of measuring cerebral blood flow with NO_2, mentioned in the previous section, was later followed by measurement of the time course of a radioactive substance. An isotope of the rare gas xenon, ^{133}Xe, dissolved in a saline solution, was injected into a carotid artery (principal artery leading to the brain) and followed with a scintillation counter (Høedt–Rasmussen et al., 1966). They were able to duplicate the Kety–Schmidt results and were thus encouraged to develop more sophisticated apparatus with many scintillation counters. In this way they were able to follow the relative dynamic blood flow in different regions of the brain. Olesen (1971) focused counters of 1.2 cm diameter on 35 small regions of the brain after ^{133}Xe injection. He showed that if a hand is exercised, the logarithm of the clearance rate from the hand portion of the somatosensory region contralateral to the hand increased up to 54%. Note that a diffusion rate generally has an exponential form, and its is therefore appropriate to use the logarithm for rate measurement, the slope of the logarithm being proportional to the exponent. (As an example of the exponential form in diffusion see Vol. I, Fig. 3.7.)

A bank of 254 scintillation counters arrayed in an area the size of the cross section of the cranium was constructed by Sveinsdottir et al. (1977). The counters were lead shielded to a collimation of 0.7 cm. The investigators were able to make simultaneous measurements of ^{133}Xe clearance rates during a variety of physical and even mental activities. For example, reading silently activates four areas: the frontal eye field, the visual association area, the supplementary motor area, and Broca's speech center. Reading aloud activates two additional centers: the mouth area and the auditory cortex. Lassen et al. (1978) has excellent color photographs of the screen display of the scintillation counters during a variety of activities. The disadvantage of the ^{133}Xe technique is that this isotope emits a low-energy γ ray. Rays emitted from the inner portions of the brain undergo a considerable amount of Compton scattering (Vol. I, p. 254), which leads to about 20% spurious counts. Therefore the camera is almost blind to small interior areas.

The technique of single and multiple scintillation counters was used by Raichle et al. (1975, 1976) to follow radioisotope-labelled [^{15}O]oxyhemoglobin to locate regions of increased blood flow during muscular movements. Among other things they found that muscular movement of the hand correlated with increased blood supply to the appropriate brain region in normal subjects but not with one suffering from mild dementia. This result is not understood at present.

COINCIDENCE COUNTING OF POSITRON ANNIHILATION

The problems of collimation and appropriate tissue absorption correction for scintillation detection that are encountered in external probe measurements do not arise with a technique that has been under development since the early 1960s. The technique is coincidence counting of positron emission, and it is both depth independent and self-collimating. The rapidity of development of this technique, called PETT, PET, or ECAT, and the rapid development of an operating device seems to have bypassed the pioneering efforts discussed above.

A number of radionucleides decay by the emission of a positive charge with the mass of an electron. This is called a *positron*. A brief summary of the origin of this particle is given in Vol. I, p. 255. When the positron encounters an electron, they mutually annihilate and produce two γ rays, each with energy 0.511 MeV. These γ rays may have any direction in space but, from momentum conservation, they must have directions 180° from each other. In other words, they are emitted at the same time, but in opposite directions. They are detected by a technique called coincidence counting. If two γ-ray counters facing each other register a count within nanoseconds, limited only by the electronics of the counter, then the coincidence conditions are met for simultaneity of two γ rays 180° apart.

The technology of such coincidence counting has been in use by physicists for several decades, and, since algorithms for computerized tomography have been developed, the transition to positron-emission transaxial tomography (PETT) required only the detailed study of its application to physio-

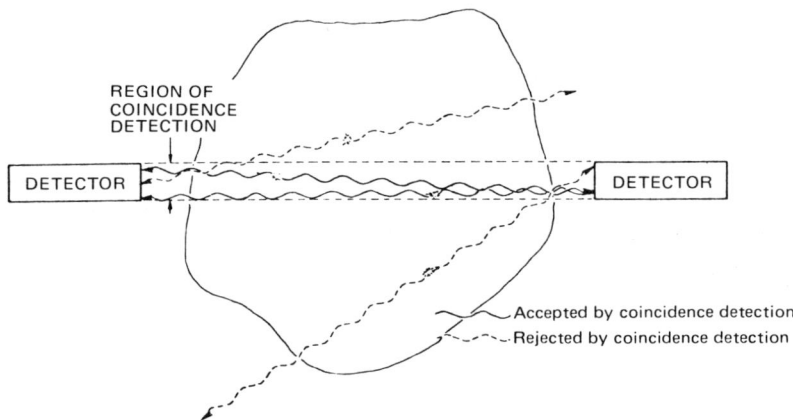

FIG. 7.2 Positron-annihilation photons are detected only in coincidence between the two counters and therefore an automatic collimation exists without the use of absorbers. [From Phelps et al. (1975).]

logical systems. [Note that this technique is sometimes referred to as positron-emission tomography (PET) and emission computerized axial tomography (ECAT).] A few positron-emitting radionuclides, namely, ^{15}O, ^{13}N, ^{11}C, and ^{18}F, can be substituted into a number of biological or pharmaceutical compounds without changing the chemical properties, and they are therefore suitable for the study of biological processes.

There are a number of distinct advantages in the use of positron-emission tomography, and these will be mentioned briefly.

1. The two γ-ray photons are emitted at 180°. This property establishes an automatic collimation and eliminates the need for absorption collimators. A schematic of this is shown in Fig. 7.2, where the solid lines represent emissions detected by the counters and the dashed lines those rejected because only one counter is activated.

2. Coincidence counting of annihilation radiation is independent of the depth of the source. This is readily seen from the following argument. The probability of survival of a pair of γ-ray photons in passing through a medium of average absorption coefficient μ is the product of the probability of survival for each γ ray separately [Vol. I, Eq. (9.40)]. If the distance $S_1 - S_3 = D$ is the total pathlength, say the width of the head at the region of detection of an emission pair, where S_1 and S_3 are two points on the exterior of the head, and S_2 is the interior point of emission, then $S_1 - S_2$ and $S_2 - S_3$ are the two pathlengths. The joint probability of survival is then

$$\exp\left(-\int_{S_1}^{S_2} \mu\, ds\right) \exp\left(-\int_{S_2}^{S_3} \mu\, ds\right) = \exp\left(-\int_{S_1}^{S_3} \mu\, ds\right) \quad (7.4)$$

Since this integral is independent of point S_2, where the emission took place, the net attenuation is constant. The attenuation for different pathlengths D is determined by placing a band on the outside of the head with a standardized positron emitter such as ^{68}Ge. This preliminary calibration directly measures the attenuation coefficients μ for positron annihilation photons at each point in a scanner slice.

3. The solid-angle efficiency is constant as a function of depth because it results from the combined efficiency of the two detectors. If the solid angle for one detector increases because of the displacement of the event from a central position between the two detectors, the solid angle at the other detector decreases. Therefore the absolute sensitivity can be calculated from only the total length D and the attenuation coefficient of the total material traversed.

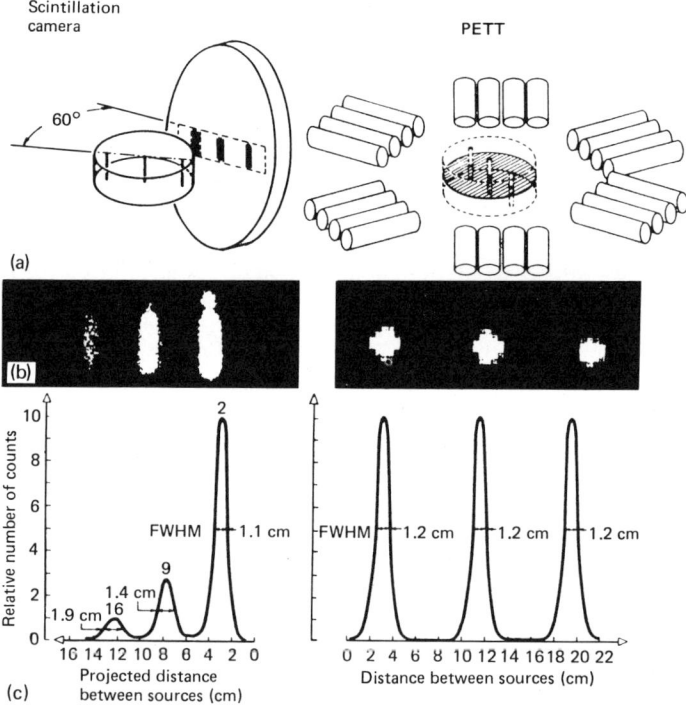

FIG. 7.3 Comparison of line spread functions and full width at half maximum (FWHM) from three positron-emitting sources in water as detected by a single scintillation camera and by PETT. (a) Physical arrangements, (b) film exposure, and (c) densitometer trace. These experiments illustrate the independence of the depth of the source for coincidence detection. [From Ter-Pogossian *et al.* (1975).]

4. Because of the well-defined depth-independent response of annihilation coincidence, the line spread functions as measured by the full width at half maximum (FWHM) is constant, whereas for a single detector it is not. Ter-Pogossian et al. (1975) demonstrated this with simulated sources. This is illustrated in Fig. 7.3. Three sources are placed at an angle of 60° from their line to the plane of a scintillation camera. The sources are in water so that there is a different penetration depth between each source and the camera. Because the absorption is depth dependent, the linespread functions have different shapes and therefore different FWHM values. In contrast, when the same system of sources is recorded by PETT, the depth independence results in identical lineshape functions and FWHM values. This result considerably improves the accuracy of the absorption factor introduced into the algorithm.

5. The efficiency of the coincidence detection increases with the square of the number of detectors, whereas that of single detectors increases only linearly with the number of detectors. Consider the simplest case illustrated in Fig. 7.4. In the upper part, if there are only two counters initially, 1 and 3, and the number is doubled by adding counters 2 and 4, the detection efficiency increases by a factor of 2. In coincidence counting if the two original detectors 1 and 3, are increased by a factor of 2 by adding detectors 2 and 4, the number of added lines of coincidence is four, as shown. This efficiency factor of coincidence counting can be generalized to the upper limit of the admittance angle of the counters. This angle is governed by the effect of radiation "spillage" of the detector crystals, which is discussed below. The configuration of detectors in some PETT designs is illustrated in Fig. 7.5. Here it is seen that coincidence detection of two opposite banks of detectors gives rise to 16 lines of detection, of which 8 are shown. Single-counter detection by these opposite banks would give only 8 lines of detection.

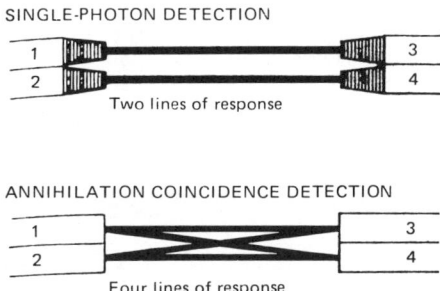

FIG. 7.4 Schematic illustrating the relative increase in number of lines of detection between single photon counters and coincidence counters with four detectors in each system. [From Phelps et al. (1975).]

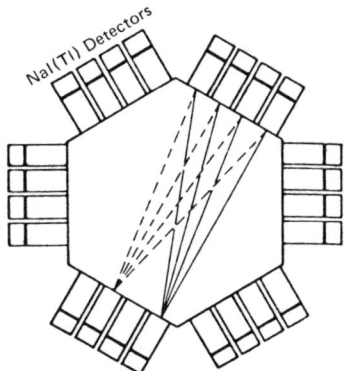

FIG. 7.5 Schematic of multiple coincidence achievable by a PETT in a hexagonal array with four detectors on each side. [From Phelps *et al.* (1975).]

DESIGN CONSIDERATIONS OF POSITRON ANNIHILATION

The detector of γ rays is a crystal of sodium iodide doped with thallium, NaI (Tl), the mechanism of which is discussed in Vol. I, p. 290. The γ ray causes a photon of wavelength 4030 Å to be emitted within the crystal, which is detected by a photomultiplier tube and recorded electronically.

The crystal scintillator must be large enough that the photons produced reach the photomultiplier tube. If it is not large enough, then some photons escape detection. This is called spillage, which is illustrated in Fig. 7.6. In the left-hand drawing the possible range of the scintillation radiation in a crystal

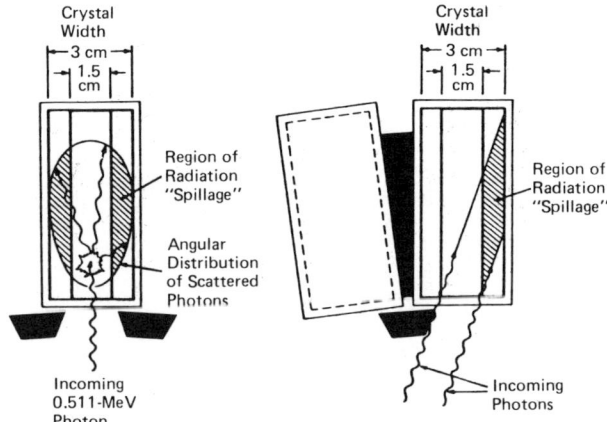

FIG. 7.6 Two types of spillage or loss of scintillation radiation to the photomultiplier if the crystal detector is too small. [From Ter-Pogossian *et al.*, *J. Comput. Assist. Tomogr.* **2**, 539. (1978a). Copyright 1978 Raven Press, New York.]

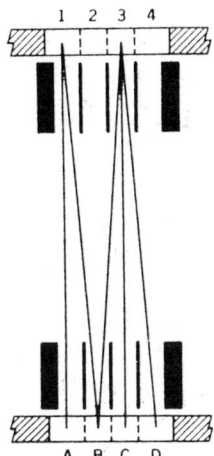

FIG. 7.7 Diagram showing the coincidence lines used in the reconstruction of the seven sections in PETT IV. [From Ter-Pogossian *et al.* (1978b).]

is shown by the oval. If the crystal is 3 cm wide, the event will be seen by the photomultiplier and recorded, whereas in a crystal only 1.5 cm wide there will be spillage. The right-hand side of Fig. 7.6 shows that a similar spillage can occur if the incident photon makes too large an angle.

Because of some slight motion of the patient during the several-minute period of data taking, the group at the Washington University School of Medicine developed a multislice tomograph that can take data on seven sections simultaneously.

The principle is shown in Fig. 7.7. The heavy lines represent lead spacers and the area numbered 1–4 and that labeled A–D represent two large detector crystals. Two photomultipliers are placed at the ends of the crystals (at positions 1 and 4, and A and D) and, from the relative intensity of a scintillation detected by each of the two detectors on a crystal, the location along the crystal can be determined. This arrangement permits not only vertical events to be recorded, e.g., 1–A, but also adjacent events, e.g., 1–B. In this way a total of seven slices can be measured simultaneously. The arrangement of the hexagonal array of detectors is shown on the left-hand side of Fig. 7.8 and the lines of events recorded on the right-hand side. The computer can then reconstruct a three-dimensional picture. The motion of the gantry, shown on the left is a rotation of 60° in 20 increments of 3° coupled with a lateral translation of 6 cm.

Although a number of PETT designs employ a hexagonal array of detectors, as in Fig. 7.8, the group at the Washington University School of Medicine has developed what they call PETT V, an array of 48 scintillation detectors

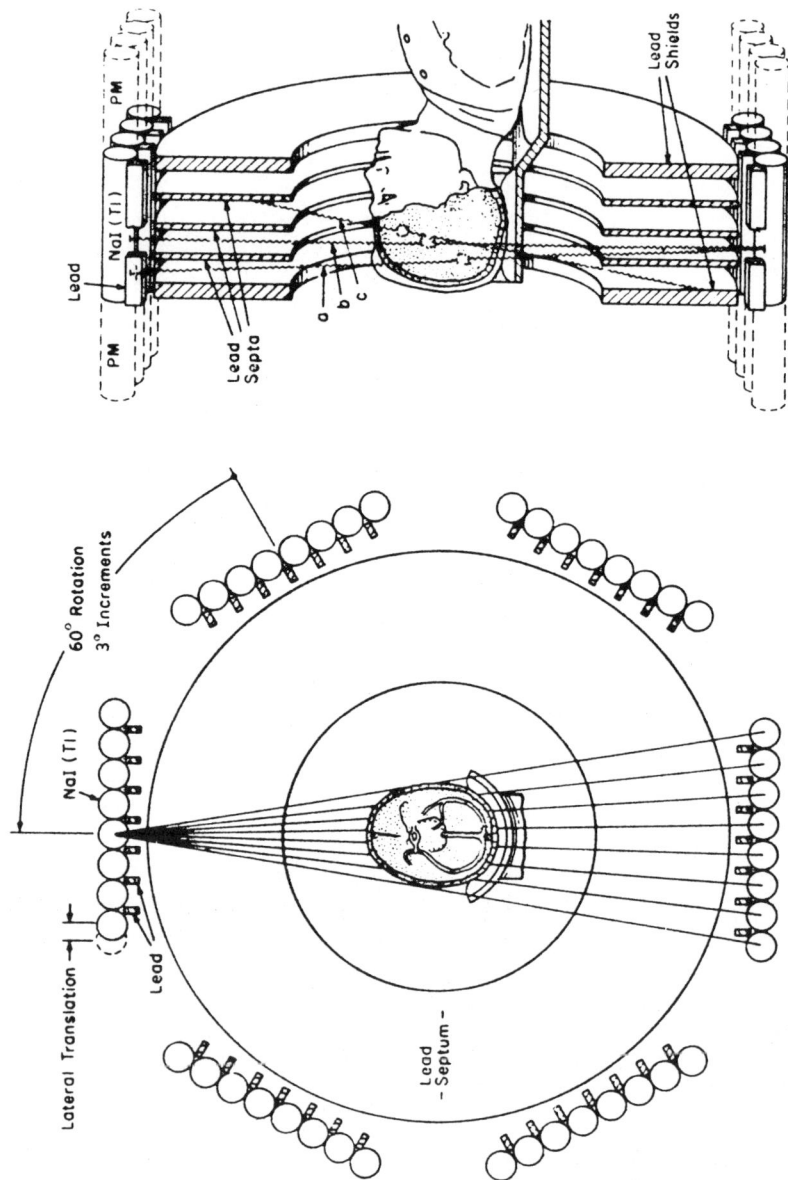

FIG. 7.8 Diagram of PETT IV showing detectors, collimators, and septa arrangement. [From Ter-Pogossian et al. (1978b).]

distributed uniformly on a circle. This arrangement gives an angle of $360°/48 = 7.5°$ between each detector. Because of the crystal width requirement, discussed above, and some shielding between crystals to eliminate counting errors, more detectors would increase the difficulties of rapid mechanical motion. This group therefore employed a twofold motion. It consists of a rotation plus a translation, called a "wobble," so that the effective rotation between counts is 3.5°. This combined motion is illustrated in Fig. 7.9, where the lower part shows the motion of the detectors.

A ring system with 95 detectors with each detector coupled in coincidence with 40 detectors on the opposite side has been developed at the Karolinske Sjukhuset in Stockholm (Bohm et al., 1978). In the conventional mode the resolution is 1.05 cm, but with a modified sampling technique they claim a resolution of 0.7 cm, with a conceivable limit of about 0.5 cm.

Attention is called at this point to the sampling theorem developed in communication theory (Appendix H): the sampling resolution must be less than or equal to half the object size to specify its size. Phelps et al. (1975) state that their positron-emission experiments showed that the necessary sampling resolution was usually from one-half to one-quarter of the object size for quantitative resolution. Thus, because of this theorem and the size constraints in construction described above, the ultimate resolution of PETT seems to be limited to about 1 cm^3, i.e., twice the resolution limit in each direction.

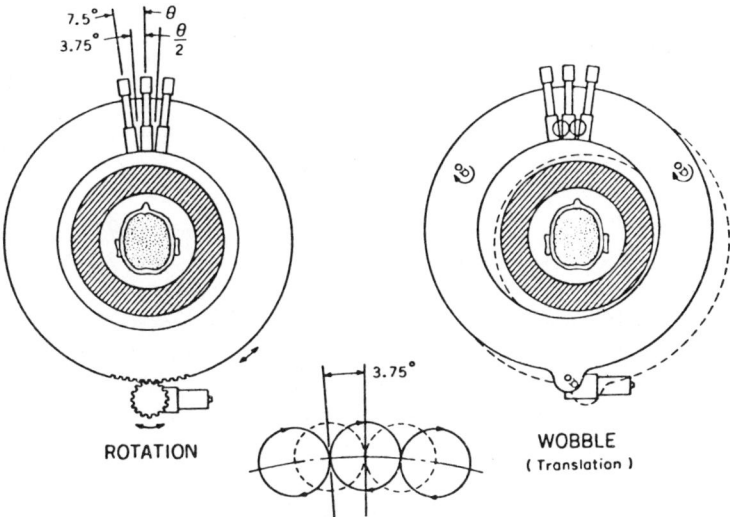

FIG. 7.9 Schematic of combined rotational and translational motion of the detector ring of PETT V. [From Ter-Pogossian et al. (1978b).]

POSITRON-EMISSION MEASUREMENT OF BRAIN METABOLISM

In an earlier section we discussed the use of labeled deoxyglucose and fluorodeoxyglucose for the measurement of regional cerebral metabolism. Their usefulness is based on the fact that their uptake by a cell is at the same rate as that of glucose, but they cannot leave the cell after phosphorylation. With the development of positron-emission tomography these same substances may be used since both ^{11}C and ^{18}F are positron emitters. The regional cerebral metabolic studies discussed earlier were repeated with PETT and yielded the same results when similar sensory stimuli and muscular motion were tested on man (Reivich et al., 1980; Greenberg et al., 1981).

When these analogs of glucose are employed, corrections must be made in the kinetic formulation for differences in membrane transport properties and enzyme affinities. There has long been uncertainty about whether these factors change for a diseased organ. The task of finding and applying correction factors for every type of disease is formidable. This can be avoided if labeled glucose or oxygen can be used because these are identical biochemically to the compound being traced. Only a known small correction in diffusion rates is necessary because of the slightly different isotopic mass. The isotopes that could be used are ^{11}C-labeled glucose or ^{15}O-labeled oxygen. If the use of these isotopes can be developed, then any pharmaceutical can be labeled without changing its biochemistry. The difficulty with these isotopes is their short half-lives (2.05 min for ^{15}O; 20.34 min for ^{11}C), therefore, experiments can only be done near a cyclotron or nuclear reactor. This requirement has limited the use of these radionuclides to a few major medical centers in the world. The number of medical cyclotrons is gradually increasing, however, and their use is being simplified. The second requirement was the development of faster emission tomographs. In the preceding section we noted the increase in the number of scintillation counters in the newer devices. This increase reduces the loss of counted emissions and thereby decreases the required counting time, currently at 5 min or less. The ^{15}O produced by the cyclotron can be inhaled by the subject and measurements can begin almost immediately. The ^{11}C-labeled glucose requires very rapid chemistry, which can now be done in 75 min (3–4 half-lives). A schematic of the chemical steps required and their times is shown in Fig. 7.10. The ^{11}C is combined with oxygen at high temperature to form CO_2. This is then fed to leaves undergoing photosynthesis, which create glucose and other carbohydrates. Standard extraction methods are then performed, and the [^{11}C] glucose is ready to be injected.

Recent measurements have been made in vivo on monkeys' brains using the labeled glucose by Raichle et al. (1978). In this experiment the collection

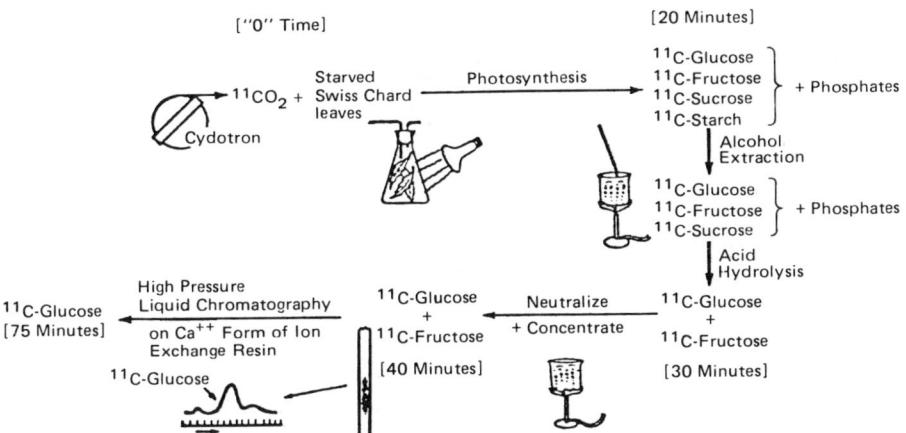

FIG. 7.10 Schematic of steps in the production and extraction of [^{11}C] glucose in 75 min. [From Raichle (1979).]

of image data began 4 min after injection and continued for 2 min. The resulting data were comparable to the glucose-analog studies on sacrifice animals. This technique indicates relative glucose uptake in cerebral regions. These investigators obtained absolute values by simultaneously monitoring the blood volume. This was done by having the monkey inhale ^{11}CO, which forms [^{11}C]carboxyhemoglobin, and sampling the blood every 15 sec. Thus, a safe, in vivo, kinetically accurate technique now exists for the study of man, both normal and diseased.

An even more attractive technique is the direct inhalation of either ^{15}O or $C^{15}O_2$ by the subject. Frackowiak et al. (1980) have worked out the kinetics of brain absorption of these (the $C^{15}O_2$ forms some $H_2^{15}O$ in the alveoli) and, coupled with blood sampling, were able to quantify the regional cerebral blood flow, the oxygen extraction, and the oxygen use in man by emission tomography. In a discussion of reservations of the technique they cite the following assumptions: (1) the water extraction rate from cells is independent of blood flow rate; (2) the partition coefficient for water is the same for both healthy and diseased tissue; and (3) the volume of water in the cranial region is equal to its fraction of body volume. Nevertheless, this technique is an attractive and simple routine chemical test using natural metabolic molecules.

This brief summary indicates the rapid advance of new techniques from only a few groups within the past few years. With the establishment of the basic technology and the commercial availability of the devices, the next few years should show startling advances in the understanding of the metabolic activities and pathways of healthy and diseased brains.

GENERAL ASPECTS OF NUCLEAR MAGNETIC RESONANCE

Nuclear magnetic resonance (NMR) experiments were first performed more than 35 years ago (Bloch et al., 1946) and have since been established as one of the more important branches of spectroscopy. While optical spectroscopy measures transitions between energy levels of orbital electrons with wavelengths in the range of 5×10^{-7} m, NMR measures transitions between alignments of the magnetic moments of nuclei with energy levels corresponding to wavelengths in the range of a few meters.

Some nuclei have a net spin, and this rotation of their internal electric charges gives rise to a magnetic moment, as does current in a loop of wire in classical physics. When these magnetic moments, which are randomly oriented, are placed in a strong magnetic field, they tend to become oriented in the direction of the field. This average orientation is a function of the strength of the field, the magnitude of the magnetic moment, and the temperature because thermal effects tend to disorient, or randomize, the magnetic order. Their orientation and disorientation in a magnetic field is quantized, that is, there are only certain allowed energy states, which may be viewed as orientation configurations. The basic method of NMR spectroscopy is to orient the magnetic moments in a strong static magnetic field and then to alter their direction of orientation with a transverse oscillating magnetic field produced by a radio-frequency (rf) coil. The rf field is then turned off, and a detector coil measures the number of the magnetic moments as they return to their former average orientation with respect to the static field. Since this is a quantized process, the magnitudes involved must be evaluated by the formalism of quantum mechanics. But just as the Bohr atom is a visual analog of quantized processes of atoms, a similar analog can be constructed for NMR once the magnitudes of the quantities involved are accepted.

Not all nuclei have net spins. Of those that do the hydrogen nucleus, the proton, is particularly suitable for NMR studies because of its frequent occurrence in numerous compounds. Nuclei in different chemical environments experience the magnetic fields of other adjacent nuclei, which adds to the static and rf fields. Therefore their resonances occur at differing frequencies of rf fields and are separately detectable. It is this property that makes NMR spectroscopy such a powerful tool in chemical structural analysis. Resonance frequencies of nuclei are generally reported as *chemical shifts*, which are proportional to the difference in the resonance frequencies of two nuclei divided by a standard reference frequency (see p. 278). NMR is generally the only technique available for the study of the local environment of certain nuclei.

The usefulness of NMR can be illustrated as follows. The magnitude of a given NMR signal is proportional to the number per unit volume of a given nucleus in a particular chemical environment. Consider ^{31}P, naturally occurring phosphorus, which has a nuclear spin and therefore can be detected by NMR. Thus the energy phosphate ATP, having three phosphorous atoms, each of which experiences a slightly different environment, exhibits three different resonance frequencies, i.e., a three-line spectrum. The magnitude of each of these ^{31}P signals can be followed in the time-dependent reactions involved and gives important information both of the reaction rate and in the identification of products.

A third technique developed in the past few years is that of NMR imaging. Although a number of techniques are evolving, the general principle is to superimpose an additional magnetic field as a linear gradient across the sample. When the rf field is swept across a frequency band a series of resonance frequencies will be observed that arise from the differing total static fields, large magnetic field plus gradient field, which the same type of nuclei experience. By rotating the gradient field through different angles, a plane may be imaged and the spatial density of the nuclei may be reconstructed by standard techniques such as those used in x-ray CT scanning (Vol. I, Chapter 10). For example, if proton resonance is used, virtually no image comes from bone, which clarifies intercranial examination. Other techniques have also been developed, such as scanning of nuclear relaxation times.

Most of these techniques will be described and discussed in the following sections, not to the extent that the reader will be able to go into the laboratory and perform the experiments, but rather in a semiquantitative descriptive way in which the reader will be able to understand the literature reporting what is going on in the laboratories. Such knowledge is extremely important because NMR is completely noninvasive, and, since the energy of an NMR photon is about 10^{-10} that of an x-ray photon, there are no injurious effects. In addition there has been no report of ill effects caused by high magnetic fields. NMR imaging is already a strong supplement to x-ray CT scanning, and within a decade it may well replace x-ray imaging.

In order to understand the physics of NMR we shall use results that may be found in most calculus-based first-year physics textbooks (e.g., Halliday and Resnick, 1981). We shall develop the concept in three steps for clarity of understanding. First, the precession of a spinning top; second, the precession of an electric dipole about the vector direction of an applied electric field; third, the precession of the magnetic dipole of a nucleus about the vector direction of an applied magnetic field. We shall then show how this precessing dipole can be made to signal its precession frequency by the application of an additional magnetic field rotating with the same frequency as the precession frequency. That is, the field rotates in resonance within the dipole; hence the name nuclear magnetic resonance.

Spinning Tops

A spinning top is observed to precess. An object spinning about its axis has an angular momentum **J** associated with it whose vector direction is given by the direction in which the thumb points when the fingers of the right hand are curled about the axis of spin with the fingers pointing in the direction of the spin. A schematic is shown in Fig. 7.11a. There are two forces acting on the top, **F**, upward at the base, and the center of mass, which in a gravity system is downward. Since **F** acts at the pivot point, it exerts no torque. The mass times the acceleration of gravity, mg, does exert a torque because it is some distance from the pivot point. This torque τ is given by the vector cross product **r** × m**g**, where **r** is the distance along the axis from the pivot point

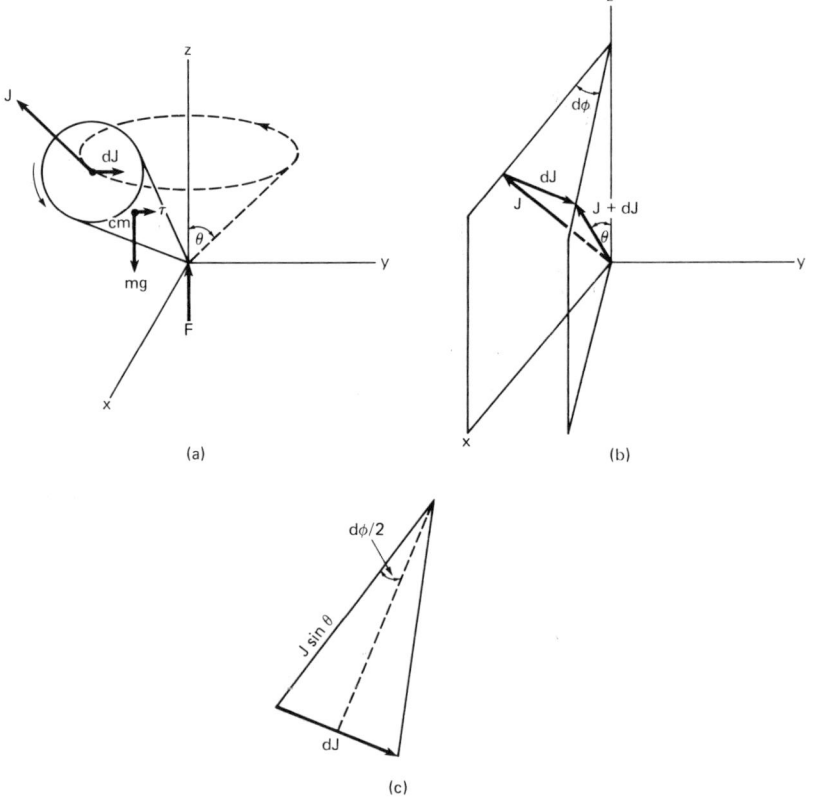

FIG. 7.11 The precession of a spinning top. **J**: angular momentum vector; cm: center of mass; θ: angle top makes with vertical; τ: torque; $d\varphi$: precession angle. (a) Precession. (b) Geometry of incremental precession angle $d\varphi$. (c) Top view of b.

GENERAL ASPECTS OF NUCLEAR MAGNETIC RESONANCE

to the center of mass. The direction of this torque, given by the right-hand rule, is shown in Fig. 7.11a. This torque causes the **J** vector, which is in the direction of the top's axis, to rotate slightly counter-clockwise to the z axis when viewed from above. From the new, slightly rotated position of the **J** vector, the same torque causes a further rotation and so on. Thus there exists a continuous torque tangential to the dashed circle, which causes the top to precess along the circle.

If the top were not spinning, the torque would simply cause the top to fall to the x axis and the torque direction would only show that it falls in a counterclockwise direction in the x–z plane about the pivot point. However, the top is spinning with a large **J** vector, and in this case $\tau = d\mathbf{J}/dt$. Thus $d\mathbf{J}$ is a vector at right angles to **J** and must be added to it, with the resultant angular momentum $\mathbf{J} + d\mathbf{J}$. If $d\mathbf{J} \ll \mathbf{J}$, however, the magnitude of **J** will remain essentially unchanged.

A schematic of the vector addition of $d\mathbf{J}$ to **J** is shown in Fig. 7.11b, greatly exaggerated. The angle $d\varphi$ is that caused by the addition of $d\mathbf{J}$ and is therefore the angle of change of the axis of the top about the z axis. We may calculate the rotational velocity precession ω_p by evaluating $d\varphi/dt$. If one views Fig. 7.11b from the top, as illustrated in Fig. 7.11c, one sees an equilateral triangle with the vector components $\mathbf{J} \sin \theta$ as the legs and $d\mathbf{J}$ as the base. The dashed line is the bisector of the base so that two right triangles are formed with $d\mathbf{J}/2$ as the base of each and $\mathbf{J} \sin \theta$ as the hypotenuse. Thus by definition

$$\sin \frac{d\varphi}{2} = \frac{d\mathbf{J}/2}{\mathbf{J} \sin \theta} \tag{7.5}$$

In the small-angle approximation $\sin \theta \sim \theta$ and, since $d\varphi/2$ is very small, we may write

$$d\varphi = \frac{d\mathbf{J}}{\mathbf{J} \sin \theta} \tag{7.6}$$

Dividing both sides by dt gives

$$\frac{d\varphi}{dt} = \frac{d\mathbf{J}/dt}{\mathbf{J} \sin \theta} \tag{7.7}$$

or

$$\omega_p = \frac{\tau}{\mathbf{J} \sin \theta} \tag{7.8}$$

It is seen from Eq. (7.8) that the larger the angular momentum due to spin **J**, the slower the frequency of precession. In this classical model for the spinning top, all angles θ are allowed. In quantum theory, energy is proportional to frequency and only certain discrete energy states are allowed. Therefore at the atomic or nuclear level not all angles θ are permitted; only quantized ones that obey certain rules of angular momenta are allowed.

We shall next develop the easiest quantized case, that of an electron moving in a Bohr orbit about a nucleus.

Larmor Precession

A loop bearing a current I has a magnetic moment μ equal to the product of the current and the cross-sectional area of the loop. If the radius of the nth Bohr orbit is r_n, then the cross-sectional area is πr_n^2 and

$$\mu = I\pi r_n^2 \tag{7.9}$$

Since there is only one electron of charge $-e$ going around the orbit, or loop, the current is the charge passing a fixed point per unit time. In this case it is the charge $-e$ times the number of revolutions per unit time. The time per revolution is the distance around the orbit $2\pi r_n$ divided by the speed v or $2\pi r_n/v$. Therefore the current I is $-ev/2\pi r_n$ and the magnetic moment arising from the orbiting electron is

$$\mu = \frac{-ev}{2\pi r_n}\pi r_n^2 = -\frac{e}{2}vr_n \tag{7.10}$$

Now multiply numerator and denominator of the right-hand side of Eq. (7.10) by the mass of the electron m_e and recall that the classical angular momentum of a particle of momentum mv at a radius r_n from a point of rotation is $J = mvr_n$. This allows Eq. (7.10) to be rewritten[†] as

$$\boldsymbol{\mu}_l = (-e/2m_e)\mathbf{J}_l \tag{7.11}$$

where the subscript l has been used to denote that this is a relation between the vector *orbital* magnetic moment and its vector orbital angular momentum. The negative sign indicates that the two vectors, magnetic moment and angular momentum, are in opposite directions for a rotating negative charge.

A useful quantity is the ratio of the magnetic moment to the angular momentum or

$$\boldsymbol{\mu}_l/\mathbf{J}_l = -e/2m_e \tag{7.12}$$

This ratio is called the *gyromagnetic ratio* for orbital motion. Another useful quantity is the magnetic moment of the hydrogen atom in its ground state, i.e., $n = 1$. It should be recalled that in the Bohr model of the atom the only

---

[†] $\mathbf{J}$ and $\boldsymbol{\mu}$ are antiparallel vectors in this equation.

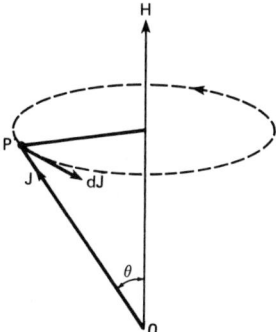

FIG. 7.12 The precession of the magnetic moment of a spinning electron about a magnetic field **H**. *OP* is the axis line along which the dipole moment lies, and **J** is the angular momentum vector.

angular momenta that are allowed are those that are integral multiples of $\hbar$, Planck's constant divided by 2π, or

$$J = m_e vr = m_e \omega r^2 = n\hbar$$

For $n = 1$ and $J = \hbar$, Eq. (7.11) written as the positive quantity becomes

$$\mu_B = e\hbar/2m_e \tag{7.13}$$

This is called the *Bohr magneton*. Although quantum mechanics shows that orbital motion of an electron in the ground state of a hydrogen atom has no magnetic moment, the Bohr magneton is useful as a unit with which to express other magnetic moments.

We now make a classical electrical analogy of the precession of the magnetic moment of an electron about the direction of an applied magnetic field to that of the precession of a spinning top. In Fig. 7.12 a magnetic field of strength **H** is applied in the direction of the z axis and line *OP* represents vector **J** making an angle θ with **H**. Note from Eq. (7.11) that the magnetic moment **μ** is proportional to the angular momentum **J**. Just as an electric field exerts a torque on an electric dipole to try to make it align with the field, so a magnetic field exerts a torque on a magnetic dipole to try to make it align with the field. The magnitude of the torque is given by the vector cross product **μ × H**. We have already mentioned that torque τ is the time derivative of angular momentum, $\tau = d\mathbf{J}/dt$. Therefore

$$\frac{d\mathbf{J}}{dt} = \boldsymbol{\mu} \times \mathbf{H} \tag{7.14}$$

Substituting Eq. (7.11) gives

$$\frac{d\mathbf{J}}{dt} = -\frac{e}{2m_e} \mathbf{J} \times \mathbf{H} \tag{7.15}$$

Since reversing the cross product reverses the sign, this equation can be rewritten

$$\frac{d\mathbf{J}}{dt} = \frac{e}{2m_e} \mathbf{H} \times \mathbf{J} \qquad (7.16)$$

This equation is similar to the equation for a classical top. The vector OP revolves in a circle of radius $J \sin \theta$ and in time dt it revolves through an angle $d\varphi$ as in the case of the classical top. But $\mathbf{H} \times \mathbf{J} = HJ \sin \theta$ and therefore

$$d\varphi = \frac{dJ}{J \sin \theta}, \qquad \omega = \frac{d\varphi}{dt} = \frac{dJ/dt}{J \sin \theta}$$

Substituting Eq. (7.16) gives

$$\omega = \frac{e}{2m_e} HJ \frac{\sin \theta}{J \sin \theta}$$

or

$$\omega_L = (e/2m_e)H \qquad (7.17)$$

which is known as the *Larmor precession frequency*. (Recall that $\omega = 2\pi v$, where v is the frequency.) What is important to note here is that the precession frequency depends only on the gyromagnetic ratio (i.e., $e/2m_e$) and the magnetic field. We shall use these concepts in the next section. The angular momentum $\mathbf{J}$ of this section will be replaced by a vector $\mathbf{I}$, which is the net nuclear spin angular momentum.

NUCLEAR MOMENTS

Nuclei contain electrical charges, and since nuclei spin about their axes, they have both an angular momentum, called spin angular momentum, and motion of charge about their axes, analogous to the motion of the electron in an orbit of the Bohr atom. This motion of charge about an axis generates a magnetic moment whose direction is parallel to the axis of rotation. Not all nuclei have a net spin and associated magnetic moment, however. When pairs of protons in the nucleus occur they align themselves with their spins in opposite directions thereby effectively canceling each other's spin. The same is true with neutrons. Although neutrons have no net charge, they do have spin. However, as with protons, pairs of neutrons cancel each other's spin. Nuclei other than hydrogen (a single proton) are assemblies of protons and neutrons. Thus, such an ensemble nucleus can have a net spin only if it has an odd number of protons or neutrons, or both. This net spin angular momentum vector $\mathbf{I}$ has either integral or half-integral values. For example, the hydrogen nucleus has a value of $I = \frac{1}{2}$, whereas deuterium, ^{2}H, has $I = 1$.

NUCLEAR MOMENTS

From quantum theory it can be shown that the magnitude of the total nuclear spin momentum of a nucleus can be written $\hbar\sqrt{I(I+1)}$. Thus, for hydrogen the total nuclear spin momentum is $\hbar\sqrt{\frac{3}{4}}$. As with electron spin, each nucleus that has a nonzero spin has a magnetic moment vector $\boldsymbol{\mu}$ coaxial with its angular momentum vector $\mathbf{I}$ and proportional to its magnitude. If the proportionality constant is γ, this relation can be written

$$\boldsymbol{\mu} = \gamma \mathbf{I} \tag{7.18}$$

The factor $\gamma = \boldsymbol{\mu}/\mathbf{I}$ is called the magnetogyric or the gyromagnetic ratio by analogy to Eq. (7.12). The magnitude of γ is an intrinsic property of each nucleus, and its value is different for each nucleus.

The energy of either an electric or magnetic dipole in a field is given by the vector dot product of the dipole vector on the field vector. In the case of a magnetic dipole of strength μ in a magnetic field of strength $\mathbf{H}_0$ the energy E is

$$E = -\boldsymbol{\mu} \cdot \mathbf{H}_0 \tag{7.19}$$

The negative sign follows since the dipole is in its minimum energy configuration when it is aligned with the field. This is because the angle in the dot product $(\mu H_0 \sin \theta)^\dagger$ is taken as 90° when the field and dipole direction are the same. The definition of work done to reorient such a dipole is $dw = \tau \, d\theta$, where τ is the torque that must be exerted by an external agent to turn the dipole from its zero energy position ($\theta = 90°$) to any other position θ. Thus, the negative sign follows from the general definition that if a system does work the energy is positive but if work is done on the system it is negative.

It is known from quantum theory that only certain energies are allowed. This discreteness in energies is introduced into Eq. (7.19) by allowing μ to have only certain fixed orientations in an applied field $\mathbf{H}_0$. Note that in the absence of an external field, $\mathbf{H}_0 = 0$, any orientation is permissible. Generally, the z axis is taken as the $\mathbf{H}_0$ field direction. Then a nucleus with nuclear spin I can have magnetic moment values in the direction of the field of the form

$$\mu = \gamma m_I \hbar \tag{7.20}$$

where m_I is called the magnetic quantum number and takes on values

$$m_I = -I, -I+1, \ldots, I \tag{7.21}$$

Thus for a nucleus with $I = \frac{1}{2}$

$$m_I = -\tfrac{1}{2}, +\tfrac{1}{2}$$

---

† Note here that the measurement of angle θ has been changed from that of Figs. 7.11 and 7.12. It is now the angle between the μ axis and the x–y plane; see Fig. 7.14a.

256   7. NEW TECHNIQUES OF BRAIN STUDIES

For a nucleus with $I = 1$

$$m_I = -1, 0, +1$$

For a nucleus in $I = \frac{3}{2}$

$$m_I = -\tfrac{3}{2}, -\tfrac{1}{2}, +\tfrac{1}{2}, +\tfrac{3}{2}$$

It is seen from these examples that the number of different nuclear state orientations in a field $\mathbf{H}_0$ is $2I + 1$.

If Eq. (7.20) is substituted into Eq. (7.19), we obtain

$$E_m = m_I \gamma \hbar H_0 \tag{7.22}$$

where the m_I values are given by Eq. (7.21). The negative sign in Eq. (7.19) has been dropped because the energy will be positive or negative depending

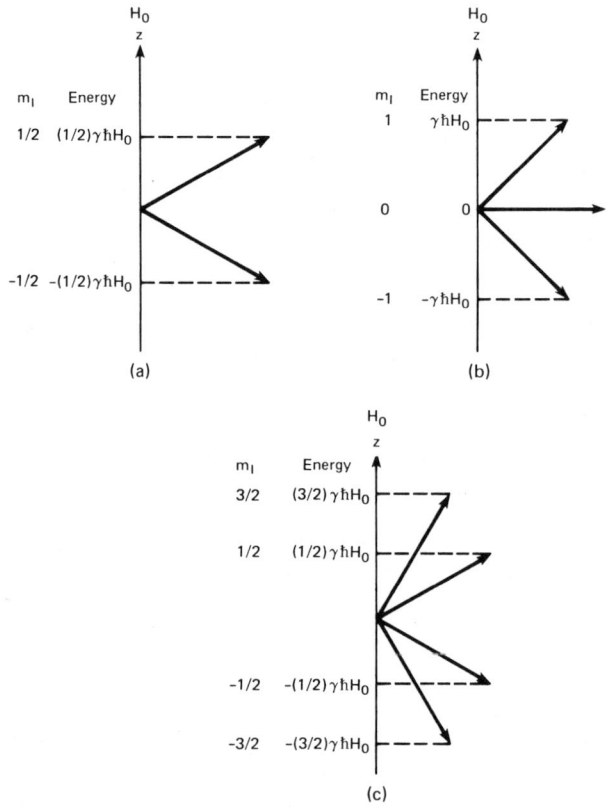

FIG. 7.13 The quantized orientations of the nuclear magnetic moment μ with respect to an applied magnetic field $\mathbf{H}_0$ in the z direction. (a) $I = \tfrac{1}{2}$; (b) $I = 1$; (c) $I = \tfrac{3}{2}$.

EFFECT OF A RADIO-FREQUENCY FIELD

upon whether the nuclear magnetic moment vector is in the direction of $\mathbf{H}_0$ or opposite to it and this is given by the value of m_I, which may be positive or negative. For the three cases given above, the allowed energies will be

$$I = \tfrac{1}{2}: \quad \tfrac{1}{2}\gamma\hbar H_0, \quad -\tfrac{1}{2}\gamma\hbar H_0$$
$$I = 1: \quad \gamma\hbar H_0, \quad 0, \quad -\gamma\hbar H_0 \quad (7.23)$$
$$I = \tfrac{3}{2}: \quad \tfrac{3}{2}\gamma\hbar H_0, \quad \tfrac{1}{2}\gamma\hbar H_0, \quad -\tfrac{1}{2}\gamma\hbar H_0, \quad -\tfrac{3}{2}\gamma\hbar H_0$$

The spatial orientation of these states and the corresponding energies, which are calculated by their projections on the $\mathbf{H}_0$ vector from the dot product relation, are shown in Fig. 7.13.

We have shown on p. 254 that a magnetic dipole will precess about a magnetic field axis with a precession frequency given by the product of the gyromagnetic ratio and the magnitude of the field. If a similar calculation is performed for the precession of the nuclear magnetic moment in an external magnetic field, a similar result is obtained, namely, the precessional frequency is

$$\omega_n = \gamma H \quad \text{or} \quad v_n = (\gamma/2\pi)H \quad (7.24)$$

where the subscript n denotes nuclear precession even though it is still called Larmor precession. The gyromagnetic ratio γ can be either positive or negative, and therefore Eqs. (7.17) and (7.24) can differ in sign. These differences are unimportant since the magnitude of the angular velocity comes into our consideration rather than the direction of rotation. Our ultimate goal is to measure the frequency of rotation v given by $v = \omega/2\pi$ so direction of rotation is not explicitly involved.

EFFECT OF A RADIO-FREQUENCY FIELD

We have seen that if a nucleus with a net spin oriented at random in space is placed in a static magnetic field $\mathbf{H}_0$ in the z direction, the magnetic moment will assume one of the allowed orientations and precess about the vector direction of the field. Since the energy of a dipole in a magnetic field is a function of the angle between the dipole and the field, it follows that the energy of the system will remain constant during the precessional motion and, furthermore, Eq. (7.24) shows that the precessional frequency is independent of the orientation of the dipole with respect to the applied magnetic field. Such a system is spectroscopically uninteresting because there is no mechanism available to extract energy from the system which can be used to measure the frequency of precession.

Consider now what will happen if a second magnetic field $\mathbf{H}_1$ is superimposed at right angles to the first field H_0, i.e., in the x–y plane. The magnetic

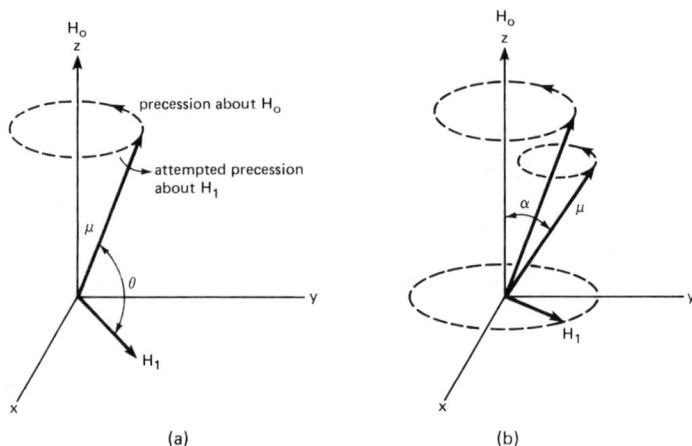

FIG. 7.14 (a) Vector diagram showing the nuclear magnetic dipole moment μ precessing about static field H_0 and attempting to precess about static field H_1 in the x–y plane at the same time. (b) When H_1 rotates in the x–y plane at the Larmor frequency, μ precesses about H_0 and simultaneously about the vector sum $H_0 + H_1$. This causes a periodic variation of the angle α between μ and H_0. Note the change in the measure of angle θ from that employed in Figs. 7.11 and 7.12.

dipole of the nucleus that is precessing about H_0 will also try to precess about H_1. This is indicated in Fig. 7.14a. It can be seen in this figure that if H_1 is fixed, the attempt at precession will be short lived because μ, in its precession about H_0, will leave the field direction of H_1 and encounter it 180° later and attempt to precess in the opposite direction when it is oppositely oriented to the vector direction of H_1. Thus there will be a net cancellation of any effect. However, suppose H_1 is rotated in the x–y plane at the Larmor frequency of the precession of μ about H_0 and in the same direction as the precession. The dipole will then precess about the field H_0 and also about the vector sum of the two fields $H_0 + H_1$. This is indicated in Fig. 7.14b. The angle α between the dipole and the H_0 direction will vary sinusoidally as μ precesses about H_0. This behavior is called *nutation*. It can be seen that this nutational orbit has a projection on the x–y plane that varies with rotation of the dipole. This varying field projection of the dipole also gives rise to an induced signal at the Larmor frequency. This will be considered shortly.

The simplest way to achieve a rotating magnetic field is with a coil having its axis in the x–y plane and carrying an oscillating current. The oscillating current will create a magnetic field along the axis of the coil, which reverses direction with the frequency of oscillation. Suppose the magnetic field is produced by a coil oriented in the x direction and has a magnitude of

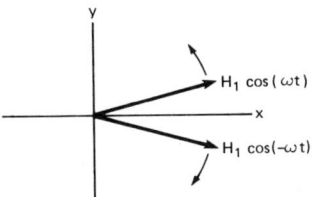

FIG. 7.15 Two vector magnetic fields of amplitude H_1 rotating in opposite directions with the same angular frequency ω.

$H_x = 2H_1 \cos \omega t$, where $\omega = 2\pi f$ and f is the frequency of oscillation. Such a field is mathematically equivalent to two fields of half the amplitude rotating in opposite directions. Figure 7.15 illustrates two such field vectors rotating in opposite directions. Components in the x direction add and components in the y direction cancel. Therefore, $H_x = 2H_1 \cos \omega t$ may be considered to consist of two fields rotating in opposite directions. One rotates in the direction of the precessing dipole whereas the other rotates in the opposite direction and may be neglected. Therefore an oscillator connected to a coil whose magnetic field lies in the x–y plane will result in the rotation of H_1. Frequencies in the megacycle range are required to match the nuclear Larmor frequencies.

As we have noted earlier, the angle θ (Fig. 7.14a) of precession about $\mathbf{H}_0$ is quantized and, therefore, not all angles with respect to $\mathbf{H}_0$ can be occupied by the magnetic moment vector. For example, for a nucleus with $I = \frac{3}{2}$, these allowed energy states are given by Eq. (7.23) and illustrated in Fig. 7.13. Transitions between allowed energy states occur by the absorption or emission of energy, although there is a quantum restriction that only transitions between adjacent energy states are allowed. The frequencies associated with these transitions are readily obtained from Eq. (7.22) and Planck's energy relation,

$$E = h\nu \tag{7.25}$$

where h is Planck's constant and ν is the frequency.

A transition energy ΔE between quantum states m_I and m_I' is given by

$$\Delta E = h\nu_m - h\nu_{m'} = m_I \gamma \hbar H_0 - m_I' \gamma \hbar, H_0 \tag{7.26}$$

(see Fig. 7.13) and, using Eq. (7.22), we may write

$$\Delta E = \gamma \frac{h}{2\pi} H_0 |m_I - m_I'| \tag{7.27}$$

It is seen in the examples of Eq. (7.23) that the positive difference $|m_I - m_I'|$ between any two adjacent states is unity, and it follows that

$$\nu = \gamma H_0 / 2\pi \tag{7.28}$$

TABLE 7.1

Isotope	I	ν in 10^4 G field (Hz)[a]	μ[b]
^{1}H	$\frac{1}{2}$	42.58×10^6	2.79
^{2}H	1	6.54×10^6	0.86
^{31}P	$\frac{1}{2}$	17.24×10^6	1.13

[a] Modern experiments express the magnetic field in teslas (T) where $1\,T = 10^4\,G$. Therefore the magnetic moment of the proton μ_p is 1.41×10^{-26} J/T. From note b the nuclear magneton μ_N has the value of 5.0509×10^{-27} J/T and thus $\mu_p = 2.793\mu_N$.

[b] In units of the nuclear magneton $\mu_N = e\hbar/2m_p c$, where m_p is the mass of the proton.

This equation is identical to the Larmor precession frequency, Eq. (7.24). The significance of this is that if the applied oscillation of the H_1 field is at the frequency of the dipole precession about H_0, absorption and emission of energy $h\nu$ from the two adjacent oriented dipole energy states will have the same frequency as the precession frequency. Equation (7.28) shows that NMR spectroscopy is similar to other types of spectroscopy in that energy absorption from an applied electromagnetic field corresponds to transitions between energy levels. Some examples of nuclear constants of use to physiological studies are shown in Table 7.1.

POPULATION AND RELAXATION

Let us consider the simplest case of $I = \frac{1}{2}$, for example, the proton, and ask what is the relative population of spins between the upper and lower states in a magnetic field 10 kG (see Fig. 7.16). Define $n(2)$ and $n(1)$ as the population in the upper and lower states, respectively. When the spin system is in thermodynamic equilibrium, the population will obey Boltzmann statistics in that the population ratio is given by the relation

$$n(2)/n(1) = \exp(-\gamma\hbar H_0/kT) \qquad (7.29)$$

where k is Boltzmann's constant and T is absolute temperature. At room temperature for the proton in a static magnetic field of 10^4 G this ratio is approximately $1 + 7 \times 10^{-6}$, which means that for every 1,000,000 nuclei in the upper-energy level, there will be 1,000,007 in the lower-energy level (the energy state in which the nuclei are oriented in the field direction). This slight difference in population gives rise to a slight macroscopic magnetization of the specimen.

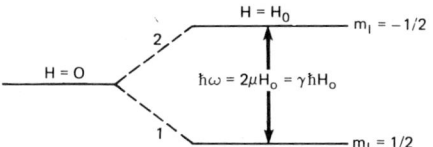

FIG. 7.16 Splitting of energy levels of a nucleus with spin $I = \frac{1}{2}$ when placed in a magnetic field.

We are accustomed to thinking about electronic transitions in spectroscopy in which an electron excited to a higher-energy state spontaneously emits energy and decays to a lower-energy state. Because of the prompt decay of the excited electrons, absorption spectroscopy measurements may be performed. Although there are quantum transition probabilities for such decays, the varying electric fields caused by the motions of neighboring atoms or molecules introduce perturbations that permit the rapid decay from excited energy states. This is not the case in nuclear resonance spectroscopy. The nuclei do not always decay promptly to the lower state, but remain in the upper state until acted upon by other magnetic fields of surrounding nuclei, which, if they satisfy the Larmor frequency condition of Eq. (7.28), will stimulate decay to a lower-energy state. Such interaction is a relatively infrequent occurrence because nuclear magnetic interactions are extremely small, in general. Many nuclei have no magnetic moment, so a nuclear magnetic dipole's immediate environment may contain only small numbers of nuclei with magnetic moments. Still, under certain conditions to be described, interaction does occur, which permits the transitions. These interactions that cause transitions back to the lower-energy state are called *relaxation processes*. The type just described, which causes stimulated emission to a lower-energy state, is called *spin–lattice relaxation*. It promotes essentially an exponential decay from higher-energy states with a characteristic exponential relaxation time called T_1. There is another phenomenon in which the stray magnetic fields cause the individual moments to assume a wide variety of Larmor frequencies and thereby spread the observed signal over a band of frequencies. This process is called *spin–spin relaxation*. This relaxation also obeys an essentially exponential decay, and its characteristic relaxation time is called T_2.

So far, we have discussed the behavior of a single magnetic dipole in static plus rf magnetic fields. Once a particular isotope and its associated Larmor frequency are selected, millions of this moment respond. The ensemble of magnetic moments which is actually measured results in a composite magnetization vector called **M**, where

$$\mathbf{M} = \sum_{\text{unit volume}} \boldsymbol{\mu}_i \qquad (7.30)$$

When the rf field is applied individual moments are transferred to higher-energy states thereby causing **M** to change. The stronger the rf field and the longer it is applied, the more individual moments are transferred into the higher-energy state until a saturation magnetization is achieved. A pulse long enough and strong enough to tip **M** from its initial position until it is rotating in the x–y plane is called a 90° or $\pi/2$ pulse. If the rf field is applied for still longer periods of time, the net spins of the system will invert, that is, they will align with their magnetic moments in a direction opposite to that of $\mathbf{H}_0$. An rf pulse applied long enough to accomplish this is called a 180° pulse.

SPIN–SPIN RELAXATION

The imposition of a large static magnetic field $\mathbf{H}_0$ in the z direction causes all the individual moments of an ensemble to precess about the $\mathbf{H}_0$ direction with the same frequency. They may be envisioned as a cone formed by a bunch of straws in a low glass sitting at the center of a turntable. Equations (7.14) and (7.18) may be combined to show that

$$\frac{d\boldsymbol{\mu}}{dt} = \gamma \boldsymbol{\mu} \times \mathbf{H} \tag{7.31}$$

The vector $\boldsymbol{\mu}$ has components $\mu_x i$, $\mu_y j$, and $\mu_z k$ whereas $\mathbf{H}$ has only a vector direction H_z. The cross product

$$(\mu_x i + \mu_y j + \mu_z k) \times H_z$$

substituted into Eq. (7.31) then yields three equations (note that

$$\mu_x i \times H_z = \mu_y H, \quad \mu_y j \times H_z = \mu_x H, \quad \text{and} \quad \mu_z k \times H_z = 0)$$

and therefore

$$\frac{d\mu_x}{dt} = \gamma \mu_y H, \quad \frac{d\mu_y}{dt} = \gamma \mu_x H, \quad \frac{d\mu_z}{dt} = 0 \tag{7.32}$$

As expected, μ_z is constant. The coupled differential equations may be solved by standard procedures into terms involving sines and cosines. Since each moment obeys these solutions and their x and y components are randomized over a large number, there is no net change of dipole moment with time, and hence no signal even though there is a net magnetization in the z direction as in Eq. (7.30). When an rf field $\mathbf{H}_1$ is applied which rotates at the frequency of rotation of these individual moments they align with this $\mathbf{H}_1$ vector. Thus, during the time of field application they sum into a single **M** vector which has a rotating component in the x–y plane (Fig. 7.17a). This rotating **M** dipole has a varying field and therefore induces a signal at the Larmor frequency in

SPIN–SPIN RELAXATION

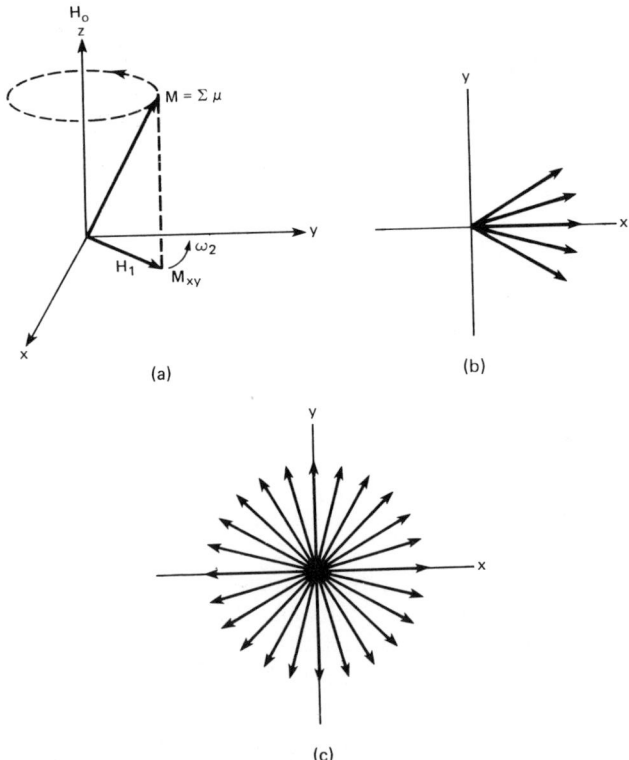

FIG. 7.17 (a) The magnetic vector **M**, the sum of the individual nuclear magnetic moments μ, rotating about $\mathbf{H}_0$ at the Larmor rotation velocity ω_L, and the projection of **M** in the x–y plane. (b) As local magnetic fields perturb the individual moments, each moment begins to precess at slightly different Larmor frequencies and their individual vectors begin to fan out in the x–y plane. (c) At a later time they have fanned out in all directions so that there is no net **M** in the x–y plane.

the detector coil. If $\mathbf{H}_0$ and $\mathbf{H}_1$ were the only magnetic fields, the **M** vector component in the x–y plane would continue to produce a signal after the rf field is turned off. However, there are other stray magnetic fields present, and since each of these individual nuclear dipole moments is at a different spatial location in the sample, **M** will experience slightly different magnetic fields. An obvious one, of course, is inhomogeneities in $\mathbf{H}_0$. There is no perfect large magnet, and the differences in total field experienced by the individual dipoles will cause each to have a slightly different Larmor precession frequency. Some will have a slightly larger frequency and some smaller. The effect is that the individual dipoles will begin to fan out in their cone of

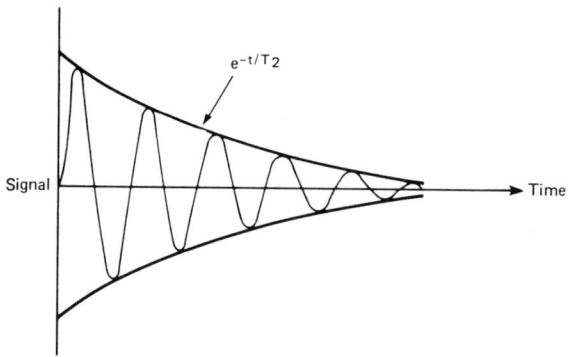

FIG. 7.18 The exponential decay of the envelope of frequencies as the frequency distribution function broadens because of spin–spin relaxation. This signal is called free-induction decay (FID).

rotation about $\mathbf{H}_0$ so that their net projection on the x–y plane will be zero. This is illustrated in Figs. 7.17b and c. An experimental method will be described shortly, called the spin–echo technique, which will allow the experimenter to compensate for magnet inhomogeneities. However, such inhomogeneities are not the only ones that the individual magnetic dipoles experience. Within the sample there are other nuclei that have magnetic moments and the fields that they contribute also perturb the local fields of the individual dipoles so that their resonance frequencies change. The change of the maximum signal when all the dipoles align until there is no net dipole signal is called the *free-induction decay* (FID). The decay is approximately exponential, and the time constant is designated as T_2 (Fig. 7.18). This decay of dipole alignment is called spin–spin relaxation. If the nuclei are in a liquid, the motion of the surrounding atoms or molecules is quite rapid and the perturbing magnetic field averages to a uniform field that all dipole moments experience equally. Thus they do not fan out very rapidly. For example, in pure water T_2 can be several seconds. In a solid, however, such averaging of the magnetic fields of other nuclei does not take place, and instead each dipole may experience a slightly different field. In this case the change in individual frequencies is very rapid. T_2 in a solid may be on the order of microseconds. A measurement of T_2 is used to characterize the environment of the nucleus under study as "liquidlike" or "solidlike." NMR was first studied in solids, and it is from these studies that the term spin–spin relaxation was named. Although the former discussion is valid, the origin of the free-induction decay may be interpreted in a quite different way.

Suppose a 90° rf pulse has been applied to a solid. A nucleus precessing at the Larmor frequency will produce a rotating magnetic field of this

SPIN–SPIN RELAXATION

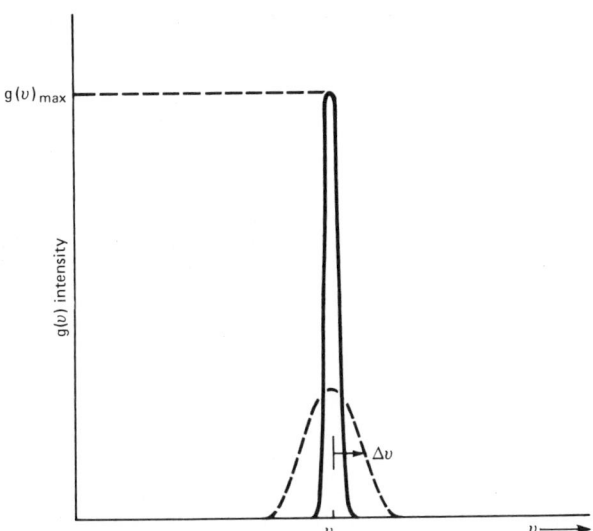

FIG. 7.19 Shape function (distribution function) $g(v)$ versus frequency. v_L is the Larmor frequency, and $g(v)_{max}$ occurs when all moments are at this frequency (solid curve). Inclusion of spin–spin interactions causes the magnetic moments to fan out, and the shape function broadens (dashed curves). $\Delta v = 1/T_2$ is the half-width at half-height of the broadened function.

frequency at the location of a similar nucleus. This added magnetic field at the other nucleus will cause a shift in its Larmor frequency and hence a different energy. The second nucleus does the same thing to the first nucleus, and thus they both flip their spins and they have exchanged energy. Although this process does not change the number of spins in each state, it is the origin of the name spin–spin relaxation. However, this process shortens the lifetime of the energy states because, by the uncertainty principle (see Vol. II, p. 141) it will broaden the frequency spectral line by an amount

$$\Delta v = 1/\Delta t = 1/T_2 \qquad (7.33)$$

This broadening of a spectral line reduces the magnitude of the observed frequency. A schematic of this effect is shown in Fig. 7.19. If one has a sharp resonance at the Larmor frequency, a magnitude of $g(v)_{max}$ is observed, where $g(v)$ is the probability distribution function of the frequencies (also called shape function or envelope function). If an uncertainty of frequency Δv is introduced through T_2, this can be represented by the half-width at half-height and the initial peak has decayed to the height of the broader peak. The characteristic decay time for this peak decay is T_2.

SPIN–LATTICE RELAXATION

When the rf field H_1 is applied the moments can absorb energy and change their quantum energy states to higher values (Fig. 7.13), which in effect shifts the population of the states from their thermal equilibrium values as expressed by Eq. (7.29). The longer the rf field is applied, the larger the population of the higher-energy state becomes. Since the transition probability for absorption and emission is the same, and since the rate of transitions is the product of the transition probability and the concentration in each state, a steady state will be achieved in which the rate of change is equal in both directions. This steady state is the maximum achievable magnetization M_0 for a given temperature and magnetic field.

Consider the case of the proton whose change in energy states in a magnetic field is illustrated schematically in Fig. 7.16. Let N_1 be the number of spins in the lower-energy state per unit volume and N_2 the number in the upper-energy state per unit volume. Then the population difference is $N_1 - N_2$ and the net magnetization in the z direction is

$$M_z = (N_1 - N_2)\mu$$

If M_0 is the equilibrium value, the rate of change of spin population due to a magnetic field in the z direction will be given by

$$\frac{dM_z}{dt} = \frac{M_0 - M_z}{T_1} \quad (7.34)$$

where T_1 is a time constant for the change. That is, at $t = 0$ there is no magnetic field and $M_z = 0$. When the field is turned on the magnetization will increase with time to the equilibrium value of $M_z = M_0$. Integrating Eq. (7.34), we obtain M_z as a function of time

$$\int_0^{M_z} \frac{dM_z}{M_0 - M_z} = \frac{1}{T_1} \int_0^t dt$$

$$\ln \frac{M_0}{M_0 - M_z} = \frac{t}{T_1} \quad (7.35)$$

$$M_z(t) = M_0[1 - \exp(-t/T_1)]$$

Equation (7.35) is an equation of exponential growth (Vol. I, p. 39), which is illustrated as M_z versus time in Fig. 7.20. At saturation, $t = \infty$ and $M_z = M_0$. If at saturation or some earlier time the magnetic field is removed, then M_z will decay to zero exponentially with the same relaxation time, as shown in Fig. 7.20. This can be seen as follows. Suppose, as in Fig. 7.20, the magnetic field is shut off when the magnetization has reached a value of M_{zt}. The rate

SPIN–LATTICE RELAXATION

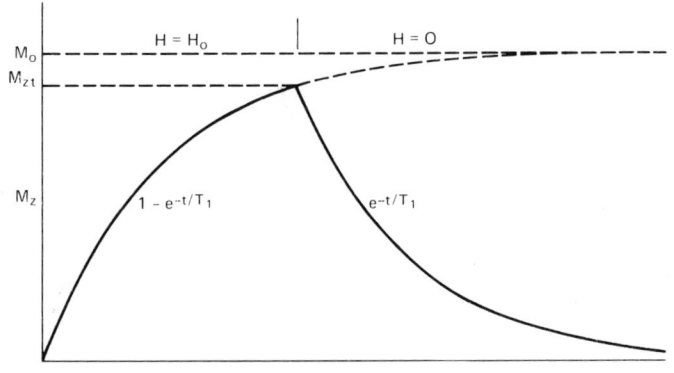

FIG. 7.20 Exponential growth of magnetization M_z as a function of time when sample is placed in a magnetic field. M_0 is the saturation magnetization governed by Eq. (7.29). If the magnetization has reached value M_{zt} at some time t and the field is turned off, this magnetization will decay exponentially to zero.

of decay with time of magnetization will be proportional to the amount of magnetization present, with the proportionality constant being $-1/T_1$

$$\frac{dM_z}{dt} = -\frac{M_z}{T_1} \tag{7.36}$$

$$\int \frac{dM_z}{M_z} = -\frac{1}{T_1} \int dt$$

$$\ln M_z = -\frac{t}{T_1} + C$$

$$M_z = C \exp(-t/T_1)$$

but when $t = 0$, $M_z = M_{zt}$ and therefore $C = M_{zt}$ or

$$M_z = M_{zt} \exp(-t/T_1) \tag{7.37}$$

Although Eqs. (7.34), and (7.36) were written for static field magnetization, they apply equally to rf field magnetization. The size of T_1 and hence the rate of return of the spins to thermal equilibrium also depends on transient magnetic fields having the appropriate frequencies to effect the transition. In a rigid lattice or in a sample at low temperatures, a broad spectrum of transient magnetic fields is rare and, in contrast to the case of T_2, T_1 is very long in a solid and very short in a liquid, which does have a broad spectrum of fields present.

SPIN–ECHO TECHNIQUE

The spin–echo technique averages out inhomogeneities in the magnetic field H_0 and permits the measurement of the spin–spin relaxation time T_2. This technique can best be described by reference to Fig. 7.21. Before the application of the rf pulse the resultant magnetization is aligned with H_0. As an rf pulse is applied at the Larmor frequency the magnetization vector **M**

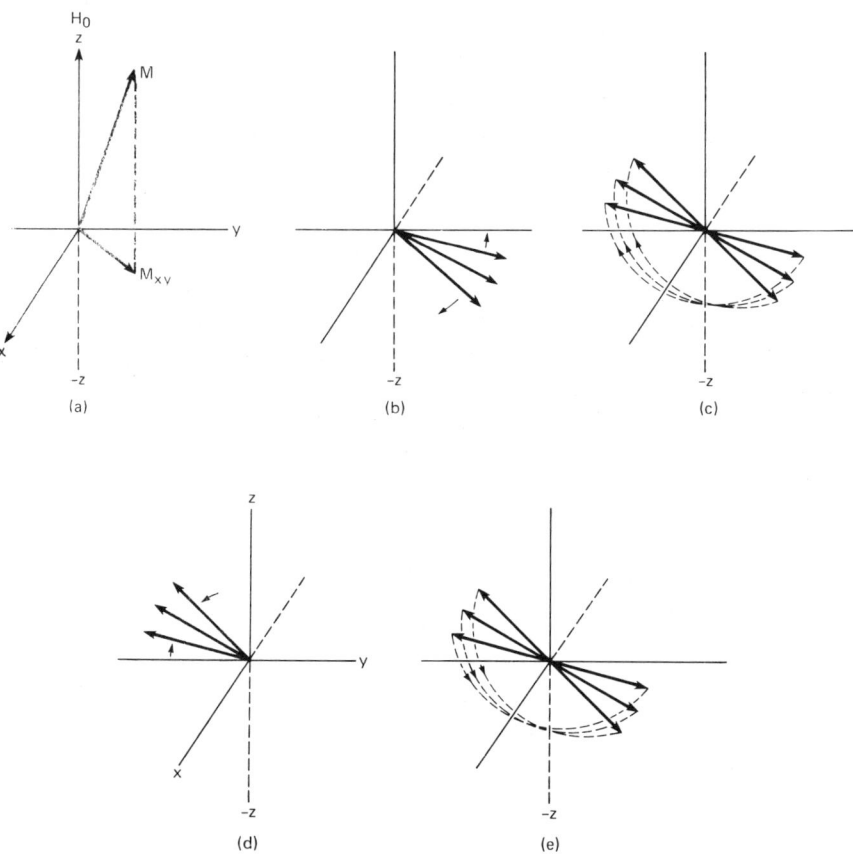

FIG. 7.21 Steps in the spin–echo technique. (a) The resultant magnetic moments align along the applied field before application of a rf field. (b) After a 90° rf pulse is applied. After termination of the pulse the individual moments begin to fan out in the x–y plane. (c) The application of a 180° pulse, which rotates the moments through the $-z$ direction and back into the x–y plane. The moments are now pointing in the opposite direction. (d) The individual moments now move in the opposite direction, thereby closing the fan. (e) Another 180° pulse is then applied. This rotates the moments through the $-z$ direction back into the x–y plane after which the moments again begin to fan out.

NUCLEAR MAGNETIC RESONANCE IMAGING

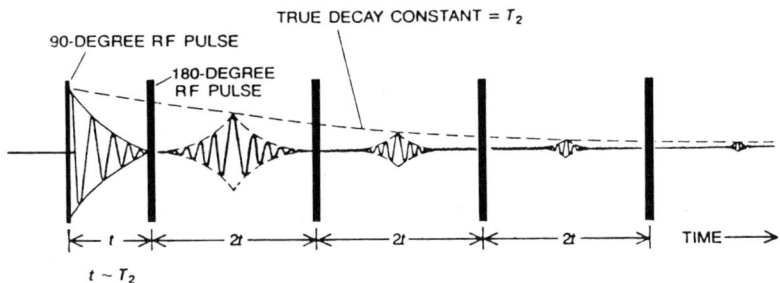

FIG. 7.22 NMR signal patterns in the spin–echo method. A 90° rf pulse is applied. After a time t a series of 180° pulses is applied at time intervals of $2t$. The decay in magnitude of the spin–spin relaxation is strongly dependent on the decay constant T_2. [From Pykett (1982). Copyright © 1982 by Scientific American, Inc. All rights reserved.]

develops a projection in the x–y plane and, as before, all of its component dipole moments will tend to align in a common direction. If the pulse is long enough or strong enough, a 90° pulse is achieved in which **M** lies totally in the x–y plane (Fig. 7.21b). Because of spin–spin relaxation, when the rf field is turned off the component dipoles begin to spread as coherence is lost. This is indicated in the figure by the fanning out of the individual magnetic moments with time. After some time t a 180° rf pulse is applied. This causes the dipoles to go through the $-z$ direction (a 90° pulse from the x–y plane) and return to the x–y plane in the reverse direction as indicated in Fig. 7.21c. After termination of the 180° pulse the direction of rotation of the individual moments remains unchanged. This motion causes them to be realigned in the reverse direction. They then begin to fan out again. As subsequent 180° pulses are applied the reversals of **M** cancel out the H_0 field inhomogeneities and only spin–spin effects cause the decrease of the FID. The selection of the time between the pulses is chosen as approximately twice the time for FID. This is shown in Fig. 7.22. The dashed line indicates the decay rate, which is of the form $\exp(-2t/T_2)$, and from this T_2 can be determined.

NUCLEAR MAGNETIC RESONANCE IMAGING

The development of x-ray computerized tomography (CT) for the brain was discussed in Vol. I, Chapter 10. In this device a collimated beam of x rays is passed through the brain in a slice about 1 cm in width from a variety of different angles, and a suitable algorithm is used with a computer to reconstruct the image. An elementary discussion of image reconstruction was given; for a more general discussion, see Brooks and DiChiro (1976). Usually, seven 1-cm contiguous slices are made so a 7-cm depth of the brain has been scanned. The total x-ray dosage in this scan is about 3 rem (Vol. I, p. 258). The maximum allowable dosage for workers in radiation environment in any 13-week period as established by the U.S. Department of Energy is 3 rem. Because of the concern of possibly creating more tumors than they cure, radiologists are naturally reluctant to follow the progress of therapy by a series of scans too closely spaced in time. Positron-emission tomography, discussed earlier, will probably require a lower dosage, but clearly there is a need for a noninvasive, nondamaging image of the brain and other internal organs.

Recent advances in NMR imaging have developed this technology so that brain scans can be produced with a spatial resolution and signal-to-noise ratio (Vol. II, pp. 184, 270) similar to the first generation of x-ray CT and without the deleterious effects. The only measurable physiological effect is a 0.3°C temperature rise in the brain from microwave heating, which is not hazardous. Three other distinct advantages of NMR imaging are (1) there are no moving parts, i.e., the control of magnetic field direction is done electronically; (2) because all directions of scanning are equally accessible, multiplane (axial, coronal, and sagittal) scans can be made and three-dimensional constructions are easily obtained; and (3) imagings of T_1 and T_2 as well as ^{31}P are possible.

As discussed earlier in connection with the sampling theorem (p. 245), if an object is to be imaged, the wavelength of the radiation field must be half the size of the smallest features that are desired to be resolved. One might expect, therefore, that the long wavelength of rf resonance frequencies used in NMR precludes any thought of using it for an imaging process. However, Lauterbur (1973, 1974) proposed a method to remove this wavelength restriction. If a second field is superimposed on the system that restricts the interaction of the object with the first field to a limited region, the resolution becomes independent of wavelength. He suggested that a small gradient magnetic field $\mathbf{H}_2$ be superimposed on the NMR system in a different direction from $\mathbf{H}_0$ and the rf field $\mathbf{H}_1$. The resulting field is the vector sum of the fields and, because the fields are coupled by the object, he proposed the name "zeugmatography," where zeugma is the Greek word for "that used for joining." Let us briefly consider how this is done. Figure 7.23 shows a schematic of two vials of water in a static magnetic field $\mathbf{H}_0$ in the upward or z direction. Suppose a static small gradient magnetic field H_2 is superimposed by small coils with typical gradient magnitudes of 10^{-4} T/cm. Each point in the vials will then experience a static field $\mathbf{H} = \mathbf{H}_0 + (\partial \mathbf{H}_2/\partial y)y$ depending on its y location. The Larmor frequency of the protons in the water is given by

$$v_L = (\gamma/2\pi)H \qquad (7.28)$$

NUCLEAR MAGNETIC RESONANCE IMAGING

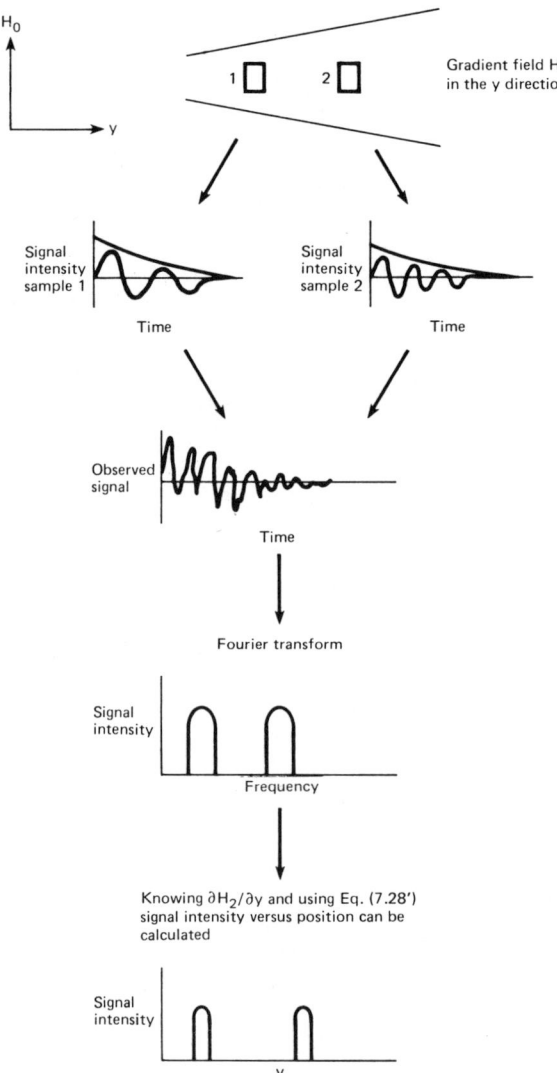

FIG. 7.23 Steps in obtaining a line image of two vials of water, 1 and 2. They are placed in a field $\mathbf{H}_0$ in the z direction and a small gradient magnetic field $\partial H_2/\partial y$ in the y direction. An rf field is applied with a broad band of frequencies around the proton resonance frequency. Each vial experiences a different field and therefore has a different Larmor frequency, which decays after the field is removed. The observed signal is a mixture of the two frequencies. A Fourier transform applied to the signal separates the signal into the two frequencies. If one knows the gradient field, the frequencies can be located in the y position, thereby permitting the location of the vials along a line to be determined.

272   7. NEW TECHNIQUES OF BRAIN STUDIES

There is no longer a single excitation frequency but a band of frequencies within the range of

$$v_L = \frac{\gamma}{2\pi}\left[H_0 + \left(\frac{\partial H_2}{\partial y}\right)y\right] \qquad (7.28')$$

where the band will be small because the gradient is small. The rf excitation pulse is then swept through this band instead of being a single frequency. The detector coil will register a band of frequency responses. This frequency band cannot be interpreted directly, but a Fourier transform (Appendix G) can be performed, giving a plot of signal intensity versus frequency. Since from Eq. (7.28) the frequency is proportional to the strength of the magnetic field and the magnetic field strength at each position is known, the spatial location of the points of maximum intensity can be calculated (see Fig. 7.23). This has produced a line projection of the objects. By rotating either the objects or the gradient, a reconstruction of the objects can be obtained by computer analysis of the one-dimensional projection taken at different angles. These angles can be achieved without any moving parts simply by having three small gradient coils in orthogonal directions. Varying their magnitudes electronically can change the direction of their resultant vector sum. Analyses of scanning projections done in this manner have been shown for both two-dimensional (Lauterbur, 1973, 1974) and three-dimensional reconstruction (Lauterbur and Lai, 1980; Lai and Lauterbur, 1980). Figure 7.24 shows

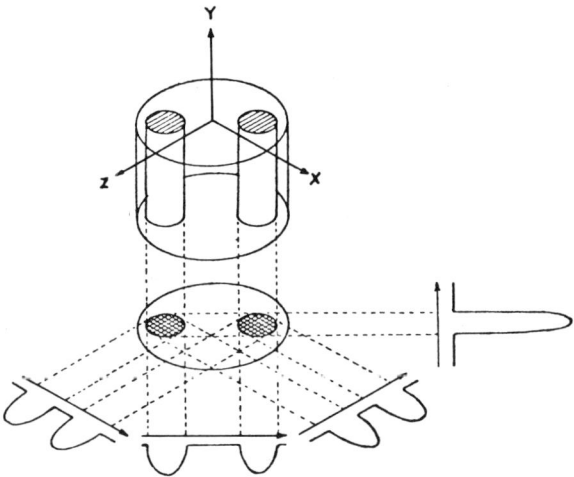

FIG. 7.24   Position versus signal intensity of two vials of water of Fig. 7.23 as a result of four different magnetic gradient directions, indicated by the arrows. [Reprinted by permission from Lauterbur, *Nature* **242**, 196 (1973). Copyright © 1973 Macmillan Journals Limited.]

sample scans in two dimensions of an object consisting of two 1-mm tubes of H_2O attached to the inside wall of a 4.2-mm-diam tube of D_2O. In this experiment the magnetic field gradient was rotated about the sample. The arrows show the direction of the four scans. The projections were used to reconstruct a two-dimensional projection of the object in a manner analogous to x-ray CT.

The requirement for reconstruction has been eliminated by what is known as the sensitive-point method, developed by Hinshaw (1976). In this method the magnetic field gradients are varied sinusoidally in time in such a way that only one point in the sample is in a stationary magnetic field, i.e., time independent. The signal from this sensitive point is selected by recording only the time-independent part of the total signal. This technique makes it possible to select one point in the sample and to measure its density, relaxation times, and macroscopic flow and the presence of NMR sensitive nuclei other than hydrogen. All of this information is capable of being imaged, not just the proton density.

This procedure can be shown in the following simple way in two dimensions (Lai et al., 1979). The principle is that linear field gradients are vectors that may be generated from orthogonal components. Suppose that the experiment is arranged so that two pairs of solenoids, a and b, are orthogonal to each other and generate magnetic field vectors **A** and **B**, respectively, which at point i produce the resultant vector **C**. Thus, at point i the vector addition is

$$\mathbf{C}_i = \mathbf{A}_i + \mathbf{B}_i \tag{7.38}$$

In general, the two sets of coils will produce a vector field

$$\mathbf{C} = \mathbf{A} + \mathbf{B} \tag{7.39}$$

and, since the **A** and **B** fields are orthogonal, vector addition is defined by the relation

$$|\mathbf{C}|^2 = |\mathbf{A}|^2 + |\mathbf{B}|^2 \tag{7.40}$$

If field **A** is produced by coil current $aI_A \sin\theta$ and **B** is produced by $bI_b \cos\theta$, where a and b are constants, then

$$|\mathbf{C}|^2 = |aI_a \sin\theta|^2 + |bI_b \cos\theta|^2$$

and

$$|\mathbf{C}| = [(aI_a \sin\theta)^2 + (bI_b \cos\theta)^2]^{1/2} \tag{7.41}$$

An equality of $aI_A = bI_b = G$, where G is a constant, may be achieved with the use of a calibrating sample by adjusting aI_A when $I_B = 0$ and bI_B when

274   7. NEW TECHNIQUES OF BRAIN STUDIES

$I_A = 0$. With this adjustment this equality may be substituted into Eq. (7.41) to obtain

$$|C(\theta)| = G (\sin^2 \theta + \cos^2 \theta)^{1/2} = G \qquad (7.42)$$

which applies to any point i.

In this way, since the amplitudes are equal, a time-independent null point will be obtained. Each gradient is produced by current through coils located symmetrically on each side of the sample region. The gradient is produced by controlling individually the current through each coil. By changing the magnitude of current through one of the coils the sensitive point will be displaced along the gradient axis and can be scanned across the object by adjusting the relative amplitude of the currents. A third orthogonal set of coils is added to give a sensitive point controllable in three dimensions. Scanning is now done automatically by microprocessors that achieve three-dimensional scanning with the senstive point. In this way scans of slices may be produced with the slice plane in any direction.

As the location of this sensitive point is systematically moved across the sample the magnitude of the signal can be recorded on successive lines of an x–y plotter or put onto the screen of a storage oscilloscope. An early example of proton resonance in water with such a series of traces of the sensitive point on an x–y plotter is shown in Fig. 7.25. The scanned object shown on the right-hand side is a glass tube containing water and a glass capillary with water in it. The rise distance from glass to water gives an estimate of the spatial resolving power of the system.

There are distinct advantages of the sensitive-point method over the projection method. The most important is that the requirement for the uniformity of the magnetic field is much less. The location of the sensitive point can be obtained by calibration scans and the resulting image corrected

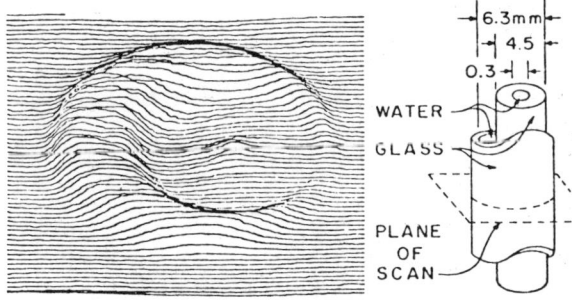

FIG. 7.25 x–y plotter traces produced by the sensitive-point method. A diagram of the object with the plane of scan is shown on the right-hand side. [From Hinshaw (1976).]

for this. Also nonuniformities in the large static magnetic field can be corrected by the same method. Second, a small region of an object may be scanned without acquiring equivalent data from the remainder. Third, reconstruction computers and algorithms are not required. In addition, as will be mentioned shortly, scans may be made of T_1 or T_2 and associated images produced. These are sometimes more useful than proton-density scans.

The sensitive-point method was shown by Hinshaw (1976) to be a function of the proton NMR response and not just the proton density. Although the response is a function of the density, it is also dependent on the T_1/T_2 ratio of relaxation times. This response was used to map planes in the cross section of a human wrist. A typical result is shown in Fig. 7.26, from Hinshaw et al. (1978, 1979). By comparison of a series of adjacent tomographs these investigators concluded that the tomograph slice thicknesses are about 5 mm and the resolution within a slice is about 0.4 mm.

This technique has been extended to imaging of the brain with initial results approaching the quality of early x-ray CT scans (Holland et al., 1979). The advantage is the ease with which coronal and sagittal cross sections are imaged.

Whole-body NMR imaging machines have been devised (Hutchison et al., 1980; Crooks et al., 1982). A cross section of the body can be imaged both by proton density and by T_1 and T_2 images. Edelstein et al. (1980) report that T_1 images generally show more interesting information than proton-density images. This is because the range of proton density values of 70–90% water in soft tissue is much less than the 100–400-msec range of T_1 values at 1.7 MHz. The measured uncertainties are not that different, about 3.5% for T_1 and 2.5% in proton densities.

A further interesting development is underway in NMR imaging. ^{19}F has a gyromagnetic ratio comparable to hydrogen and also has spin $\frac{1}{2}$. In addition, fluorine has a negligible natural occurrence in biological tissues. Therefore the same NMR instrument can be used both to map a portion of the body and to locate fluorine. Imaging of ^{19}F has been demonstrated by Holland et al. (1977). An important application is in the imaging of vascular systems and defects therein. Fluorocarbons have been developed as artificial blood substitutes when they are emulsified with a buffered aqueous solution and a nonionic surfactant. Therefore blood flow and vascular imaging by NMR can be made after an injection of this material, and disorders may be detectable even in local regions in internal organs which are currently difficult to image. Differences in blood velocity are measurable because a liquid flowing through an NMR device after the rf pulse has no nuclei in the excited state and therefore will not emit a signal. Adjustment of time between rf pulses can therefore yield a range of grays on the screen whose relative brightness can be calibrated to the fluid velocity.

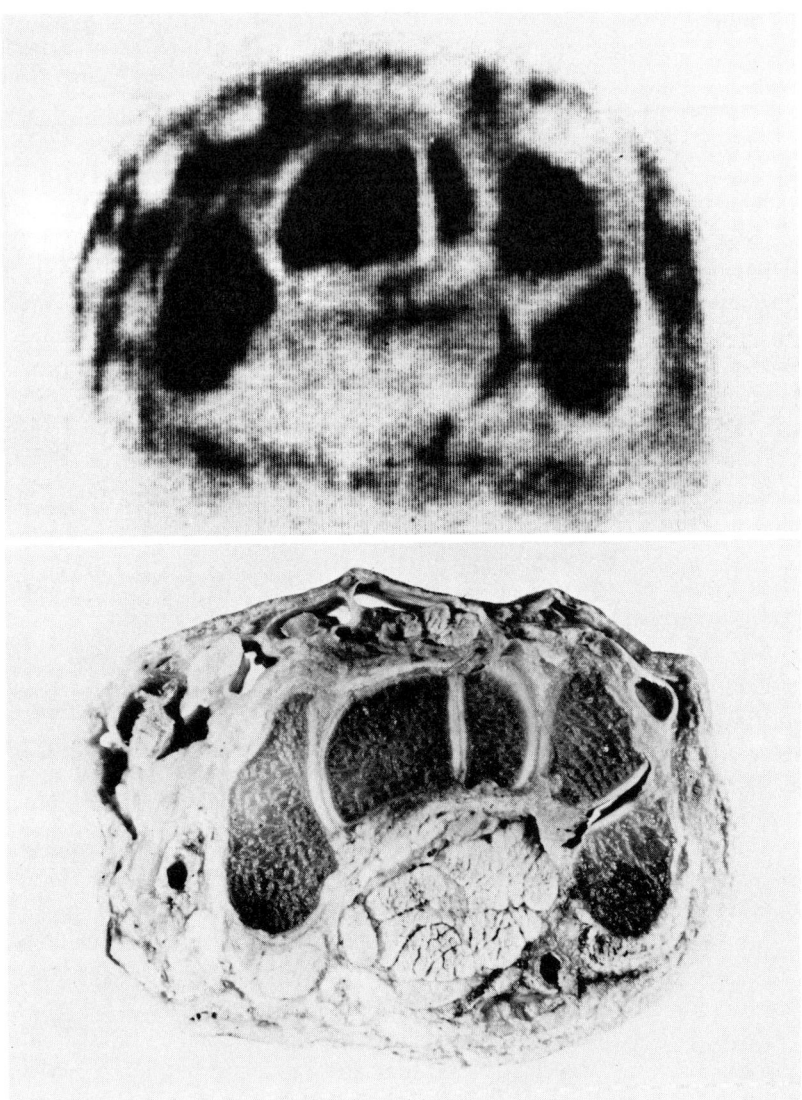

FIG. 7.26 (a) NMR image through a human forearm. (b) Anatomical drawing. [From Hinshaw et al. (1978).]

CHEMICAL SHIFTS

Very exciting research has been taking place in the study of NMR spectroscopy of ^{31}P in-ATP metabolism. Although most studies so far have been only on muscle, this technique is currently being applied to brain studies.

The usefulness of nuclear magnetic resonance resides in the fact that the effective total magnetic field acting at a particular nuclear site is the resultant of the applied magnetic fields and an additional field caused by the screening or shielding effects of the electrons surrounding the nucleus. As an example, we note that each carbon atom in the benzene ring of the ethyl vanillin

ethyl vanillin

molecule experiences a slightly different resultant magnetic field when placed in a constant uniform magnetic field, since the local electron distribution for each carbon atom in the ring is slightly different. It is this situation that makes NMR useful in molecular structural determinations and as a tool for probing metabolic pathways in cells. If we denote by H_0 the applied field, then the total magnetic field H that a given nucleus experiences can be written

$$H = H_0(1 - \sigma) \tag{7.43}$$

The parameter σ is called the *shielding* or *screening factor* and is generally a very small number, the order of 10^{-4} or less. Unfortunately the absolute values of σ cannot be determined experimentally and in only a few simple cases can the values of σ be calculated, although the screening factors can be accounted for theoretically in principle. For these reasons, reported values of σ or the related parameter, the chemical shift to be defined below, are given with reference to some agreed-upon internal reference such as tetramethylsilane (TMS) for nonaqueous solvents in which TMS is soluble. For aqueous solutions, no simple reference molecule is predominant. Thus in any publication it is essential that the reference to which other values are given be clearly stated.

Inclusion of the shielding factor modifies the Larmor frequency at which resonance absorption is obtained. Instead of Eq. (7.28), the resonance frequency when the shielding effects of the local electrons are included is given as

$$v_L = (\gamma H_0/2\pi)(1 - \sigma) \tag{7.44}$$

where H_0 is the applied magnetic field. Consider two samples, labeled r and s, and the corresponding shielding factors σ_r and σ_s, located in a field of strength H_0. If the applied field is fixed and the frequency of the applied radio frequency is varied to scan the spectrum, then the Larmor frequency of the sample (v_s) and the frequency of the reference (v_r) are

$$v_s = \frac{\gamma H_0}{2\pi}(1 - \sigma_s), \qquad v_r = \frac{\gamma H_0}{2\pi}(1 - \sigma_r) \tag{7.45}$$

where the same type of nucleus, characterized by a magnetogyric ratio γ, is probed. The dimensionless *chemical shift*, denoted by the symbol δ, and expressed in parts per million (ppm) is defined as

$$\delta \equiv \frac{v_s - v_r}{v_r} \times 10^6 \tag{7.46}$$

which, by substituting from Eq. (7.45) can be related to the shielding factors, namely,

$$\delta = \frac{\sigma_r - \sigma_s}{1 - \sigma_r} \times 10^6 \cong (\sigma_r - \sigma_s) \times 10^6 \tag{7.47}$$

The latter approximation is usually satisfactory since $\sigma_r \ll 1$. Although δ is field-independent, and therefore it is independent of the NMR spectrometer used, the actual separation between NMR resonance peaks, before being divided by v_r, increases linearly with the magnetic field. The larger the magnetic field, the greater is the separation between any two resonances, and

TABLE 7.2

Chemical Shift of Some Phosphate Metabolites[a]

Metabolite	Chemical shift (ppm)[b]
Inorganic phosphate	2.2
Phosphocreatine	−3.0
γ-phosphate of adenosine triphosphate	−5.7
β-phosphate of adenosine diphosphate	−6.1
α-phosphate of adenosine diphosphate	−10.4
α-phosphate of adenosine triphosphate	−10.8
β-phosphate of adenosine triphosphate with Mg^{2+} bound	−19.5
β-phosphate of adenosine triphosphate	−21.2

[a] From Burt et al. (1979).
[b] Chemical shifts at pH 7 and measured from the standard, 85% orthophosphoric acid.

therefore the better the resolution. In reporting NMR spectra the abscissa is expressed in units of δ ppm and the ordinate is left unspecified since usually only the ratios of areas under the NMR resonance absorption lines are meaningful.

Because theory alone is insufficient to predict the chemical shifts accurately, recourse to empirical data is essential. Table 7.2 lists a few of the known chemical shifts for phosphate molecules of interest in the experiment to be described as measured relative to the standard, 85% orthophosphoric acid. The molecular structure of some of the molecules is given in Fig. 7.27. An example of an NMR spectrum is shown in Fig. 7.28 for rat muscle. This figure illustrates a ^{31}P spectrum obtained from the leg muscle of an anesthetized rat. The chemical shifts for this particular experiment are measured with respect to phosphocreatine. Peaks I, II, and III are assigned to the β, α-, and γ-phosphate of adenosine triphosphate (ATP) molecule on the basis of their chemical shifts, -16.10, -7.54, and -2.47, respectively. Note that these are all displaced about -3.4 relative to their values listed in Table 7.2. This is because the chemical shifts in Table 7.2 are measured with respect to a different reference than those in Fig. 7.28, and, in addition, the pH is different for both scales.

The identification of $\delta = -16.10$ with β-phosphate with allowance for the -3.4 difference between values in Table 7.2 and Fig. 7.28 gives a chemical

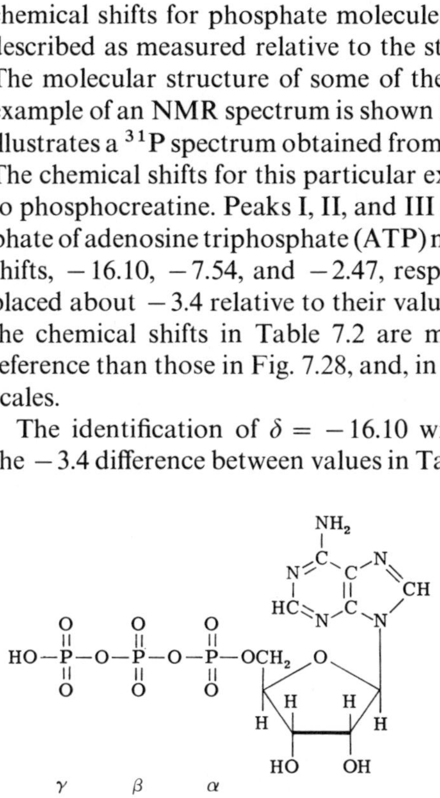

Adenosine triphosphate (ATP)

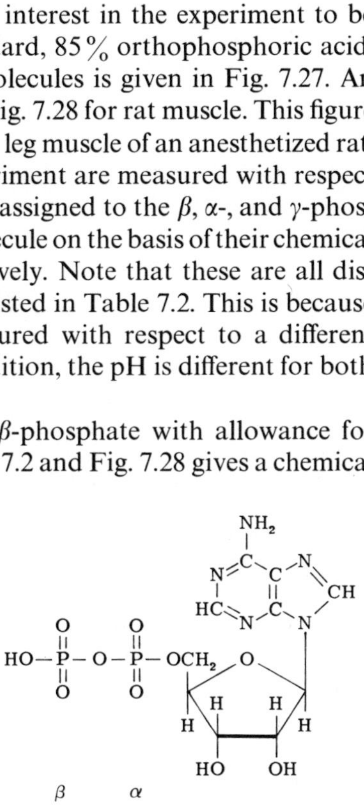

Adenosine diphosphate (ADP)

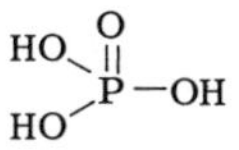

Orthophosphoric acid

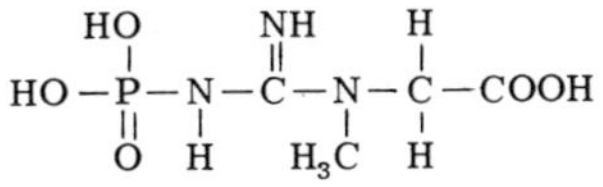

Phosphocreatine (PC)

FIG. 7.27 The molecular structures of the molecules discussed in the text. The α, β, and γ positions of the phosphate atom in the molecule are indicated in cases where the molecule has more than one phosphorus atom.

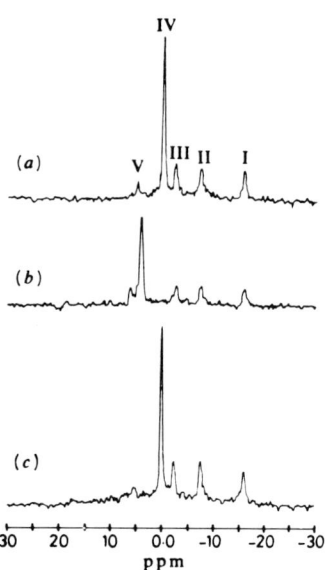

FIG. 7.28 Nuclear magnetic resonance ^{31}P spectra of the leg muscle of an intact anesthetized rat. Chemical shift assignments relative to phosphocreatine: I: β-phosphate of ATP, -16.10; II: α-phosphate of ATP, -7.54; III: γ-phosphate of ATP, -2.47; IV: phosphocreatine, -0.00; V: inorganic phosphate, 4.90. (a) Muscle below knee joint; (b) ischemic muscle from same area by application of a tourniquet above the area; (c) spectrum of muscle above the tourniquet. [From Burt et al. (1979).]

shift of $\delta = -19.5$, which, on comparison with the data in Table 7.2, indicates that the ATP in muscle is bound to a divalent cation, presumably Mg^{2+}. The areas under the resonance lines provide information on the concentration of the different molecular species. This can be seen as follows. An examination of Table 7.2 shows that γ-phosphate ATP and the β-phosphate of ADP have similar chemical shifts; their values differ by approximately -0.4, a difference below the experimental resolution of the data shown in Fig. 7.28. Hence, the signal labeled III in Fig. 7.28 may be due to contributions from both ATP and ADP. Since the area under an absorption line is proportional to the concentration of the molecules participating in forming that resonance line it follows that the ratio R may be written

$$R \equiv \frac{\text{area of peak I}}{\text{area of peak III}} = \frac{[\text{ATP}]}{[\text{ADP}] + [\text{ATP}]} \qquad (7.48)$$

where the brackets denote concentrations. Experimentally it is found that $R \approx 1$, implying that the $[\text{ADP}] \ll [\text{ATP}]$, i.e., the concentration of ADP is so small compared with ATP that its concentration is below detectability limits. In a similar manner Burt *et al.* (1979) showed that the area under peak IV implies that phosphocreatine has a concentration in muscle tissue about four times that of ATP. Figure 7.28b illustrates the effect of localized ischemia on rat leg muscle. Application of a tight tourniquet around the leg and placed just above the knee caused the NMR spectrum of the area below the knee to change significantly. Figure 7.28c shows that no change was observed in muscle tissue located above the tourniquet, indicating healthy, nonischemic muscle. This observation demonstrates that NMR spectroscopy has potential application for the determination of ischemic areas in diseased muscle. The ischemic region of the muscle showed no evidence for phosphocreatine, i.e., peak IV is absent in the NMR spectrum. There are also reduced amounts of ATP phosphates and an increased concentration of inorganic phosphate.

INTRACELLULAR PH MEASUREMENT BY ^{31}P NUCLEAR MAGNETIC RESONANCE

It is also possible to infer the intracellular pH in muscle cells from the position of the inorganic phosphate, peak V in Fig. 7.28. The equation necessary for calculating the intracellular pH from NMR data will now be developed.

pH is defined as the negative logarithm (base 10) of the hydrogen-ion concentration. It is an important biological parameter because the hydrogen-ion concentration has a regulatory effect on cellular metabolism and a modulatory effect on the excitability of neural tissue. Knowledge of pH is also important for understanding the effects of acidosis on coordinating functions of vital organs such as the heart. It should be recalled that pH = 7 is neutral, neither acidic nor basic. Therefore the ability to measure the intracellular pH in both cell suspensions and perfused organs by ^{31}P NMR constitutes an important application of nuclear magnetic resonance. The technique relies on the observation that the chemical shift (δ) of an inorganic phosphate such as K_2PO_4 is highly dependent on pH (Moon and Richards, 1973). In principle, any signal whose frequency is sensitive to proton concentration can be used; however, the phosphorous signal is most suitable for intact tissue. The method has significant advantages compared with determination of intracellular pH by dye absorption or microelectrode techniques in that it is noninvasive and relatively rapid; a measurement takes no longer than 10 sec for an adult rat heart. In special cases the method allows the measurement of pH in different locations within the cell.

The relation to be derived between the measured phosphate chemical shift δ_0 and the pH can be expressed as

$$pH = pK + \log[(\delta_a - \delta_0)/(\delta_0 - \delta_b)] \tag{7.49}$$

where pK is defined as the negative logarithm (base 10) of the reaction (*dissociation* or *ionization*) constant K. The dissociation constant is defined as (Vol. I, p. 351)

$$K = [H^+][A^-]/[HA] \qquad (7.50)$$

for the chemical reaction

$$HA \rightleftharpoons H^+ + A^- \qquad (7.51)$$

where [HA] represents the concentration of undissociated acid and [A⁻] the concentration of the salt. The chemical shifts δ_a and δ_b are the limiting chemical shifts at high and low pH. As an example, for phosphates,[†] $\delta_a = 3.29$, $\delta_b = 5.81$, and $pK = 6.90$, since biological systems are near neutral pH. These values for pK, δ_a, and δ_b are the fixed reference values.

Using Eq. (7.49) and the observed chemical shift δ_0, a value for the pH can be calculated. The origin of Eq. (7.49) is straightforward and can be derived using the very general result obtainable from NMR theory, namely, that if a given nucleus exchange rate between two different sites is very fast compared to the transition rate for absorption of rf energy, then only a single NMR line is observed and the resonance frequency is the average of the Larmor frequencies for the different sites, weighted according to the probability that a nucleus is at each site:

$$v_{obs} = p_1 v_1 + p_2 v_2 \qquad (7.52)$$

Here v_1 and v_2 are the Larmor frequencies of the two sites, and p_1 and p_2 are the probabilities of occupancy of the sites. From elementary chemistry it is known that the tribasic acid H_3PO_4 dissociates according to the equilibrium equations

$$H_3PO_4 \rightleftharpoons H^+ + H_2PO_4^- \qquad (7.53)$$

$$H_2PO_4^- \rightleftharpoons H^+ + HPO_4^{2-} \qquad (7.54)$$

$$HPO_4^{2-} \rightleftharpoons H^+ + PO_4^{3-} \qquad (7.55)$$

with pK values of approximately 2, 7, and 12, respectively. Physiological pH is near neutrality, i.e., $pH = 7$, and therefore only the second reaction is of biological importance. Since the exchange of the proton between $H_2PO_4^-$ and HPO_4^{2-} is known to be extremely rapid, estimated to be 10^9–10^{10} sec^{-1}, it follows that the observed phosphate resonance frequency is given by Eq. (7.52). With the definition of the chemical shift as given by Eq. (7.46) and the

---

[†] Strictly speaking, this is for orthophosphates.

relation between the observed resonance frequency and its intrinsic frequency components, the observed chemical shift δ_0 can be written

$$\delta_0 \equiv \frac{v_{obs} - v_r}{v_r} = \frac{p_1 v_1 + p_2 v_2 - v_r}{v_r} = \frac{p_1 v_1 + p_2 v_2 - (p_1 + p_2) v_r}{v_r}$$

$$= p_1 \left(\frac{v_1 - v_r}{v_r} \right) + p_2 \left(\frac{v_2 - v_r}{v_r} \right)$$

$$= p_1 \delta_a + p_2 \delta_b = p_1 \delta_a + (1 - p_1) \delta_b \qquad (7.56)$$

where δ_a and δ_b are the chemical shifts of $H_2PO_4^-$ and HPO_4^{-2}, respectively, and p_1 and p_2 are their mole fractions.

From the definition of pK, the negative logarithm of the equilibrium reaction constant, it follows that

$$pK \equiv -\log\{[H^+][HPO_4^{2-}]/[H_2PO_4^-]\}$$
$$= -\log[H^+] + \log\{[H_2PO_4^-]/[HPO_4^{2-}]\}$$
$$= pH + \log\{[H_2PO_4^-]/[HPO_4^{2-}]\} \qquad (7.57)$$

where the brackets denote concentration and the definition of

$$pH = -\log[H^+]$$

has been used. The concentrations of $[HPO_4^{-2}]$ and $[H_2PO_4^-]$ are, respectively, p_1 and p_2. These parameters are related to the chemical shift by Eq. (7.56). Solving this equation for p_1 and p_2 in terms of δ_0, δ_a, and δ_b and inserting into Eq. (7.57) gives the desired relation between chemical shifts, pK and pH, as given in Eq. (7.49).

As an example of the use of Eq. (7.49), note that with the chemical shift of the inorganic phosphate for nonischemic muscle (peak V in Fig. 7.28) of 4.90 and the known values of $pK = 6.90$, $\delta_a = 3.29$, and $\delta_b = 5.81$, the intracellular pH of the leg muscle is found to be 7.1. The chemical shifts of inorganic phosphate for the ischemic part of the muscle is reduced to 4.31 and this indicates a more acidic pH of 6.7. Thus, a technique is emerging that, when combined with imaging, may be able to image local ischemia and dead tissue as well as lesions, malignancies, and other degenerative diseases in which there is a lowered or accelerated metabolic rate involving ATP.

In some physiological preparations it is possible not only to measure the mean pH but also the distribution of pH_i in i compartments. The details of the spectrum analysis in this case are specific to the cell type studied. For example, an intracellular pH gradient between the cytoplasm (pH = 7.1) and mitochrondria (pH = 7.70) in suspensions of liver cells has been inferred from the splitting of the P_i NMR peak into two discernible peaks (Shulman *et al.*, 1979). Information of this nature allows for an assessment of the contribution of the pH gradient to the proton motive force. With other biochemical data this information helps elucidate electron transport and intracellular metabolic processes.

The most difficult and most complex organ to study by biochemical and perfusion techniques is the brain. With the advent of high-field superconducting magnets, which not only increase sensitivity but also allow for greater volume, it is now possible to perform NMR experiments on the intact brain. Figure 7.29 shows the ^{31}P NMR spectra of the brain of an intact rat. The phosphate signals shown in Fig. 7.29a are composite signals from both brain and bone tissue. The very broad background spectrum, shown in Fig. 7.29b, is due to the skull, and this contribution to the total signal can be eliminated by convolution techniques (Campbell *et al.*, 1973). The difference spectrum, i.e., signal (a) minus signal (b), gives only the phosphate signals from brain and is shown in Fig. 7.29c. The chemical shifts of the three ATP resonances are easily discernible (I, II, and III) and, like muscle discussed above, show that ATP is complexed with divalent cations in vivo. Similar to the results on muscles, the ADP concentration is below the limits of detectability. It is seen that there are peaks present not observed in muscle, and these are identified in the caption. Clearly, however, we are on the threshold of a new and important technique.

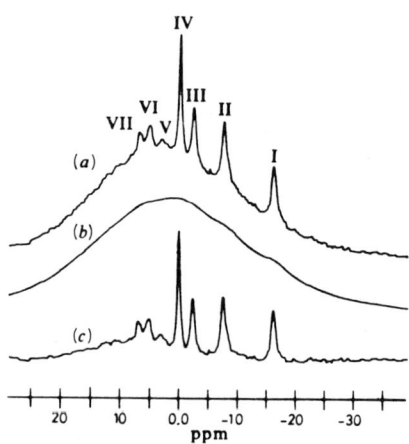

FIG. 7.29 The nuclear magnetic resonance ^{31}P spectra of the brain of an intact rat. (a) Spectrum of brain and skull; (b) spectrum of bone; and (c) difference spectrum. Chemical shift assignments: I: β-phosphate of ATP, −16.2; II: α-phosphate of ATP, −7.62; III: γ-phosphate of ATP, −2.48; IV: phosphocreatine, 0.00; V: phosphodiesters, 3.0, inorganic phosphate; VI: the 2-phosphate of 2,3-diphosphoglycerate, 5.00; VII: sugar phosphates and the 3-phosphate of 2,3-diphosphoglycerate, 6.65. [Reprinted by permission from Ackerman *et al. Nature* **283**, 170 (1980). Copyright © 1980 Macmillan Journals Limited.]

REFERENCES

Ackerman, J. J. H., Grove, T. H., Wong, G. G., Gadian, D. G., and Radda, G. K. (1980). Mapping of metabolites in whole animals by ^{31}P NMR using surface coils. *Nature (London)* **283**, 167.

Bloch, F., Hansen, W. W., and Packard, M. (1946). The nuclear induction experiment. *Phys. Rev. B* **78**, 474.

Bohm, C., Eriksson, L., Bergström, M., Litton, J., Sundman, R. and Singh, M. (1978). A computer assisted ringdector positron camera system for reconstruction tomography of the brain. *IEEE Trans. Nucl. Sci.* **NS-25**, 624.

Brooks, R. A., and DiChiro, G. (1976). Principles of computer assisted tomography (CAT) in radiographic and radioisotope imaging. *Phys. Med. Biol.* **21**, 689.

Burt, C. T., Cohen, S. M., and Bárány, M. (1979). An analysis of intact tissue with ^{31}P NMR. *Annu. Rev. Biophys. Bioeng.* **8**, 1.

Campbell, I. D., Dobson, C. M., William, R. J. P., and Xavier, A. V. (1973). Resolution enhancement of protein PMR spectra using the difference between a broadened and a normal spectrum. *J. Magn. Reson.* **11**, 172.

Crooks, L., Arakawa, M., Hoenninger, J., Watts, J., McRee, R., Kaufman, L., Davis, P. L., Margulis, A. R., and DeGroot, J. (1982). Nuclear magnetic resonance whole-body imager operating at 3.5 K Gauss. *Radiology* **143**, 169.

Edelstein, W. A., Hutschinson, J. M. S., Johnson, G., and Redpath, T. W. (1980). Spin warp NMR imaging and applications to human whole body imaging. *Phys. Med. Biol.* **25**, 751.

Frackowiak, R. S. J., Lenzi, G. L., Jones, T., and Heather, J. D. (1980). Quantitative measurement of regional cerebral blood flow and oxygen metabolism in man using ^{15}O and positron emission tomography: Theory, procedure and normal values. *J. Comput. Assist. Tomogr.* **4**, 727.

Greenberg, J. H., Reivich, M., Alavi, A., Hand, P., Rosenquist, A., Rintelmann, W., Stein, A., Tusa, R., Dann, R., Christman, D., Fowler, J., MacGregor, B., and Wolf, A. (1981). Metabolic mapping of functional activity in human subjects with the [^{18}F] fluorodeoxyglucose technique. *Science* **212**, 678.

Halliday, D., and Resnick, R. (1981). "Fundamentals of Physics," 2nd ed., ext. vers. Wiley, New York.

Hinshaw, W. S. (1976). Image formation by nuclear magnetic resonance: The sensitive point method. *J. Appl. Phys.* **47**, 3709.

Hinshaw, W. S., Andrew, E. R., Bottomley, P. A., Holland, G. N., Moore, W. S., and Worthington, B. S. (1978). Internal structural mapping by nuclear magnetic resonance. *Neuroradiology* **16**, 607.

Hinshaw, W. S., Andrew, E. R., Bottomley, P. A., Holland, G. N., Moore, W. S., and Worthington, B. S. (1979). An *in vivo* study of the fore-arm and hand by thin section NMR imaging. *Br. J. Radiol.* **52**, 36.

Høedt-Rasmussen, Sveninsdottir, E., and Lassen, N. A. (1966). Regional cerebral blood flow in man deteremined by intra-arterial injection of radioactive inert gas. *Circ. Res.* **18**, 237.

Holland, G. N., Bottomley, P. A., and Hinshaw, W. S. (1977). ^{19}F magnetic resonance imaging. *J. Magn. Reson.* **28**, 133.

Holland, G. N., Hawkes, R. C., and Moore, W. S. (1979). Nuclear magnetic resonance (NMR) tomography of the brain: Coronal and sagittal sections. *J. Comput. Assist. Tomogr.* **4**, 429.

Hutchison, J. M. S., Edelstein, W. A., and Johnson, G. (1980). A whole-body NMR imaging machine. *J. Phys. E* **13**, 947.

Kennedy, C., DesRosiers, M. H., Sakuroda, O., Shinohara, M., Reivich, M., Jehle, J. W., and Sokoloff, L. (1976). Metabolic mapping of the primary visual system of the monkey by means of autoradiographs [^{14}C] deoxyglucose technique. *Proc. Natl. Acad. Sci. U.S.A.* **73**, 4230.

Kety, S. S., and Schmidt, C. F. (1945). The determination of cerebral blood flow in man by the use of nitrous oxide in low concentrations. *Am. J. Physiol.* **143**, 53.
Kuhl, D. E., and Edwards, R. Q. (1970). The Mark III scanner. A compact device for multiple-view and section scanning of the brain. *Radiology* **96**, 563.
Lai, C.-M., and Lauterbur, P. C. (1980). A gradient control device for complete three-dimensional nuclear magnetic resonance, Zeugmatographic imaging. *J. Phys. E* **13**, 747.
Lai, C.-M., Shook, J. W., and Lauterbur, P. C. (1979). Microprocessor-controlled reorientation of magnetic field gradients. *Chem. Biomed. Environ. Instrum.* **9**, 1.
Lassen, N. A., Ingvar, D. H., and Skinhøj, E. (1978). Brain function and blood flow. *Sci. Am.* **239** (10), 62.
Lauterbur, P. C. (1973). Image formation by induced local interactions: Examples employing nuclear magnetic resonance. *Nature (London)* **242**, 190.
Lauterbur, P. C. (1974). Magnetic resonance Zeugmatography. *Pure Appl. Chem.* **40**, 149.
Lauterbur, P. C., and Lai, C.-M. (1980). Zeugmatography by reconstruction from projections. *IEEE Trans. Nucl. Sci.* **NS-27**, 1227.
Moon, R. B., and Richards, J. H. (1973). Determination of intracellular pH by ^{31}P magnetic resonance. *J. Biol. Chem.* **248**, 7276–7278.
Olesen, J. (1971). Contralateral focal increase of cerebral blood flow in man during arm work. *Brain* **94**, 635.
Phelps, M. E., Hoffman, E. J., Mullani, N. A., and Ter-Pogossian, M. M. (1975). Application of annihilation coincidence detection to transaxial reconstruction tomography. *J. Nucl. Med.* **16**, 210.
Pykett, J. L. (1982). NMR imaging in medicine. *Sci. Am.* **246** (5), 78.
Radda, G. K., and Seeley, P. J. (1979). Recent studies on cellular metabolism by nuclear magnetic resonance. *Annu. Rev. Physiol.* **41**, 749–769.
Raichle, M. E. (1979). Quantitative *in vivo* autoradiography with positron emission tomography. *Brain Res. Rev.* **1**, 47.
Raichle, M. E., Larson, K. B., Phelps, M. E., Grubb, R. L., Jr., Welch, M. J., and Ter-Pogossian, M. M. (1975). *In vivo* measurement of brain glucose transport and metabolism employing glucose—^{11}C. *Am. J. Physiol.* **228**, 1936.
Raichle, M. E., Grubb, R. L., Jr., Gado, M. H., Eichling, J. O., and Ter-Pogossian, M. M. (1976). Correlation between regional cerebral blood flow and oxidative metabolism. *Arch. Neurol. (Chicago)* **33**, 523.
Raichle, M. E., Welch, M. J., Grubb, R. L., Jr., Higgins, C. S., Ter-Pogossian, M. M., and Larsen, K. B. (1978). Measurement of regional substrate utilization rates by emission tomography. *Science* **199**, 986.
Reivich, M., Kuhl, D., Wolf, A., Greenberg, J., Phelps, M., Ido, T., Casella, V., Fowler, J., Hoffman, E., Alavi, A., Som, P., and Sokoloff, L. (1979). The [^{18}F] fluorodeoxyglucose method for the measurement of local cerebral glucose utilization in man. *Circ. Res.* **44**, 127.
Reivich, M., Greenberg, J. H., Alavi, A., Fowler, J. S., Christman, D. R., Rosenquist, A., Rintelmann, W., Hand, P. J., MacGregor, R. R., and Wolf, A. P. (1980). The [^{18}F] fluorodeoxyglucose method for measuring lGMR$_{gl}$ in man: Effects of physiological stimuli. *In* "Cerebral Metabolism and Neural Function" (J. V. Passonneau, R. A. Hawkins, W. D. Lust, and F. A. Welsh, eds.), p. 398. Williams & Wilkins, Baltimore, Maryland.
Shulman, R. G., Brown, T. R., Ugurbil, K., Ogawa, S., Cohen, S. M., and den Hollander, J. A. (1979). Cellular applications of ^{31}P and ^{13}C nuclear magnetic resonance. *Science* **205**, 160–166.
Sokoloff, L., Reivich, M., Kennedy, C., Des Rosiers, M. H., Patlak, C. S., Pettigrew, K. D., Sakurada, O., and Shinohara, M. (1977). The [^{14}C] deoxyglucose method for the measurement of local cerebral glucose utilization: Theory, procedure, and normal values in the conscious and anesthetized albino rat. *J. Neurochem.* **28**, 897.

REFERENCES

Sveinsdottir, E., Larsen, B., Rommer, P., and Lassen, N. A. (1977). A multidetector scintillation camera with 254 channels. *J. Nucl. Med.* **18**, 168.

Ter-Pogossian, M. M., Phelps, M. E., Hoffman, E. J., and Mullani, N. A. (1975). A positron-emission transaxial tomograph for nuclear imaging (PETT). *Radiology* **114**, 89.

Ter-Pogossian, M. M., Mullani, N. A., Hood, J. T., Higgins, C. S., and Ficke, D. C. (1978a). Design considerations for a positron emission transverse tomograph (PETT V) for imaging of the brain. *J. Comput. Assist. Tomogr.* **2**, 539.

Ter-Pogossian, M. M., Mullani, N. A., Hood, J., Higgins, C. S., and Currie, M. C. (1978b). A multislice positron emission computed tomograph yielding transverse and longitudinal images. *Radiology* **128**, 477.

APPENDIX A

Open-Channel Distribution Function

Suppose the total number of identical independent membrane channels is N and that each channel has only two states, open and closed. Under these conditions there are 2^N possible configurations for the N-channel system. For example, if $N = 3$, there are eight possible configurations of the complete system, namely, ooo, ooc, oco, coo, cco, coc, occ, and ccc, where o and c denote open and closed states and the first, second, and third terms in the triplets label the states of channels 1, 2, and 3, respectively. Each of these configurations does not necessarily occur with equal probability, and, as discussed below, the probability of a given configuration depends on the assigned probability for the open and closed states. In the general case we seek an analytical expression for the probability P_K that K channels of the N channels are open, irrespective of which K channels they are, given that the probability an individual channel is open is $p \leq 1$. If K channels are open, then of the 2^N configurations only $C_K^N = N!/K!(N-K)!$ have K indistinguishable open channels. This is evident since C_K^N is simply the number of ways in which K indistinguishable objects and another $N - K$ indistinguishable objects can be ordered. If the channels are noninteracting, then the probability of any channel being open does not depend on the state of any other channel. This assumption is generally summarized by saying that the channels are *statistically independent* of each other. When this condition does not hold the channel system is said to exhibit cooperativity. With the assumption of statistically independent channels, the probability for a given sequence of K open and $N - K$ closed channels is $p^K(1 - p)^{N-K}$. This follows since the probability of a composite event composed of statistically independent events equals the product of the individual probabilities of each component. Combining the probability factor $p^K(1 - p)^{N-K}$ with the number of possible sequences having K open channels gives the net probability P_K that K channels are open:

$$P_K = \frac{N!}{K!(N-K)!} p^K(1-p)^{N-K} \qquad (A.1)$$

A distribution function of this form is called a binomial distribution since P_K is the $(K + 1)$th term in the binomial expansion of $(p + (1 - p))^N$:

$$1 = (p + (1-p))^N = \sum_{K=0}^{N} \frac{N! p^K(1-p)^{N-K}}{K!(N-K)!} = \sum_{K=0}^{N} P_K \qquad (A.2)$$

The average number of open channels, denoted by $\langle K \rangle$, is by definition equal to the product of the probability that K channels are open times the number K summed over all K. Thus

$$\langle K \rangle \equiv \sum_{K=0}^{N} K P_K \qquad (A.3)$$

Evaluation of $\langle K \rangle$ is most easily performed by defining the function

$$f(\lambda) = (\lambda p + (1-p))^N \qquad (A.4)$$

where λ is an arbitrary parameter. Taking the first derivative of $f(\lambda)$, using the expansion of $(\lambda p + (1-p))^N$ given in Eq. (A.2), and setting $\lambda = 1$ gives

$$f'(\lambda)|_{\lambda=1} = Np(\lambda p + (1-p))^{N-1}|_{\lambda=1} = Np \qquad (A.5)$$

and

$$f'(\lambda)|_{\lambda=1} = \sum_{K=0}^{N} \frac{N! K p^K(1-p)^{N-K}}{K!(N-K)!} = \sum_{K=0}^{N} K P_K \qquad (A.6)$$

Thus

$$\langle K \rangle \equiv \sum_{K=0}^{N} K P_K = Np \qquad (A.7)$$

In a similar manner the second derivative of $f(\lambda)$ at $\lambda = 1$ gives

$$f''(\lambda) = N(N-1)(\lambda p + (1-p))^{N-2} p^2 |_{\lambda=1}$$

$$= N(N-1)p^2 = \sum_{K=0}^{N} K(K-1) P_K \qquad (A.8)$$

From Eqs. (A.7) and (A.8) it is straightforward to calculate the variance $\langle (\delta K)^2 \rangle$ of the distribution function given by Eq. (A.1). By definition

$$\langle (\delta K)^2 \rangle = \langle (K - \langle K \rangle)^2 \rangle = \langle K^2 \rangle - \langle K \rangle^2 \qquad (A.9)$$

where

$$\langle K^2 \rangle = \sum_{K=0}^{N} K^2 P_K \qquad (A.10)$$

and

$$\langle K \rangle = \sum_{K=0}^{N} K P_K \qquad (A.11)$$

Substitution of the results of Eqs. (A.7) and (A.8) into Eq. (A.9) gives

$$\langle (\delta K)^2 \rangle = Np(1-p) \qquad (A.12)$$

The binomial distribution has several properties worth noting. (1) For $p = \frac{1}{2}$, the distribution is symmetric about Np. (2) For $p \neq \frac{1}{2}$, the distribution is asymmetric about Np, although the asymmetry decreases as N increases. (3) For large N and small p, the binomial distribution reduces to the Poisson distribution function. This last property can be seen as follows. Let $m = Np$; then upon substituting $p = m/N$ into Eq. (A.1) and rearranging the terms

$$P_K = \frac{N!}{(N-K)! K!} \left(\frac{m}{N}\right)^K \left(1 - \frac{m}{N}\right)^{N-K} \qquad (A.13)$$

$$P_K = \frac{m^K}{K!} \left(\frac{N(N-1) \cdots (N-K+1)}{N^K}\right) \left(1 - \frac{m}{N}\right)^{N-K} \qquad (A.14)$$

For N large, $N \gg K$,

$$\frac{N(N-1) \cdots (N-K+1)}{N^K} \approx 1 \qquad (A.15)$$

and

$$(1 - m/N)^{N-K} \approx e^{-m} \qquad (A.16)$$

TABLE A.1
Comparison of Binomial and Poisson Distribution[a]

m	$K = 0$	$K = 1$	$K = 2$	$K = 3$
0.418	0.646	0.294	0.0537	0.0059
	0.658	0.275	0.056	0.006
0.158	0.851	0.138	0.009	—
	0.85	0.135	0.016	—

[a] First row of each m lists the binomial values for $N = 5$. Second row lists the Poisson distribution values.

Equation (A.16) is just the definition of the exponential function. (For additional details, see Vol. I, p. 362.) Hence, for large N and small p, the distribution function P_K reduces to

$$P_K(m) = e^{-m} m^K / K! \qquad (A.17)$$

the Poisson distribution with a mean of m. Note that from Eq. (A.12) it follows that the variance of the Poisson distribution is m. A comparison of the binomial distribution for $N = 5$ and the Poisson distribution for two values of m is given in Table A.1. This table illustrates the important point that both distributions are almost identical even for rather small values of N.

APPENDIX **B**

Rall's Branching Rule

To derive Rall's three-half sum rule, we consider for simplicity a dendrite trunk that branches into only two daughter segments. Generalization to branching that is more complex will be evident. Let λ_0 denote the space constant characterizing the initial (parent) dendritic segment and let L_0 and d_0 represent its physical length and diameter. Suppose furthermore that at $x = x_1$ the dendrite branches into two segments and that these segments have physical lengths L_{11} and L_{12} and diameters d_{11} and d_{12}, respectively. Figure B.1 illustrates these notations.

The steady-state solution, Eq. (5.17'), for the initial L_0 segment, $0 \le x \le x_1$, can be written with $L_0 = x_1$ and $V(L_0) \equiv V(x_1) \equiv V_1$ (the potential at the branch point) as

$$V(x) = V_1 \cosh[(x_1 - x)/\lambda_0] + V_1 B_1 \sinh[(x_1 - x)/\lambda_0] \quad \text{(B.1)}$$

B_1 is a constant determined by the boundary conditions. We may determine the axial current at any point in the segment $i(x)$ by differentiating Eq. (B.1)

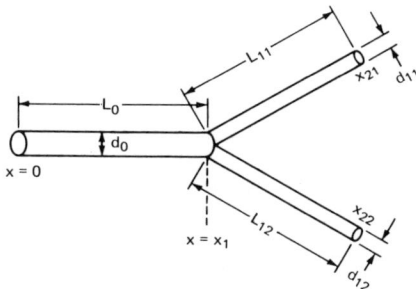

FIG. B.1 Arbitrary dendritic branching diagram illustrating notation used to define the boundary conditions at branch points. The total dendritic network is not shown. Physical lengths are denoted by L_0, L_{11}, and L_{12} and the corresponding diameters of the branches are given by d_0, d_{11}, and d_{12}, respectively.

and substituting the resulting expression into

$$i(x) = -\frac{1}{r_{i1}} \frac{\partial V(x)}{\partial x} \tag{5.2}$$

Here r_{i1} is the core resistance per unit length for the initial dendritic segment; thus

$$i(x) = \frac{V_1}{r_{i1}\lambda_0} \left[\sinh\left(\frac{x_1 - x}{\lambda_0}\right) + B_1 \cosh\left(\frac{x_1 - x}{\lambda_0}\right) \right] \tag{B.2}$$

It can be seen from Eq. (B.2) that B_1 is related to the amount of axial current flowing at the branch point $x = x_1$ for at this point the hyperbolic sine term is zero and the hyperbolic cosine term is unity, i.e., $i(L_0) = V_1 B_1/r_{i1}\lambda_0$. At the soma ($x = 0$) the axial current is

$$i(0) = \frac{V_1}{r_{i1}\lambda_0} \left[\sinh\left(\frac{x_1}{\lambda_0}\right) + B_1 \cosh\left(\frac{x_1}{\lambda_0}\right) \right] \tag{B.3}$$

whereas from Eq. (B.1) the electric potential at the soma is given by

$$V(0) \equiv V_0 = V_1 \left\{ \cosh\left(\frac{x_1}{\lambda_0}\right) + B_1 \cosh\left(\frac{x_1}{\lambda_0}\right) \right\} \tag{B.4}$$

Dividing Eq. (B.3) by Eq. (B.4) one obtains the relation

$$i(0) = \frac{V_0}{r_{i1}\lambda_0} \left\{ \frac{\sinh(x_1/\lambda_0) + B_1 \cosh(x_1/\lambda_0)}{\cosh(x_1/\lambda_0) + B_1 \sinh(x_1/\lambda_0)} \right\} \tag{B.5}$$

RALL'S BRANCHING RULE

Dividing numerator and denominator of the right-hand side of Eq. (B.5) by $\cosh(x_1/\lambda_0)$ and substituting $x_1 = L_0$ results in the expression

$$i(0) = \frac{V_0}{r_{i1}\lambda_0} \left\{ \frac{B_1 + \tanh(L_0/\lambda_0)}{1 + B_1 \tanh(L_0/\lambda_0)} \right\} \tag{B.6}$$

If the term in brackets is denoted by the symbol B_0, then Eq. (B.6) can be written in a shorthand notation as

$$i(0) = V_0 B_0 / r_{i1} \lambda_0 \tag{B.7}$$

Two limiting values of B_0 in Eq. (B.7) are of particular interest, namely, $B_0 = 0$ and $B_0 = 1$ because they are values for very short and for very long dendritic trunk lengths. For very short trunk lengths compared with λ_0, i.e., for small values of L_0/λ_0, the function $\tanh(L_0/\lambda_0)$ is approximately zero and therefore $B_0 \approx B_1$. For very long trunk lengths, i.e., $L_0 \gg \lambda_0$, $\tanh(L_0/\lambda_0)$ is close to unity and therefore $B_0 \approx 1$ regardless of the value of B_1. For the special case of $B_1 = 1$, it is evident from Eq. (B.6) that $B_0 = 1$. For intermediate trunk lengths, the value of B_0 always lies between B_1 and unity. For long dendritic trunk lengths, where $B_1 = 1$, Eq. (B.1) simplifies and yields an exponential solution. This is seen by introducing the relations $2 \sinh u = e^u - e^{-u}$ and $2 \cosh u = e^u + e^{-u}$ into Eq. (B.1); thus

$$V(x) = V_1 \exp[(x_1 - x)/\lambda_0] \tag{B.8}$$

At $x = 0$ Eq. (B.8) becomes

$$V_0 = V_1 \exp(x_1/\lambda_0)$$

and therefore Eq. (B.8) may be written

$$V(x) = V_0 \exp(-x/\lambda_0)$$

This is just the exponential decay solution given by Eq. (5.18) (p. 161) for an infinitely long cylinder. Note that in this particular case the cylinder need not be infinitely long.

Consider the case in which a dendritic trunk of intermediate length has a sealed end, i.e., negligible current can flow out of it. In this limit, at $x = x_1$, $\sinh u = 0$ and $\cosh u = 1$ it follows that if $i(x_1) = 0$, then B_1 must be zero. As discussed by Rall (1959), the sealed-end boundary condition constitutes an excellent approximation to a trunk whose end is sealed with a disk having the same electrical membrane properties as the cylinder wall.† For the special case where $B_1 = 0$, Eq. (B.1) becomes

$$V(x) = V_1 \cosh[(L_0 - x)/\lambda_0] \qquad (B.9)$$

and

$$V(0) = V_1 \cosh[L_0/\lambda_0] \qquad (B.10)$$

Dividing Eq. (B.9) by (B.10) and setting $V(0) = V_0$ gives

$$V(x) = \frac{V_0 \cosh[(L_0 - x)/\lambda_0]}{\cosh[L_0/\lambda_0]} \qquad (B.11)$$

Note that when L_0 becomes very large the special solution ($B_1 = 0$) reduces to the exponential solution given by Eq. (5.18). Examples of potentials obeying the relation given by Eq. (B.11) are shown in Fig. 5.11.

Consider now the general case where $B_1 \neq 0$ and where axial current is allowed to flow into the daughter branches. By examining Fig. B.1, it can be seen that if one wishes to calculate the electric potential or current in either of the two branches, the same procedure given for a single dendrite cylinder

---

† This can be seen as follows. From Eq. (B.2) it follows that the axial current at the terminal end ($x = x_1$) is given by

$$i(x_1) = B_1 V_1/\lambda_0 r_{i1}$$

This current is also equal to the voltage (V_1) divided by the resistance of the disk to current flow, $R_m/\pi a^2$, where a is the radius of the cylinder and R_m is the passive membrane resistance for a unit area. Equating these two equivalent expressions for the axial current gives

$$i(x_1) = \frac{B_1 V_1}{\lambda_0 r_{i1}} = \frac{V_1}{R_m/\pi a^2}$$

Thus for the case where cylinder terminal is a disk with the same membrane properties as the cylinder's walls

$$B_1 = \lambda_0 (r_{i1} \pi a^2)/R_m$$

Using Eq. (5.11), this relation may be written

$$B_1 = \frac{\lambda_0 R_i}{R_m} = \sqrt{\left(\frac{R_i}{R_m}\right) \frac{a}{2}}$$

For typical values,

$$R_i = 50 \, \Omega \, \text{cm}, \qquad R_m = 1250 \, \Omega \, \text{cm}^2, \qquad \text{and} \qquad a = 2 \, \mu\text{m}$$

this equation gives $B_1 = 2 \times 10^{-3}$, a value differing negligibly from zero, the limit assumed in the sealed-end boundary conditions.

is followed except that the zero position is now x_1 and the terminal position is x_{21} or x_{22}, which may be written x_{2k} where $k = 1$ or 2. Thus it is evident from the form of the potential given by Eq. (B.1) that the potential in these branches can be written

$$V(x) = V_{2k} \cosh[(x_{2k} - x)/\lambda_{1k}] + V_{2k} B_{2k} \sinh[(x_{2k} - x)/\lambda_{1k}] \quad (B.12)$$

where $x_1 \leq x \leq x_{2k}$, and where λ_{1k} are the space constants for the branches under consideration, e.g., λ_{11}, λ_{12}, and the B_{21} and B_{22} constants are determined by the boundary conditions at the ends of these two branches, x_{11} and x_{22}, respectively. V_{21} and V_{22} are the corresponding potentials at these positions. It follows from these equations that the input currents i_k into these branches, i.e., currents at x_1, if the core resistances are r_{i11} and $r_{i12} = r_{ilk}$, may be written

$$i_k = \frac{V_{2k}}{r_{i1k} \lambda_{1k}} \left[\sinh\left(\frac{L_{1k}}{\lambda_{1k}}\right) + B_{2k} \cosh\left(\frac{L_{1k}}{\lambda_{1k}}\right) \right] \quad (B.13)$$

Analogously with the first trunk derivation, we evaluate $V(x)$ of Eq. (B.12) at the branch point $x = x_1$. Calling this potential V_1 and recognizing that $x_{2k} - x_1 = L_{1k}$ and that the potential must be continuous at the branch point allows Eq. (B.12) to be written

$$V_1 = V_{2k} \cosh(L_{1k}/\lambda_{1k}) + V_{2k} B_{2k} \sinh(L_{1k}/\lambda_{1k}) \quad (B.14)$$

where $k = 1$ or 2. Dividing Eq. (B.13) by Eq. (B.14) gives the axial current in the kth branch at $x = x_1$ as

$$i_k = \frac{V_1}{r_{i1k} \lambda_{1k}} \left\{ \frac{\sinh(L_{1k}/\lambda_{1k}) + B_{2k} \cosh(L_{1k}/\lambda_{1k})}{\cosh(L_{1k}/\lambda_{1k}) + B_{2k} \sinh(L_{1k}/\lambda_{1k})} \right\} \quad (B.15)$$

Dividing numerator and denominator of the right-hand side of Eq. (B.15) by $\cosh(L_{1k}/\lambda_{1k})$ results in an equation similar to Eq. (B.6):

$$i_k = \frac{V_1}{r_{i1k} \lambda_{1k}} \left\{ \frac{B_{2k} + \tanh(L_{1k}/\lambda_{1k})}{1 + B_{2k} \tanh(L_{1k}/\lambda_{1k})} \right\} \quad (B.16)$$

This expression can be further simplified by letting the term in braces be represented by the symbol B_{1k}; thus

$$i_k = V_1 B_{1k}/r_{i1k} \lambda_{1k}, \quad k = 1, 2 \quad (B.17)$$

This relation can be generalized to other branches by setting $1 = j$, where j labels an arbitrary branch point. By applying Kirchhoff's rule for the sum of currents at a branch point, namely, that the current flowing into point x_1

equals the net current flowing into all branches emanating from $x = x_1$, it follows that

$$i(L_0) = \frac{V_1 B_1}{r_{i1}\lambda_0} = \sum_{k=1}^{2} i_{1k} \qquad (B.18)$$

Using Eq. (B.17), this condition can be written

$$B_1(r_{i1}\lambda_0)^{-1} = \sum_{k=1}^{2} B_{1k}(r_{i1k}\lambda_{1k})^{-1} \qquad (B.19)$$

Although the derivation has been restricted to only two daughter branches arising from the initial dendritic trunk, it should be apparent that the condition is valid for any number of branches emanating from a general branch point, provided B_1 and B_{1k} are defined appropriately. Equation (B.19) can be cast into a particularly simple form if the electrical parameters R_i (the volume resistivity of the intracellular medium) and R_m (the resistance across a unit area of passive membrane) are independent of spatial variables. The resistance r_i was defined in Eq. (5.11) as $r_i = R_i/\pi a^2$ and therefore $r_i \propto a^{-2}$, where a is the radius of the cylinder. The space constant λ was defined in Eq. (5.9) as $\lambda^2 = r_m/(r_i + r_e)$. For a conducting external bath, $r_e \ll r_i$, and therefore we may make the approximation $\lambda^2 = r_m/r_i$. With the definitions given in Eq. (5.11) it follows that $\lambda^2 = (R_m/2\pi a)/(R_i/\pi a^2)$, and therefore $\lambda \propto a^{1/2}$. Thus the product $r_i\lambda_i \propto a^{-3/2}$ and $r_i\lambda_i$ has the same proportionality to the cylinder's diameter d. On substituting the relation $r_i\lambda_i \propto d_i^{-3/2}$ into Eq. (B.19), the branching condition assumes the more compact form

$$B_1 d_0^{3/2} = \sum_K B_{1k} d_{1k}^{3/2} \qquad (B.20)$$

where d_0 and d_{1k} are the diameters of the dendrite segments. The relation given by Eq. (B.20) is generally referred to as Rall's three-half sum rule. For the first two branches, Eq. (B.20) may be written

$$B_1 = B_{11}(d_{11}/d_0)^{3/2} + B_{12}(d_{12}/d_0)^{3/2} \qquad (B.21)$$

where

$$B_{11} = \frac{B_{21} + \tanh(L_{11}/\lambda_{11})}{1 + B_{21}\tanh(L_{11}/\lambda_{11})} \qquad (B.22)$$

and

$$B_{12} = \frac{B_{22} + \tanh(L_{12}/\lambda_{12})}{1 + B_{22}\tanh(L_{12}/\lambda_{12})} \qquad (B.23)$$

In idealized cases where B_1 and B_{1k} are unity, Eq. (B.20) reduces to a purely geometrical relation,

$$d_0^{3/2} = \sum_K d_{1k}^{3/2} \tag{B.24}$$

For convenience, we shall call the relation expressed by Eq. (B.24) the *restricted* three-half power branching law.

REFERENCE

Rall, W. (1959). Branching dendrite trees and motoneuron membrane resistivity. *Exp. Neurol.* **1**, 491.

APPENDIX C

Proof of the Equivalent Cylinder Approximation

To find the appropriate conditions necessary for an entire dendrite tree to be equivalent to a cylinder of finite length, we consider for simplicity a tree composed only of a trunk and two branches. Generalization to cases involving extensive branching will be evident from this special case. The notation given in Fig. B.1 will be followed. The two daughter branches may or may not have equal physical lengths or equal diameters; however, if they have the same boundary condition at the terminals, then the B constants [see Eq. (B.12)] will have the same value:

$$B_{21} = B_{22} = B_2 \tag{C.1}$$

If the branches also have the same electrotonic lengths, $L_{12}/\lambda_{12} = L_{11}/\lambda_{11}$, then it follows from Eqs. (B.16) and (B.17) that

$$B_{11} = B_{12} = \frac{B_2 + \tanh(L_{12}/\lambda_{12})}{1 + B_2 \tanh(L_{12}/\lambda_{12})} \tag{C.2}$$

Thus, for equal electrical distances along the branches, the membrane potentials in each branch are identical to each other. This can be illustrated

PROOF OF THE EQUIVALENT CYLINDER APPROXIMATION

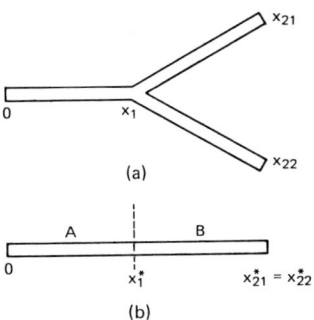

FIG. C.1 (a) A simple dendritic tree consisting of a trunk and two branches. (b) Equivalent compartment model of the tree in (a) if the terminal points have the same boundary conditions and are at the same electrotonic distance to the branch point x_1^*.

graphically by saying that the tree in Fig. C.1a is equivalent to the tree in Fig. C.1b. The asterisk denotes the corresponding electrotonic distances. The perpendicular dashed line in the figure passes through the branch point and denotes the boundary between the two compartments. The potential $V_A(x^*)$ in compartment A is given by

$$V_A(x^*) = V_0 \left\{ \frac{\cosh(x_1^* - x^*) + B_1 \sinh(x_1^* - x^*)}{\cosh(x_1^*) + B_1 \sinh(x_1^*)} \right\} \quad \text{(C.3)}$$

as can be seen by dividing Eq. (B.1) by Eq. (B.4), whereas the potential $V_B(x^*)$ for electrotonic distances x^* within compartment B is

$$V_B(x^*) = V_{21}\{\cosh(x_{21}^* - x^*) + B_{21} \sinh(x_{21}^* - x^*)\}$$

as is evident by rewriting Eq. (B.1) in electrotonic units. Suppose $(d_{11}/d_0)^{3/2} + (d_{12}/d_0)^{3/2} = 1$; then using

$$B_1 = B_{11}(d_{11}/d_0)^{3/2} + B_{12}(d_{12}/d_0)^{3/2} \quad \text{(B.21)}$$

it follows from Eq. (C.2) that

$$B_1 = B_{11} = B_{12} \quad \text{(C.4)}$$

With this relation and Eq. (C.2) and using common trigonometric identities, it can be shown that the potential $V(x^*)$ for all x^* between $x^* = 0$ and $x^* = x_{21}^*$ is the same as that calculated for an equivalent cylinder of finite electrotonic length x_{21}^* with the same terminal boundary conditions, i.e., the same B_2 as appears in Eq. (C.1). That is, the potential

$$V(x^*) = V_0 \left\{ \frac{\cosh(x_{21}^* - x^*) + B_2 \sinh(x_{21}^* - x^*)}{\cosh(x_{21}^*) + B_2 \sinh(x_{21}^*)} \right\} \quad \text{(C.5)}$$

reduces to $V_A(x^*)$ for $0 \leq x^* \leq x_1^*$ and to $V_B(x^*)$ for $x_1^* \leq x \leq x_{21}^*$. In the special case where the endpoints of the tree satisfy the sealed end conditions, namely,

$$\left.\frac{\partial V}{\partial x^*}\right|_{x^*=x_{21}^*} = 0 \qquad (C.6)$$

the constant $B_2 = 0$ and Eq. (C.5) reduce to

$$V(x^*) = V_0\{\cosh(x_{21}^* - x^*)/\cosh(x_{21}^*)\} \qquad (C.7)$$

where x_{21}^* is the total electrotonic length of the tree. It is easy to see that the proof for a tree with only one branch point given above can be generalized to more complicated dendritic tree patterns provided the following conditions are valid: (1) all terminal points satisfy the same boundary condition; (2) all terminal points are equally distant electrically from the cell body; and (3) the restricted $\frac{3}{2}$ power branching rule, Eq. (B.24), is valid at each branch point.

APPENDIX **D**

Solution to the Cable Equation for a Cylinder with Sealed Ends

To find the solution to the cable equation for a cylinder of physical length L requires assumptions regarding the boundary conditions. The approximation assumed here is that the resistance at the terminal ends is so large that the axial current vanishes. As mentioned in Appendix B, this condition is a good approximation if the terminal end membrane has the same membrane properties as the cylinder wall. The vanishing of the axial current, at the cylinder terminals is called the *sealed-end* boundary condition. Since the axial current is proportional to the first derivative of the potential (see Eq. (5.2)), the boundary conditions can be written mathematically

$$\frac{\partial V}{\partial x} = 0, \quad x = 0, \quad t > 0 \qquad (D.1)$$

$$\frac{\partial V}{\partial x} = 0, \quad x = L, \quad t > 0 \qquad (D.2)$$

The general solution to the cable equation subject to these conditions can be found by the method of separation of variables. This technique assumes a solution of the form

$$V(x, t) = f(x)g(t) \tag{D.3}$$

where f is a function of x only and g is a function of t only. Substituting Eq. (D.3) into Eq. (5.10), one obtains

$$\lambda^2 \frac{d^2f}{dx^2} g - \tau f \frac{dg}{dt} - fg = 0 \tag{D.4}$$

Rearrangement of the terms and dividing by $f(x)g(t)$ gives

$$\frac{1}{f} \lambda^2 \frac{d^2f}{dx^2} = \frac{\tau}{g} \frac{dg}{dt} + 1 \tag{D.5}$$

where the right-hand side depends only on the variable t and the left-hand side is a function only of x. For Eq. (D.5) to be valid for all x and t, both sides of the equation must be equal to some constant yet to be determined, which will be given the symbol $-\alpha^2$. The constant $-\alpha^2$ is called the *separation constant*. Thus, the partial differential, Eq. (5.10), reduces to two ordinary differential equations, namely,

$$\lambda^2 \frac{d^2f}{dx^2} + \alpha^2 f = 0 \tag{D.6}$$

and

$$\tau \frac{dg}{dt} + (1 + \alpha^2)g = 0 \tag{D.7}$$

Solutions of Eqs. (D.6) and (D.7) can be shown by direct substitution to be of the form

$$f(x) = A \sin(\alpha x/\lambda) + B \cos(\alpha x/\lambda) \tag{D.8}$$

$$g(t) = \exp(-(1 + \alpha^2)t/\tau) \tag{D.9}$$

A solution to the cable equation is therefore

$$V(x, t) = \left[A \sin\left(\frac{\alpha x}{\lambda}\right) + B \cos\left(\frac{\alpha x}{\lambda}\right) \right] \left[\exp\left(-\frac{(1 + \alpha^2)t}{\tau}\right) \right] \tag{D.10}$$

where the constants A, B, and α are determined from the boundary conditions as given by Eqs. (D.1) and (D.2). The derivative of Eq. (D.10) with respect to x is

$$\frac{\partial V}{\partial x} = \left[A \frac{\alpha}{\lambda} \cos\left(\frac{\alpha x}{\lambda}\right) - B \frac{\alpha}{\lambda} \sin\left(\frac{\alpha x}{\lambda}\right) \right] \left[\exp\left(-\frac{(1 + \alpha^2)t}{\tau}\right) \right] \tag{D.11}$$

SOLUTION TO THE CABLE EQUATION

Equation (D.1) requires $\partial V/\partial x = 0$ at $x = 0$. Substitution of $x = 0$ above shows that Eq. (D.11) can equal zero only if

$$(\alpha A/\lambda) \exp(-(1 + \alpha^2)t/\tau) = 0$$

which implies that $A = 0$. Substitution of $A = 0$ into Eq. (D.11) and using the boundary condition $\partial V/\partial x = 0$ at $x = L$ show that Eq. (D.11) equals zero only when

$$(B\alpha/\lambda) \sin(\alpha L/\lambda) = 0 \tag{D.12}$$

For a nontrivial solution, $B \neq 0$, this condition requires that $\sin(\alpha L/\lambda) = 0$; this value occurs when

$$\alpha = n\pi\lambda/L, \quad n = 0, 1, 2, \ldots \tag{D.13}$$

Therefore, particular solutions of the cable equation in the form given by Eq. (D.3) are

$$V_n(x, t) = B_n \cos\left(\frac{n\pi x}{L}\right) \exp\left\{-\left(1 + \frac{n^2\pi^2\lambda^2}{L^2}\right)\frac{t}{\tau}\right\} \tag{D.14}$$

one solution for each nonnegative value of n. Negative values of n are not included, since they correspond to solutions that are not linearly independent of solutions with positive n values. Since the cable equation is a linear differential equation, it follows that the sum of any set of solutions is also a solution. Hence, the most general solution of a cylinder with both ends sealed is

$$V(x, t) = \sum_{n=0}^{\infty} B_n \cos\left(\frac{n\pi x}{L}\right) \exp\left\{-\left(1 + \frac{n^2\pi^2\lambda^2}{L^2}\right)\frac{t}{\tau}\right\} \tag{D.15}$$

$$V(x, t) = \sum_{n=0}^{\infty} B_n \cos\left(\frac{n\pi x}{L}\right) \exp\left\{-\frac{t}{\tau_n}\right\} \tag{D.16}$$

where the time constants τ_n are defined as

$$\tau_n = \frac{R_m C_m}{(1 + n^2\pi^2\lambda^2/L^2)} = \frac{\tau_0}{(1 + n^2\pi^2\lambda^2/L^2)} \tag{D.17}$$

Although this relation was derived for a cylinder of physical length L, the τs may be written alternatively in terms of electrotonic length by the substitution $L = \lambda L^*$. The zero-order time constant τ_0 is assigned the value $R_m C_m$, which follows from the Eq. (5.29), and in Eq. (D.10) it is the value of τ_n when $n = 0$. The constants B_n are still arbitrary, and their values can be evaluated only when the constraints produced by the initial conditions are used. This aspect of the problem need not concern us. Equation (D.16) demonstrates that the transient response of the membrane potential due to injecting current located at $x = 0$, the position of the cell body and the recording electrode, is a simple sum of exponentials,

$$V(0, t) = B_0 \exp(-t/\tau_0) + B_1 \exp(-t/\tau_1) + B_2 \exp(-t/\tau_2) + \cdots \quad \text{(D.18)}$$

From the form of Eq. (D.17) it is seen that τ_n is long for small values of n and short for large values of n. In practice only two or three terms in Eq. (D.18) are used, with τ_0 as the longest decay constant because other τ values occur with such short times that they cannot be extracted from the experimental data.

APPENDIX **E**

Autocorrelation Function

Suppose at two different time intervals of a recording of data there result two patterns as shown in Fig. E.1. Although these records appear to be completely dissimilar, each when analyzed may be found to have the same statistical properties. The *autocorrelation function* is actually a self-correlation of records taken at two different times, and it provides a measure of their statistical independence. When the autocorrelation function is zero, they are completely independent.

The mean (time-average) value of a random function $N(t)$ was defined by Eq. (6.1) as

$$\langle N \rangle = \lim_{T \to \infty} \frac{1}{T} \int_0^T N(t)\, dt \qquad (E.1)$$

FIG. E.1 Recording of data at two different time intervals.

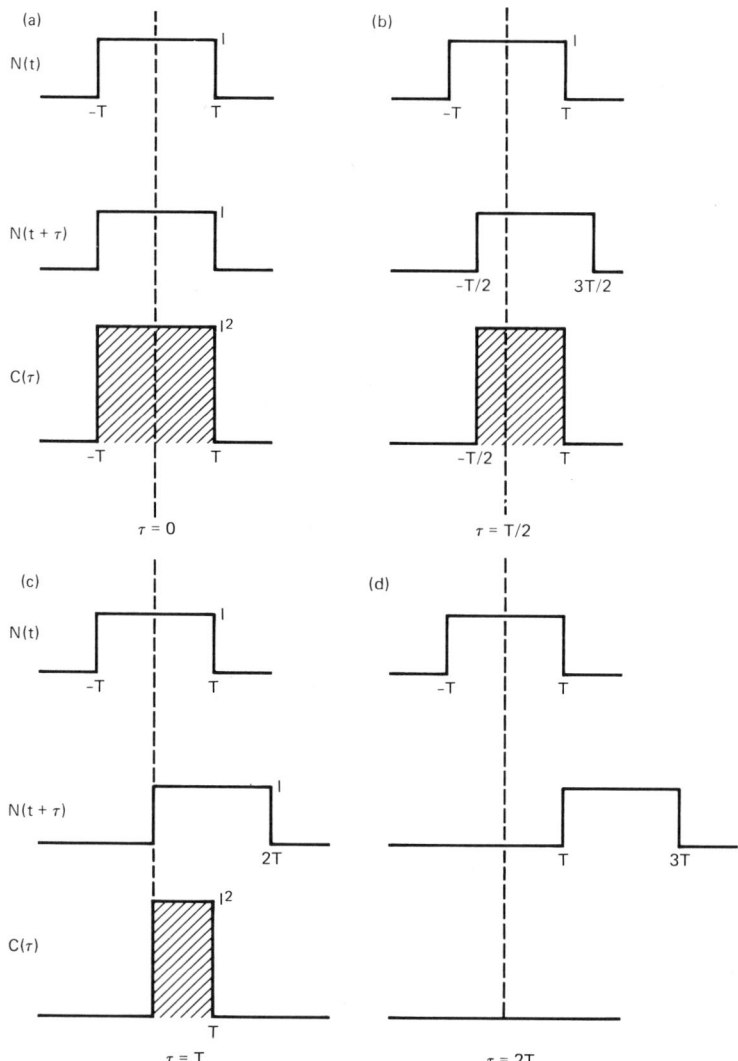

FIG. E.2 Graphical determination of the autocorrelation function of a single square wave as a function of time τ.

AUTOCORRELATION FUNCTION

We wish to find $C(\tau)$, the autocorrelation function, which is the mean value of the product, i.e., the joint probability, of the function N at two different times t_1 and t_2, where $t_2 = t_1 + \tau$. If the time sample of the function is taken from $-T$ to T, the time over which it is averaged is $2T$. Therefore the expectation value of the joint probability, which is the autocorrelation function $C(\tau)$, can be written

$$C(\tau) = \langle N(t_1)N(t_1 + \tau) \rangle = \lim_{T \to \infty} \frac{1}{2T} \int_{-T}^{T} N(t_1)N(t_1 + \tau) \, dt_1 \qquad (\text{E.2})$$

The equation as written assumes that $C(\tau)$ is independent of the selection of t_1, a condition that holds when the random process is stationary.

As an elementary example, consider the autocorrelation of the square wave in Fig. E.2 for different values of τ. Each τ value has three diagrams associated with it. The upper diagram is $N(t)$ versus time, with a value of either I or zero; the middle diagram is the square wave at time τ later; and the lower diagram (shaded) is the product of the overlapping areas whose height is I^2 but whose width, and therefore the overlap product, changes with τ. It is evident from Eq. (E.2) that because $N(t)$ has a constant value I in the domain $-T$ to T,

$$C(\tau) = \frac{1}{2T}(I)(I)\Delta = \frac{1}{2T}I^2(2T - \tau) \qquad (\text{E.3})$$

where Δ is the region of overlap of the functions $N(t)$ and $N(t + \tau)$ for which both functions are nonzero. Figure E.3 shows a plot of the autocorrelation function for a single square wave determined from Fig. E.2. In this example the autocorrelation decreases linearly with τ as seen in Fig. E.3. If the square

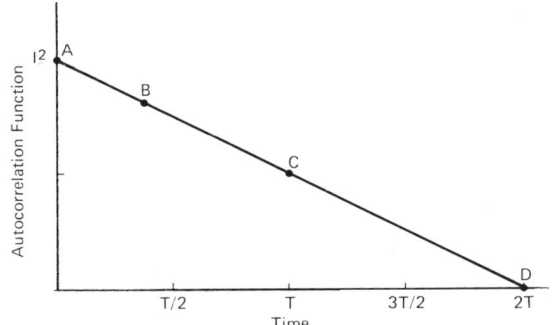

FIG. E.3 Autocorrelation function for a single square pulse versus time constructed from graphs shown in Fig. E.2. Points A, B, C, and D correspond to the positions of the square waves of Fig. E.2.

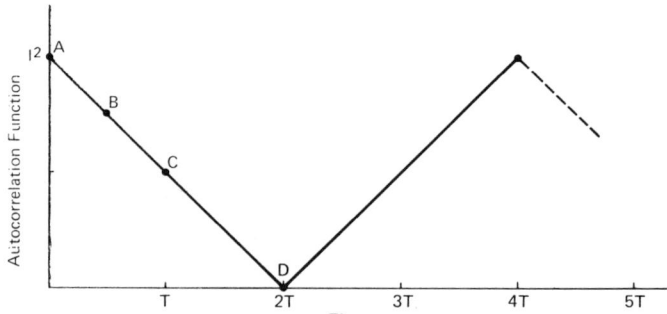

FIG. E.4 Autocorrelation function for a square wave as illustrated in Fig. E.2 but with a periodicity of 4T. Dotted line segment signifies that the function repeats for large values of time.

wave is repeated in time with a periodicity of 4T, then the autocorrelation function will repeat every 4T units. The autocorrelation function in this case decreases linearly for τ between 0 and 2T and then increases linearly between 2T and 4T. This functional dependence is illustrated in Fig. E.4. In most cases of biological interest, the autocorrelation function decreases exponentially with τ.

APPENDIX F

Fourier Coefficients

Given the Fourier series

$$f(x) = \frac{a_0}{2} + \sum_{n=1}^{N} [a_n \cos nx + b_n \sin nx] \tag{F.1}$$

the constants may be determined in the following way.

To determine a_0, multiply Eq. (F.1) by dx and integrate the sum term by term from 0 to 2π. Since for all n

$$\int_0^{2\pi} \cos nx \, dx = \int_0^{2\pi} \sin nx \, dx = 0 \tag{F.2}$$

it follows that all terms in the summation equal zero and there remains only the term

$$\int_0^{2\pi} f(x) \, dx = \frac{a_0}{2} \int_0^{2\pi} dx = a_0 \pi$$

Hence

$$a_0 = \frac{1}{\pi} \int_0^{2\pi} f(x)\, dx \qquad (F.3)$$

The coefficients a_n may be determined by multiplying both sides of Eq. (F.1) by cos $mx\, dx$ and integrating term by term from 0 to 2π. The values of the integrals are

$$\int_0^{2\pi} \sin nx \cos mx\, dx = 0 \qquad \text{for all} \quad m \text{ and } n$$

and

$$\int_0^{2\pi} \cos nx \cos mx\, dx = 0 \qquad \text{for} \quad m \neq n$$

Thus only nonzero term in the infinite sum is

$$\int_0^{2\pi} f(x) \cos mx\, dx = a_n \int_0^{2\pi} \cos^2 nx\, dx = a_n \pi \qquad \text{when} \quad m = n$$

Hence

$$a_n = \frac{1}{\pi} \int_0^{2\pi} f(x) \cos nx\, dx \qquad (F.4)$$

Note that Eq. (F.4) reduces to Eq. (F.3) when $n = 0$.

In a similar manner, multiplying both sides of Eq. (F.1) by sin $nx\, dx$ and integrating term by term yields

$$b_n = \frac{1}{\pi} \int_0^{2\pi} f(x) \sin nx\, dx \qquad (F.5)$$

In the case of Eqs. (6.10) and (6.11) the variable x has been changed to $2\pi t/T$ and the corresponding limits of integration are 0 to T.

APPENDIX **G**

Fourier Transforms

Consider the function $f(t)$ expanded in an infinite Fourier series where T is the period:

$$f(t) = \frac{a_0}{T} + \frac{2}{T} \sum_{n=1}^{\infty} [a_n \cos n\omega_0 t + b_n \sin n\omega_0 t]$$

It is convenient to substitute the exponential forms of the sine and cosine, namely,

$$\sin y = \frac{e^{iy} - e^{-iy}}{2i}, \qquad \cos y = \frac{e^{iy} + e^{-iy}}{2}$$

Thus

$$\begin{aligned} f(t) &= \frac{a_0}{T} + \frac{2}{T} \sum_{n=1}^{\infty} \left[\frac{a_n}{2} [(\exp(in\omega_0 t) + \exp(-in\omega_0 t)] \right. \\ &\quad \left. + \frac{b_n}{2i} [\exp(-in\omega_0 t) - \exp(-in\omega_0 t)] \right] \\ &= \frac{a_0}{T} + \frac{2}{T} \sum_{n=1}^{\infty} \left[\frac{a_n + ib_n}{2} \exp(-in\omega_0 t) + \frac{a_n - ib_n}{2} \exp(-in\omega_0 t) \right] \quad (G.1) \end{aligned}$$

313

Define C_n as the complex quantity

$$C_n = a_n - ib_n, \qquad C_{-n} = a_n + ib_n$$

Substitute these in Eq. (G.1) and extend the summation to $-\infty$ in order to include the $-n$ terms. The result is the simplification of Eq. (G.1), namely,

$$f(t) = \frac{1}{T} \sum_{n=-\infty}^{\infty} C_n \exp(in\omega_0 t) \tag{G.2}$$

The constant C_n may be evaluated by multiplying both sides of Eq. (G.2) by $\exp(-in\omega_0 t)\, dt$ and integrating between $-T/2$ and $T/2$. All terms vanish except when $n = m$, in which case

$$\int_{-T/2}^{T/2} f(t) \exp(-in\omega_0 t)\, dt = \frac{1}{T} \int_{-T/2}^{T/2} C_n\, dT = C_n$$

or

$$C_n = \int_{-T/2}^{T/2} f(t) \exp(-in\omega_0 t)\, dt \tag{G.3}$$

Is it possible to perform a Fourier analysis of a nonperiodic function? This can be done by using a Fourier transform. Suppose a disturbance consists of an envelope of frequencies in which to describe it properly the harmonic spectrum of frequencies requires spacing so small that it is best represented by a continuous rather than a discrete frequency spectrum. If the harmonic spacing is $\Delta f = 1/T$, then Eq. (G.2) can be rewritten

$$f(t) = \sum_{n=-\infty}^{\infty} C_n \exp(in\omega_0 t)\, \Delta f \tag{G.4}$$

and, recalling that $\omega = 2\pi f$ or $\omega_0 = 2\pi\, \Delta f$ allows Eq. (G.3) to assume the form

$$C_n = \int_{-T/2}^{T/2} f(t) \exp(-in2\pi t\, \Delta f)\, dt \tag{G.5}$$

Now let $T \to \infty$ in such a way that $n/T = n\, \Delta f$ approaches the continuous variable f, where n is the number of harmonics. Under these conditions the Fourier transform of the function $f(t)$ is defined as

$$F(f) = \lim_{T \to \infty} C_n \tag{G.6}$$

Therefore changing the limits on Eq. (G.5) and letting $n\, \Delta f \to f$ permits us to write

$$F(f) = \int_{-\infty}^{\infty} f(t) e^{-i2\pi f t}\, dt \tag{G.7}$$

FOURIER TRANSFORMS

From the definition of an integral as the limit of a sum, i.e.,

$$\int f(x)\,dx = \lim_{\substack{n \to \infty \\ \Delta_i x \to 0}} \sum_{i=1}^{n} f(x_i)\,\Delta_i x \qquad (G.8)$$

and using the relation in Eq. (G.6) for C_n allows Eq. (G.4) in the continuous spectral limit to become

$$f(t) = \int_{-\infty}^{\infty} F(f)e^{i2\pi ft}\,df \qquad (G.9)$$

Equations (G.7) and (G.9) are called Fourier transforms of each other. Equation (G.9) indicates that a nonperiodic time function can have a continuous frequency spectrum whose magnitude at frequency f is $F(f)$, and therefore $F(f)$ indicates the relative magnitude of each of the frequency components.

APPENDIX H

Sampling Theorem

Theorem: If $f(t)$ contains no frequency components larger than f_{max}, then it can be completely determined by finding its ordinates at a series of points spaced every $1/2f_{max}$ or less.

Before proving this theorem we note that if $F(f)$ is periodic, then a Fourier analysis may be made in the interval $-f_{max}$ to f_{max}. This is evident since Eqs. (G.2) and (G.3) are

$$f(t) = \frac{1}{T} \sum_{n=-\infty}^{\infty} C_n \exp(in\omega_0 t) \qquad \text{(G.2)}$$

$$C_n = \int_{-T/2}^{T/2} f(t) \exp(-in\omega_0 t)\, dt \qquad \text{(G.3)}$$

and by letting $t = f$ and $\omega_0 = 2\pi/2f_{max}(2f_{max} = T)$ allows the function $F(f)$ to be written

$$F(f) = \frac{1}{2f_{max}} \sum_{n=-\infty}^{\infty} x_n \exp(in\omega/2f_{max})$$

where

$$x_n = \int_{-f_{max}}^{f_{max}} F(f) \exp(-in\omega/2f_{max}) \, df$$

or, since $\omega = 2\pi f$

$$x_n = \frac{1}{2\pi} \int_{-2\pi f_{max}}^{2\pi f_{max}} F(f) \exp(-in\omega/2f_{max}) \, d\omega \tag{H.1}$$

Proof of the sampling theorem: The Fourier transform of $f(t)$ was given by

$$F(f) = \int_{-\infty}^{\infty} f(t) e^{-i\omega t} \, dt \tag{G.7}$$

where $f(t)$ in terms of the inverse transform is

$$f(t) = \frac{1}{2\pi} \int_{-\infty}^{\infty} F(f) e^{i\omega t} \, d\omega \tag{G.9}$$

Since $F(f)$ is taken as band limited, it follows that $F(f)$ exists only between $-f_{max}$ and f_{max}; thus

$$f(t) = \frac{1}{2\pi} \int_{-2\pi f_{max}}^{2\pi f_{max}} F(f) e^{i\omega t} \, d\omega \tag{H.2}$$

If the sampling points are taken at integral multiples of $1/2f_{max}$, i.e., $t = +n/2f_{max}$, then Eq. (H.2) can be written in the form

$$f\left(+\frac{n}{2f_{max}}\right) = \frac{1}{2\pi} \int_{-2\pi f_{max}}^{2\pi f_{max}} F(f) \exp(-in\omega/2f_{max}) \, d\omega \tag{H.3}$$

The left-hand side of Eq. (H.3) is $f(t)$ at the sampling instances whereas the right-hand side is identical to Eq. (H.1). Thus the value of the function at the sampling points are sufficient to uniquely define the coefficients x_n, and therefore $F(f)$ is completely specified.

REFERENCE

Hancock, J. C. (1961). "An Introduction to the Principles of Communication Theory," p. 18. McGraw-Hill, New York.

APPENDIX I

One-Dimensional Random Walk

Equation (6.26), $\overline{x^2} = 2\mu kTt$, can be derived for one-dimensional random motion as follows. Consider a single particle whose position is initially taken to be the origin, $x = 0$. Assume furthermore that it suffers impacts by molecules composing the surrounding medium at a steady rate of λ sec^{-1} and that each impact moves the particle a very small distance in either the positive or negative direction along the x axis. Let d_n denote the nth collision step displacement. For simplicity, we shall assume each displacement has identical magnitude, that is, $d_n = -d$ or $+d$, depending on whether the particle moves in the negative or positive x-axis direction. In more refined treatments this simplifying assumption can be removed. After m collisions, corresponding to a time $t_m = m/\lambda$, the particle has moved a total displacement

$$x = d_1 + d_2 + \cdots + d_m = \sum_{l=1}^{m} d_l \qquad (I.1)$$

For a given particle at time t_m, x takes on values between $-md$ and $+md$. Note, however, that not all values are possible, e.g., if $m = 3$, then x can not

ONE-DIMENSIONAL RANDOM WALK

equal $-2d$, 0, or $2d$. From Eq. (I.1) it is not possible to predict the outcome of a single experiment. However, if a large number of identical particles are considered, then prediction can be made about the ensemble behavior. In particular

$$\bar{x} = \sum_{l=1}^{m} \bar{d}_l = 0 \tag{I.2}$$

where the bar signifies the ensemble average. This follows since the nth displacement is equally likely to be either positive or negative, i.e., $\bar{d}_l$ is zero for all n. Equation (I.2) does not give any information about how large in magnitude a particular particle displacement is likely to be after a time t. This information is provided by the ensemble mean square deviation or variance:

$$\overline{(x - \bar{x})^2} = \overline{x^2} = \overline{\left(\sum_{l=1}^{m} d_l\right)^2} \tag{I.3}$$

$$\overline{(x - \bar{x})^2} = \sum_{l=1}^{m} \overline{d_l^2} + \sum_{l \neq n} \overline{d_l d_n} \tag{I.4}$$

$$\overline{(x - \bar{x})^2} = \sum_{l=1}^{m} \overline{d_l^2} = md^2 \tag{I.5}$$

Since the magnitude of all displacements is the same, the sum of the average of the squared values is the same as the sum of the squares, which, in turn, is a multiple of the square of a single displacement. Terms $\overline{d_n d_m}$ ($m \neq n$) are zero because the nth displacement is independent of the mth displacement and there are equal numbers of positive and negative products. The total number of steps m taken in a time t is λt, where λ is the collision frequency. Thus, from Eq. (I.5) it follows that

$$\overline{x^2} = \lambda t d^2 = 2Dt \tag{I.6}$$

where $D = \lambda d^2/2$ is defined as the diffusion coefficient for random motion in one dimension. Equation (6.26) follows immediately from Eq. (I.6) because of another relation derived by Einstein [Vol. I, Eq. (3.12)], namely,

$$D = \mu k T$$

where μ is the particle's mobility, k Boltzmann's constant, and T temperature. An important point to note about Eq. (I.6) is that for unpredictable, irreversible motion the variance varies linearly with time. This should be contrasted with reversible motion, namely, motion describable by Newton's laws of motion, where the particle displacement varies as the square of the time interval.

APPENDIX J

End-Plate Current Spectral Function

The spectral function $S(f)$ for the end-plate current fluctuations is most conveniently calculated from the autocorrelation function using the Wiener–Khintchine theorem. Since ion channels are assumed to be noninteracting, it follows that the end-plate autocorrelation function $C(\tau)$ is equal to N, the total number of end-plate channels, times the autocorrelation function $C_1(\tau)$ for a single channel. By definition [see Eq. (6.6)]

$$C_1(\tau) = \langle (i(t + \tau) - \langle i \rangle)(i(t) - \langle i \rangle) \rangle \tag{J.1}$$

where $i(t)$ denotes the current at time t transported by a single channel and $\langle i \rangle$ represents the mean current for a single channel. The angle brackets indicate the expected (ensemble) average. Equation (J.1) can be rewritten

$$C_1(\tau) = \langle i(\tau)i(0) \rangle - \langle i(\infty) \rangle^2 \tag{J.2}$$

since for stationary random processes $C_1(\tau)$ is independent of the variable t. Defining $g(\tau)$ as the conductance of a single channel at time τ, then

$$C_1(\tau) = (V - E_s)^2 \{ \langle g(\tau)g(0) \rangle - \langle g(\infty) \rangle^2 \} \tag{J.3}$$

END-PLATE CURRENT SPECTRAL FUNCTION

where Eq. (4.5) written for single channels,

$$i(\tau) = g(\tau)(V - E_s) \tag{J.4}$$

has been used. E_s is the membrane reversal potential, and V is the membrane potential. By definition, the ensemble averages in Eq. (J.3) are given by

$$\langle g(\tau)g(0)\rangle = \sum_i \sum_k \gamma(j)\gamma(k)p(j, k, \tau) \tag{J.5}$$

$$\langle g(\infty)\rangle = \sum_j \gamma(j)p(j, \infty) \tag{J.6}$$

where the sum is over all states a given channel can assume. $p(j, k, \tau)$ denotes the joint probability of finding the ion channel in state j initially ($\tau = 0$) and in state k at time τ. $p(j, \infty)$ is the probability of finding the ion channel in state j at time $\tau = \infty$. The quantities $\gamma(j)$ and $\gamma(k)$ are the conductances for states j and k, respectively.

Neuromuscular junction endplate channels are assumed to have only two states: *open*, with a time-independent conductance γ, and *closed*, with a conductance zero. Thus the only nonzero terms in Eqs. (J.5) and (J.6) are when the channels are open. That is,

$$\langle g(\tau)g(0)\rangle = \gamma^2 p(\text{open, open}, \tau) \tag{J.7}$$

and

$$\langle g(\infty)\rangle = \gamma p(\text{open}, \infty). \tag{J.8}$$

The probability $p(\text{open, open}, \tau)$ is by definition the product of the probability $p(0)$ that the channel is open at time zero times the conditional probability $p(0|\tau)$ that it will be open at time τ if it is open at time zero. That is,

$$p(\text{open, open}, \tau) = p(0)p(0|\tau). \tag{J.9}$$

Similarly, since we are considering membrane current fluctuations under the application of a constant application of iontophoretic current, it follows that the random process of current fluctuations is *stationary*. Thus

$$p(\text{open}, \infty) = p(0) \tag{J.10}$$

With the results given in Eqs. (J.7)–(J.10) it follows that Eq. (J.3) can be rewritten

$$C_1(\tau) = \gamma^2(V - E_s)^2\{p(0)p(0|\tau) - p^2(0)\} \tag{J.11}$$

The conditional probability $p(0|\tau)$ satisfies an equation similar to that given for the total membrane conductance [see Eq. (4.31)], namely,

$$\frac{dp(0|\tau)}{d\tau} = -\alpha p(0|\tau) + \beta \frac{[T(\tau)]^n}{K} \tag{J.12}$$

provided $\beta[T(\tau)]^n/K \ll \alpha$. Here K is the equilibrium constant defined by Eq. (4.29), α the rate at which the channel closes, β the rate at which channels open, and $T(\tau)$ the concentration of transmitter at the postsynaptic membrane at time τ. The solution to Eq. (J.12) can be found by methods identical to those discussed on p. 127. Hence for a constant concentration c of transmitter at the postsynaptic membrane, a condition that is valid if a constant iontophoretic current is maintained,

$$p(0|\tau) = e^{-\alpha\tau}(1 - \beta c^n/\alpha K) + \beta c^n/\alpha K \tag{J.13}$$

which, since $\beta c^n/\alpha K$ has been taken to be much less than unity, can be approximated by

$$p(0|\tau) \approx e^{-\alpha\tau} + \beta c^n/\alpha K \tag{J.14}$$

It is also evident [see Eq. (J.10)] that

$$p(0) \equiv p(\text{open}, \infty) = \beta c^n/\alpha K \tag{J.15}$$

Inserting Eqs. (J.14) and (J.15) into Eq. (J.11) gives the single-channel autocorrelation function

$$C_1(\tau) = \gamma^2 (V - E_s)^2 (\beta c^n/\alpha K) e^{-\alpha\tau} \tag{J.16}$$

The correlation time τ_c as defined on p. 190 for a single channel is therefore equal to the reciprocal of the closing channel rate, i.e., $\tau_c = \alpha^{-1}$.

The spectral density function $S(f)$ can be obtained from Eq. (J.16) by multiplying it by the total number of independent channels N and using the Wiener–Khintchine relation given by Eq. (6.20). Thus

$$S(f) = 4N \int_0^\infty C_1(\tau) \cos 2\pi f \tau \, d\tau$$

$$= 4N\gamma^2(V - E_s)^2 \frac{\beta c^n}{\alpha K} \int_0^\infty e^{-\alpha\tau} \cos 2\pi f \tau \, d\tau$$

$$= 4\gamma^2 N(V - E_s)^2 \left(\frac{\beta c^n}{\alpha K}\right) \frac{1/\alpha}{1 + (2\pi f/\alpha)^2} \tag{J.17}$$

This expression for $S(f)$ can be simplified using

$$\langle I \rangle = N\langle g(\infty) \rangle(V - E_s) = (N\beta\gamma c^n/\alpha K)(V - E_s) \tag{J.18}$$

which relates the mean total end-plate current $\langle I \rangle$ to the mean total conductance $N\langle g(\infty) \rangle$. Substituting Eq. (J.18) into Eq. (J.17) gives the desired relation

$$S(f) = \frac{4\gamma \langle I \rangle (V - E_s)/\alpha}{1 + (2\pi f/\alpha)^2} \tag{J.19}$$

END-PLATE CURRENT SPECTRAL FUNCTION

Straightforward integration of $S(f)$ over all *positive* frequencies yields the variance σ_I^2 of the total end-plate current, i.e.,

$$\int_0^\infty S(f)\,df = 4\gamma\langle I\rangle \frac{(V-E_s)}{\alpha} \int_0^\infty \frac{df}{1+(2\pi f/\alpha)^2}$$

$$= \gamma\langle I\rangle(V-E_s) \qquad (\text{J.20})$$

The difference of a factor of 2 between Eq. (J.19) and Eq. (6.46) is that the equation in the text has introduced negative frequencies through the definition of $S(-f) = S(f)$.

APPENDIX **K**

Intracellular Trapping of 2-Deoxy-D-glucose

Glycolysis is a series of sequential steps in many biological systems in which glucose is converted to pyruvate and where there is a net production of two molecules of adenosine triphosphate (ATP) per glucose molecule consumed. Subsequent metabolic reactions result in the ultimate conversion to waste products CO_2 and H_2O. Of this series of complex steps in the glycolytic pathway an understanding of only the first two is necessary to appreciate the usefulness of 2-deoxy-D-glucose in autoradiology of cells. These two steps are

1. glucose + ATP → glucose 6-phosphate + ADP + H^+
2. glucose 6-phosphate ⇌ fructose 6-phosphate

The open-chain structure of glucose cyclizes into a ring structure in solution. Figure K.1a shows the open chain on the left curling into a ring shape in the middle, and on the right the double-bonded oxygen becomes singly bonded to two carbons, closing the ring, while the OH shifts to the carbon atom, which was initially double bonded to the oxygen. In the figure on the right the carbon atoms are not included in the drawing but are in the

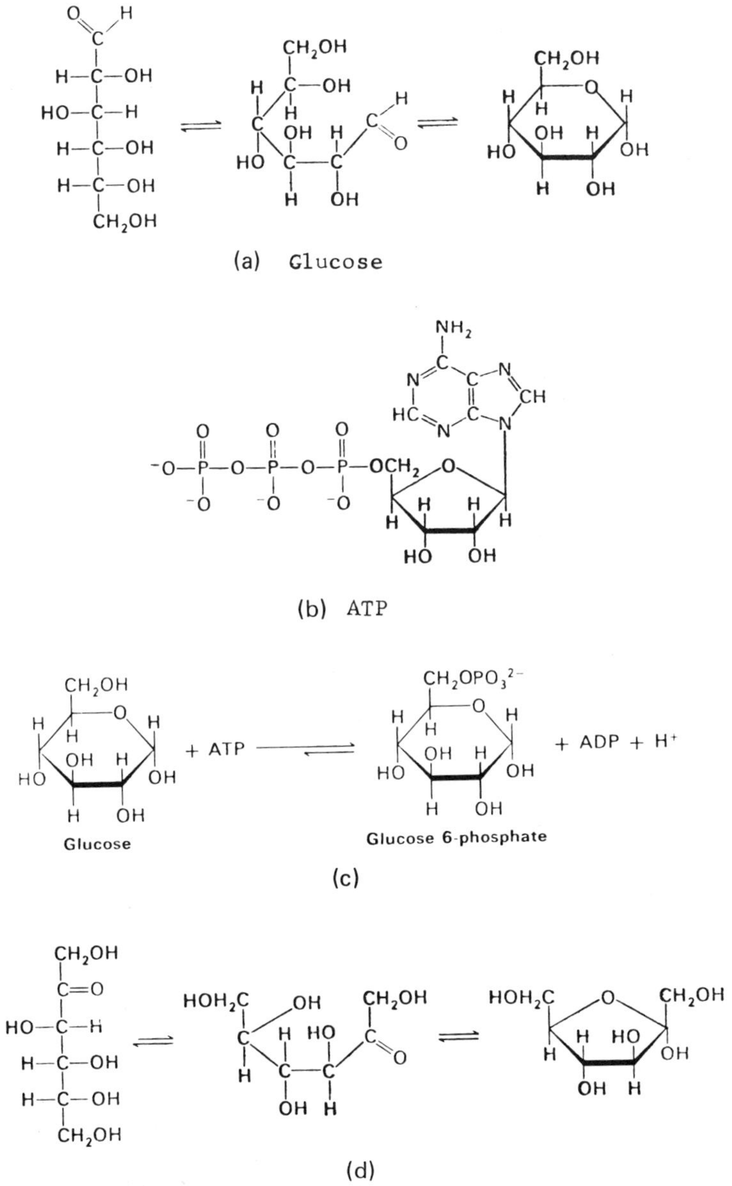

INTRACELLULAR TRAPPING OF 2-DEOXY-D-GLUCOSE   325

(a) Glucose

(b) ATP

Glucose + ATP ⇌ Glucose 6-phosphate + ADP + H⁺

(c)

(d)

FIG. K.1 (a) Open chain and ring form of glucose. (b) Structure of adenosine triphosphate (ATP). (c) Phosphorylation of glucose. (d) Open-chain and ring form of fructose. (e) Isomerization of glucose 6-phosphate—open-chain drawing. (f) Same as (e), ring drawing. (g) Open-chain drawing of 2-deoxy-D-glucose. [From "Biochemistry," 2nd ed. by L. Stryer, W. H. Freeman and Company. Copyright © 1981. All rights reserved.]

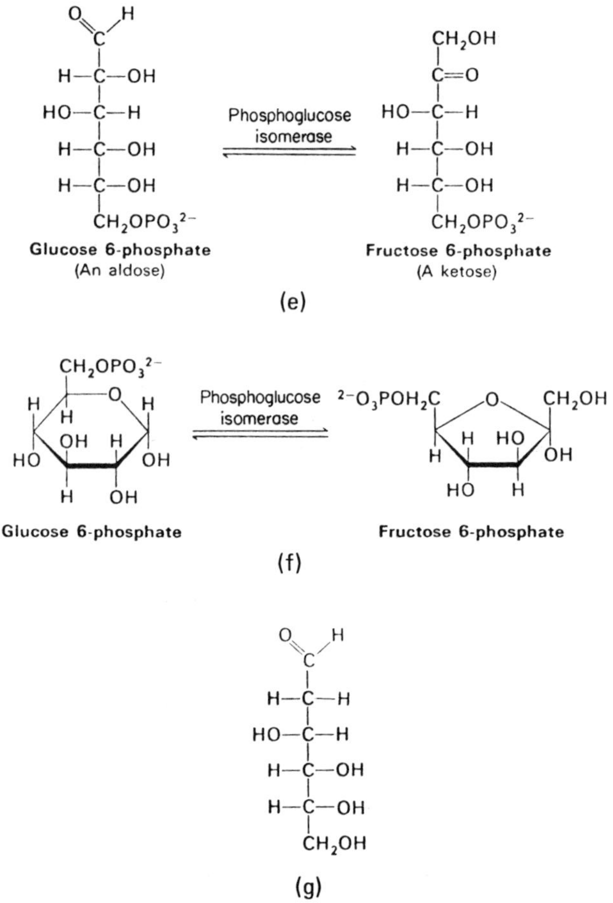

(e)

Glucose 6-phosphate (An aldose) ⇌ [Phosphoglucose isomerase] ⇌ Fructose 6-phosphate (A ketose)

Glucose 6-phosphate:
$$\begin{array}{c} \text{CHO} \\ \text{H—C—OH} \\ \text{HO—C—H} \\ \text{H—C—OH} \\ \text{H—C—OH} \\ \text{CH}_2\text{OPO}_3^{2-} \end{array}$$

Fructose 6-phosphate:
$$\begin{array}{c} \text{CH}_2\text{OH} \\ \text{C=O} \\ \text{HO—C—H} \\ \text{H—C—OH} \\ \text{H—C—OH} \\ \text{CH}_2\text{OPO}_3^{2-} \end{array}$$

(f)

Glucose 6-phosphate ⇌ [Phosphoglucose isomerase] ⇌ Fructose 6-phosphate

(cyclic structures shown with $CH_2OPO_3^{2-}$ and $^{2-}O_3POH_2C$ substituents)

(g)

$$\begin{array}{c} \text{CHO} \\ \text{H—C—H} \\ \text{HO—C—H} \\ \text{H—C—OH} \\ \text{H—C—OH} \\ \text{CH}_2\text{OH} \end{array}$$

FIG. K.1 (*Continued*)

positions shown in the middle figure. The ring structure is drawn, for clarity, so that its plane is perpendicular to the plane of the paper but tilted slightly toward the reader with the heavy bond toward the reader.

The structure of adenosine triphosphate (ATP) is shown in Fig. K.1b in which the ring is ribose, the upper two rings are adenine (also called 6-aminopurine), and the chain of oxygens and phosphors is triphosphate. The negative signs on the oxygens represent unsatisfied bonds. Adenosine diphosphate (ADP) is the same as ATP except the triphosphate chain has one fewer link.

Step 1 above occurs as follows. The cyclic structure of glucose reacts with ATP in the presence of the enzyme *hexokinase* and removes a PO_3^{2-} from

ATP and attaches it to the CH_2O in the place of hydrogen, which is released. The ATP becomes ADP. The product is glucose 6-phosphate, as shown in Fig. K.1c.

The open-chain drawing of the fructose molecule is shown on the left-hand side of Fig. K.1d. In comparison with the open-chain drawing of the glucose molecule of Fig. K.1a, both similarities and differences are evident. First, it should be noted that each has 5 carbons, 6 oxygens, and 12 hydrogens. While the lower parts of both chains are identical, the upper parts differ. The structure at the top of the glucose molecule, the double-bonded oxygen and single-bonded hydrogen to the end carbon, has the generic name of *aldose*. The double-bonded oxygen to a carbon atom within the chain of the fructose has the generic name of *ketose*.

Fructose cyclizes in solution in a way quite different from that of glucose, as seen in Fig. K.1d. The most important difference is that the ring is a pentagon instead of the hexagon of glucose because two carbons remain out of the ring instead of one, as is the case in glucose.

Step 2, the conversion of glucose 6-phosphate into fructose 6-phosphate, is the isomerization of an aldose into a ketose, which is catalyzed by the enzyme *isomerase*. This isomerization is readily seen in the open-chain drawing of Fig. K.1e. Note that no atoms are either added or removed and the phosphate group remains in place. Figure K.1f illustrates the ring drawing of the same process.

2-Deoxy-D-glucose differs from glucose in that a hydrogen has been substituted for the OH on the second carbon. This is shown in Fig. K.1g. It is seen in Fig. K.1e that with a hydrogen in this position instead of an OH the isomerization of aldose to ketose cannot take place. Therefore, although 2-deoxy-D-glucose is sufficiently similar to glucose to be absorbed by the cell and can be phosphorylated in step 1, step 2 cannot occur, and the 2-deoxy-D-glucose remains trapped in the cell.

APPENDIX **L**

Statistical Properties of Single-Channel Events

Proof that the distribution of open (and closed) channel lifetimes is exponential can be seen by the following argument. The opening and closing of a single channel as a function of time is a random signal. This is illustrated schematically in Fig. L.1, where the current $i(t) = 0$ when the channel is closed and $i(t) = i_0$ when the channel is open. Let $P_o(t)$ denote the probability that the channel is open at time t given that it was open at $t = 0$ and has not closed during the time interval t. Assume the probability of a channel making a transition from the open configuration to the closed state during the time t to $t + \Delta t$ is proportional to Δt, where Δt is very small increment of time, and that this probability is independent of events prior to time t. The latter assumption implies that

(probability of being open at $t + \Delta t$)

= (probability of being open at t)

× (probability of not closing during the time t to $t + \Delta t$) (L.1)

STATISTICAL PROPERTIES OF SINGLE-CHANNEL EVENTS

FIG. L.1 Single-channel current $i(t)$ record as a function of time.

If there are only two states of the channel, either open or closed, then the following relation holds:

(probability of not closing during time t to $t + \Delta t$)
$$\equiv 1 - \text{(probability of closing during time } t \text{ to } t + \Delta t)$$
$$= 1 - \alpha \, \Delta t \tag{L.2}$$

where α is the rate of channel closing (see p. 126). Combining Eqs. (L.1) and (L.2) gives

$$P_0(t + \Delta t) = P_0(t)(1 - \alpha \, \Delta t) \tag{L.3}$$

Rearranging terms and letting Δt go to zero gives

$$\frac{dP_0}{dt} = -\alpha P_0(t) \tag{L.4}$$

which upon integration and using the boundary condition $P_0(0) = 1$ yields the desired relation

$$P_0(t) = e^{-\alpha t} \tag{L.5}$$

Hence the open-channel lifetimes are exponentially distributed and the mean open-channel lifetime (see p. 222) equals the reciprocal of the closing-channel rate. A similar argument obviously holds for the distribution of closed-channel lifetimes. In that case, however, the decay constant of the exponential function equals the transition rate (β) from the closed- to open-channel state. Thus measurement of the single-channel closed and open lifetime distributions provides a means of inferring the conformation rates α and β that appear in Eq. (4.2).

To calculate the autocorrelation function of the random signal $i(t)$ we note that the probability p of a channel being open is $\alpha^{-1}/(\alpha^{-1} + \beta^{-1}) = \beta/(\alpha + \beta)$. This follows since α^{-1} and β^{-1} are the mean lifetimes of the open- and closed-channel states. The autocorrelation function $C(T)$ was defined in Eq. (6.6) by the relation

$$C(T) = \langle i(t)i(t + T) \rangle \tag{6.6}$$

where the angle brackets denote an appropriate time (or ensemble) average. Thus

$$C(T) = \sum_{n,m} i_n i_m \{\text{probability that } i(t) = i_n\}$$
$$\times \{\text{probability that } i(t+T) = i_m$$
$$\text{given that } i(t) = i_n\} \quad \text{(L.6)}$$

where n and m vary such that i_n and i_m sample all possible amplitudes of the random signal $i(t)$. There are four possible terms in Eq. (L.6): closed–open, closed–closed, open–closed, and open–open; however, there is only a single nonvanishing term in the sum expression, namely, the term $i_n = i_m = i_0$ since $i(t) = 0$ if the channel is closed. Hence

$$C(T) = i_0^2 \{\text{probability that } i(t) = i_0\}$$
$$\times \{\text{probability that } i(t+T) = i_0$$
$$\text{given that } i(t) = i_0\}$$
$$= i_0^2 p P_e(T) \quad \text{(L.7)}$$

where $P_e(T)$ is the probability of an even number of open and closed transitions in time T if the system starts in the open state at time 0. Figure L.2 illustrates two time segments from the single-channel current record. In case (a) there were an *odd* number of transitions during the time interval from A to B, thereby leaving the system at B in a closed state. In case (b) the system

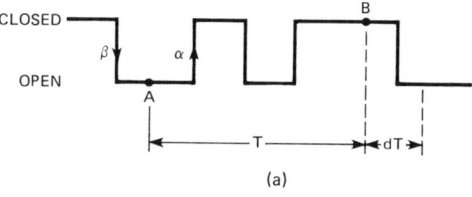

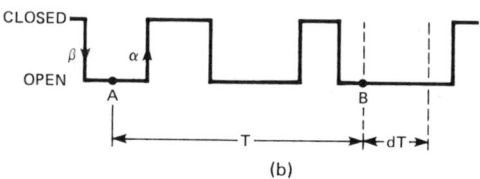

FIG. L.2 Two different time segments of the random signal $i(t)$. (a) Three transitions have occurred between points A and B. (b) Four transitions have occurred between the two open states A and B.

in going from A to B has experienced an even number of transitions so that both A and B are open states. A differential equation for $P_e(T)$ can be constructed by noting that the probability $P_e(T + dT)$ of an even number of transitions in $T + dT$ units of time, where dT is a very small increment of time, is the sum of the probabilities of two mutually exclusive events: (1) an odd number of transitions in time T and one transition in dT units of time (illustrated in Fig. L.2a); (2) an even number of transitions in time T and no transitions in dT (illustrated in Fig. L.2b). Hence

$$P_e(T + dT) = \underbrace{(1 - P_e(T))\beta\, dT}_{\text{case (a)}} + \underbrace{P_e(T)(1 - \alpha\, dT)}_{\text{case (b)}} \quad \text{(L.8)}$$

where the cases refer to those in Fig. L.2. Rearranging terms and passing to the limit $dT \to 0$ one obtains the differential equation for $P_e(T)$, i.e.,

$$\frac{dP_e(T)}{dT} + (\alpha + \beta)P_e(T) = \beta \quad \text{(L.9)}$$

The solution of this equation can be verified by direct substitution to be

$$P_e(T) = \frac{\alpha}{\alpha + \beta} e^{-(\alpha + \beta)T} + \frac{\beta}{\alpha + \beta} \quad \text{(L.10)}$$

where $P_e(0)$ has to be set equal to unity since in zero time no transitions can occur. Substituting this expression for $P_e(T)$ into Eq. (L.7) and using the relation that the probability p that a channel is open is given by $p = \beta/(\alpha + \beta)$ yields the desired expression for the single-channel autocorrelation function:

$$C(T) = i_0 p(1 - p)\, e^{-(\alpha + \beta)T} + (i_0 p)^2 \quad \text{(L.11)}$$

This equation reduces to Eq. (6.58) if we identify $\alpha + \beta$ with τ_0^{-1}.

Index

A

Acetylcholine (ACh), 4
Acetylcholinesterase, 4
Action potential, 2
Adenosine triphosphate (ATP), 279, 326
Adrenergic receptor, 134
Adrenergic synapse, 131
Afferent, 35
Agonist, 131
Aliasing errors, 196, 209
Alkaloid, 133
   muscarine, 133
   nicotine, 133
γ-aminobutyric acid (GABA), 93
Amytal test, 26
Antagonists, 133
   competitive, 133
      curare, 133
   noncompetitive, 133
Antidromic stimulation, 97
Aphasia, 26
Auditory evoked potentials, 73
Autocorrelation, 50

Autocorrelation function, 189, 222, 307
   digitized, 189
Autonomic nervous system, 19
Axon, 1, 34, 85, 86, 89, 155, 166, 172
   collaterals, 172
   hillock, 86
   terminals, 89
      knobs, 89
   unipolar, 89
Axoplasm, 155

B

Basal ganglia, 138
Binomial distribution, 115, 291
Biogenetic law, 10
Bird song, 28
Blocked channel, 227
Boltzmann statistics, 260
Botulinum toxin, 114, 132
Bouton, 35, 90
Brain
   alpha rhythm, 44, 49
   asymmetry, 26

Brain *(cont.)*
    autoradiography, 232
    brain death, 44
    Broca's area, 26
    cephalization, 23
    corpus callosum, 21, 27
    cortex, 21
    cortex map, 19
    development of, 12
    electrical properties of, 41
        EEG, 41
        evoked potentials, 41, 62
        power spectrum, 49
    evolution of, 7
    gray matter, 14
    gray-white matter, 37
    gyrus, 7
    human stages of growth, 153
    limbic, 21
    lobe, 8
    metabolism, 234, 237
        autoradiograph, 234
        dynamic radiographic studies, 237
    morphology of, 7
    neocortex, 21
    nuclear magnetic resonance, 232, 233
    positron annihilation, 232, 233, 238
    readiness potential, 20
    sulci, 7
    triune brain, 20
    two-sided brain, 9
    Wernicke's area, 26
    white matter, 14
Brownian motion, 199

C

Cable equation, 156, 159
    sealed end, 162
    sealed-end boundary condition, 295
    solutions to, 160
        compartmental model, 169
        equivalent cylinder approximation, 167
Cell body, 86
Cerebral dominance, 31
Channel closing rate, 130
Cholinergic synapse, 131
Cisternae, 86, 125
Coated vesicle, 125
Collateral, 86
Conductance fluctuations, 225
    local anesthetic, 225
    neuronal activity, 225
Conductance noise, 202
Corner frequency, 198, 210
Correlation time, 190
Cortical neuron, 32, 36
    pyramidal, 32, 36
    stellate, 32, 36
Curare, 4
Cytoplasm, 86

D

Dale's principle, 95
Dendrite, 1, 36, 85, 147, 155, 166
    dendritic spines, 164
    dendritic structure, 149
    electrotonus, 155
    environment, effects of, 149, 154
    malnutrition effects of, 149, 151, 154
    number of dendrite branches, 149
    synaptic inputs, 147
    2-deoxy-D-glucose, 234, 324
        brain metabolism studies, 234
        intracellular trapping of, 324
Dissociation constant, 282
Dopamine, 140
Drug effects, 131, 135, 225
    CNS neurons, 135
    local anesthetics, 225
    synapse, 131
Dyslexia, 57

E

Eccles model, 145
Efferent, 35
Electrical coupling, 96
    conductive, 96
    inductive, 96
Electric dipole moment, 129
    proteins, 130
Electroencephalogram (EEG), 6, 42, 48
    BEAM, 56
    cellular origin, 57
    computer diagnosis, 48
Electroencephalography, *see* Electroencephalogram
Electrotonic distance, 162
Electrotonic junction, 97
Electrotonic spreading, 96
Endoplasmic reticulum, 86
End plate, 4
    curarized, 106

INDEX

End-plate current, 105, 106, 126
   monophasic, 109
   spectral function, 320
End-plate potential, 104, 106, 116
Ensemble, 187
   ensemble average, 188
   equivalent ensemble, 190
   statistically equivalent, 188
Epilepsy, 47
Equalization times, 178
Equilibrium constant, 127
Equilibrium potential, 104
Evoked potentials, 6, 62, 63, 72, 79
   amblyopia, 68
   auditory, 63, 73
   mental chronometry, 79
   origin, 72
   somatosensory, 63
   visual evoked cortical potential, 63
      optic neuritis, 69
Excitatory equilibrium potential, 102
Excitatory postsynaptic potential (EPSP), 5, 94
Exocytosis, 121

F

Fast Fourier transforms, 209
Filament, 86
Flicker noise, 202
   $1/f$ noise, 202
Fourier coefficients, 311
Fourier series, 192
Fourier transforms, 313

G

Gap junction, 96
Gaussian distribution, 118
Glial cells, 34

H

Haeckel's biogenetic law, 10
Helix, 95
Holding potentials, 207
Hyperpolarize, 5

I

Inhibitory equilibrium potential, 102
Inhibitory neurons, 136
Inhibitory postsynaptic potential (IPSP), 6, 94, 146

Inhibitory synapse, 5
Input resistance, 177, 178, 181
Interneurons, 104
Intracellular recording, 113

J

Johnson noise, 199

K

Kirchhoff's current rule, 157

L

Larmor precession frequency, 260, 265, 282
L-Dopa, 140
Left-handedness, 29
Length constant, 162, 163
Linear cable equation, 155

M

Mean value, 186
Membrane capacitance, 159
Membrane noise, 184
Meyer–Overton rule, 226
Miniature end-plate potential (mepp), 112
Mitochondria, 90
Motoneuron, 2
Myasthenia gravis, 137
   anticholinesterases, 138
   curare test, 138
Myelin, 3

N

Nernst equilibrium potential, 108
Neuromuscular junction, 4, 212
   channels at, 212
   pore structure, 216
Neuron, 2
   information content, 88
   state of polarization, 88
Neuronal integration, 145
Neurotransmitter, 91
Nissl bodies, 86
Nodes of Ranvier, 3
Noise, 199
   conductance, 202
   end-plate, 203
   flicker, 202
   shot, 201
   thermal, 199

Notochord, 11
Nuclear magnetic resonance, 248
  adenosine triphosphate (ATP), 279
  chemical shifts, 277
  free-induction decay (FID), 264
  gyromagnetic ratio, 252, 275
  imaging, 269
  intracellular pH, 281
  Larmor precession, 252
  nuclear moments, 254
  population and relaxation, 260
  ^{31}P spectra, 284
    brain, 284
    ischemia, 283
  radio-frequency field, 257
  spin–echo technique, 268
  spin–lattice relaxation, 261, 266
  spinning tops, 250
  spin–spin relaxation, 261, 262
Nutation, 258
Nyquist interval, 197

O

One-dimensional random walk, 318
Open-channel distribution function, 289
Optic chiasma, 27
Optic neuritis, 69

P

P300 wave, 80
Parkinson's disease, 137
Passive time constant, 177
Perikaryon, 32
Picrotoxin, 135
Poisson distribution, 115, 117, 291
Polyribosome, 86
Positron-emission tomography, 239
  coincidence counting, 238
  design considerations, 242
Postsynaptic conductance, 106
Postsynaptic cytoplasm, 88
Postsynaptic density, 88
Postsynaptic membrane, 93, 131
  desensitization, 131
Postsynaptic potential (PSP), 98, 145
  excitatory postsynaptic potential (EPSP), 100
  graded response, 98
  inhibitory postsynaptic potential (IPSP), 100
  temporal summation, 99
Postsynaptic process, 88
Power spectrum, 49, 192, 195, 196, 222
  conformation model, 211
  frequency composition, 208
  fundamental frequency, 192
  Lorentzian, 198
Preganglionic fiber, 89
Presynaptic membrane, 122
Presynaptic process, 88
Probability density function, 204

Q

Quantum leakage, 120

R

Radionuclide tomography, 233
Rall model, 164
Rall theory, 145, 293
  cable equation, 303
  equivalent cylinder, 300
  Rall's branching rule, 166, 293
  sealed-end boundary condition, 295
  sealed ends, 303
  shape indices, 171, 173
Random processes, 185
  ergodic process, 188, 209
  Gaussian, 206
  random fluctuations, 184
  standard deviation, 186
  stationary random process, 188, 209
  statistically independent, 290
  stochastic process, 185
Rapid eye movement (REM), 46
Receptor–channel complex, 94, 125
Renshaw cells, 135, 136
Renshaw interneuron, 147
Reversal potential, 104, 108, 175

S

Saltatory conduction, 4
Sampling rate, 195
Sampling theorem, 316
Secretion, 121
Secretory protein, 86
Shot noise, 201

INDEX

Single-channel events, 219, 328
  open-channel lifetimes, 221
  patch-clamp method, 219
  statistical properties of, 328
Soma, 86
Soma potential, 146
  dendritic synaptic inputs, 169
  shape and time behavior of, 170
  transmembrane potential, 86
Spectral density, 195
Spinal motoneurons, 135
Spine, 89
Strychnine, 135, 136
Substantia nigra, 139
Subsynaptic membrane, 93
Sympathetic ganglion neuron, 89
Synapse, 2, 4, 85, 88
  axodendritic, 172
  chemical properties of, 85, 88, 90
    postsynaptic potentials, 98, 125
    presynaptic potentials, 110
  conductance, 128
  dendrodendritic, 88, 147, 149
  diseases of, 137
  disynaptic, 102
  drug action, 131
  electrical properties of, 85, 96
    crayfish giant motor synapse, 97
    squid giant synapse, 97, 123
  junctions, 88
  monosynaptic, 102, 136
  neuromuscular, 112, 120, *see also* Neuromuscular junction
  Ω-shaped profiles, 122
  polysynaptic, 102
  presynaptic membrane, 122
  synaptic transmission, 88
  synapto, 35
  vesicles, 90, 111, 114
Synaptic cleft, 35
Synaptic vesicle, *see* Synapse
Synaptosome, 90

T

Time average, 186
Time constant, 161
Transient passive membrane responses, 176
Transmembrane potential, 86
Transmitter–receptor complex, 107
Tubule, 86

U

Use–disuse hypothesis, 150

V

Variance, 186
Visual evoked potentials, *see* Evoked potentials
Voltage-clamp, 130

W

Wiener–Khintchine theorem, 197, 222